ANIMAL CONFLICT

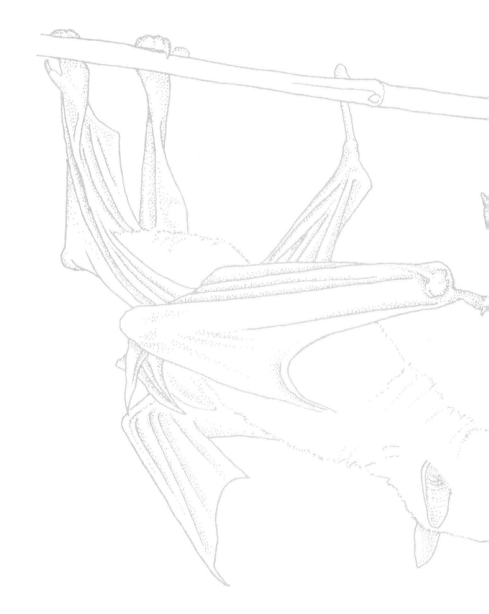

Illustrated by Leslie M. Downie
Department of Zoology, University of Glasgow

ANIMAL CONFLICT

Felicity A. Huntingford and **Angela K. Turner**

Department of Zoology, University of Glasgow

London   New York
CHAPMAN AND HALL

Chapman and Hall Animal Behaviour Series

SERIES EDITORS

D.M. Broom
Colleen Macleod Professor of Animal Welfare, University of Cambridge, UK

P.W. Colgan
Professor of Biology, Queen's University, Canada

Detailed studies of behaviour are important in many areas of physiology, psychology, zoology and agriculture. Each volume in this series will provide a concise and readable account of a topic of fundamental importance and current interest in animal behaviour, at a level appropriate for senior undergraduates and research workers.

Many facets of the study of animal behaviour will be explored and the topics included will reflect the broad scope of the subject. The major areas to be covered will range from behavioural ecology and sociobiology to general behavioural mechanisms and physiological psychology. Each volume will provide a rigorous and balanced view of the subject although authors will be given the freedom to develop material in their own way.

To Tim, Joan and Jessica, with thanks,
and to all the children who provided inspiration

First published in 1987 by
Chapman and Hall Ltd
11 New Fetter Lane, London EC4P 4EE
Published in the USA by
Chapman and Hall
29 West 35th Street, New York, NY 10001

Printed in Great Britain at the University Press, Cambridge

ISBN 0 412 25920 6 (Hb) 0 412 28750 1 (Pb)

British Library Cataloguing in Publication Data

Huntingford, Felicity
Animal conflict.
1. Aggressive behaviour in animals
I. Title   II. Turner, Angela K.
591.51   QL758.5

ISBN 0–412–25920–6
ISBN 0–412–28750–1 Pbk

Library of Congress Cataloging in Publication Data

Huntingford, Felicity.
Animal conflict.

Bibliography: p.
Includes index.
1. Agonistic behaviour in animals.   2. Conflict
(Psychology)   I. Turner, Angela K., 1954–
II. Title.   III. Title: Animal conflict.
QL758.5.H86   1987   591.51   86–20751
ISBN 0–412–25920–6
ISBN 0–412–28750–1 (pbk.)

Contents

Contents

Preface

In the past twenty years there have been many new developments in the study of animal behaviour: for example, more sophisticated methods of neurophysiology; more precise techniques for assessing hormonal levels; more accurate methods for studying animals in the wild; and, on the functional side, the growth of behavioural ecology with its use of optimality theory and game theory. In addition, there has been a burgeoning number of studies on a wide range of species.

The study of aggression has benefited greatly from these developments; this is reflected in the appearance of a number of specialized texts, both on behavioural ecology and on physiology and genetics. However, these books have often been collections of papers by specialists for specialists. No one book brings together for the non-specialist all the diverse aspects of aggression, including behavioural ecology, genetics, development, evolution and neurophysiology. Neither has there been a comparative survey dealing with all these aspects. Therefore one of our aims in writing this book was to fill in these gaps.

Another of our aims was to put aggression into context with respect to other aspects of an animal's lifestyle and in particular to other ways in which animals deal with conflicts of interest. Aggressive behaviour does not occur in a biological vacuum. It both influences and is influenced by the animal's ecological and social environment, so we consider both the complex antecedent conditions in which aggressive behaviour occurs, and its ramifying consequences in the ecosystem.

Throughout the text common names are used wherever they are available. Scientific names are given in the species index.

Acknowledgements

We are grateful to: Jon Barnes, Pat Bateson, Don Broom, Barry Everitt, John Goss-Custard, Mike Hansell, Alasdair Houston and Marilyn Ramenofsky for constructive and very helpful criticism of early drafts; M. E. Archer, T. Bakker, P. Bateson, P. Bronstein, K. D. Browne, T. Caro, D. A. Collins, D. A. Dewsbury, E. G. Dunning, V. Geist, T. Groothuis, T. R. Halliday, I. R. Inglis, J. Lazarus, P. J. Murphy, E. Nevo, P. W. Sherman, P. Thompson, G. A. Van Oortmersson and J. M. Williams for permission to use unpublished data, for papers seen in press and for recent reprints; P. W. Sherman for assistance with references; Neil Metcalfe for permission to use a photograph and H. Dingle, M. Enquist, Charlie Evans, H. W. Fricke, V. Geist, E. B. Keverne and A. Samuels for use of illustrations; Liz Denton and Peter Rickus for preparation of figures, Clarice Swan and Lynn Mackenzie for help in typing figure legends and tables; and Alan Crowden for being a patient and helpful editor.

PART ONE
Patterns of animal conflict

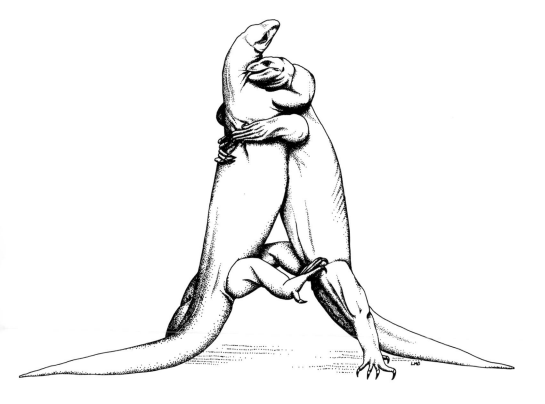

1
Conflict in the animal world

1.1 ANIMAL FIGHTS

In a pond full of breeding toads, a large male clings onto the back of a female, waiting for her to lay her eggs so that he can fertilize them. A smaller male approaches but moves away when his rival lets out a loud, low pitched croak (Fig. 1.1(a)). On an evening in late May, an adult male hare persistently chases a female. Whenever he catches up with her she attacks him, standing upright on her hind legs and repeatedly and forcefully beating his head with her fore paws (Fig.1.1(b)). The male eventually leaves without having mated. Shortly after they emerge in the spring, foraging ants from two neighbouring nests meet and start to fight, chasing and threatening members of the rival colony and seizing, biting and dragging them about (Fig.1.1(c)). Thousands of reinforcement workers arrive from the two nests, joining in a full-scale battle which is visible from a considerable distance as a seething dark mass and which continues on and off for several weeks. By the end of the battle hundreds of thousands of workers have been killed.

In each of these episodes the animals are behaving in a way which is usually called *aggressive*, borrowing a word from our everyday language. There are many thousands of other examples which we

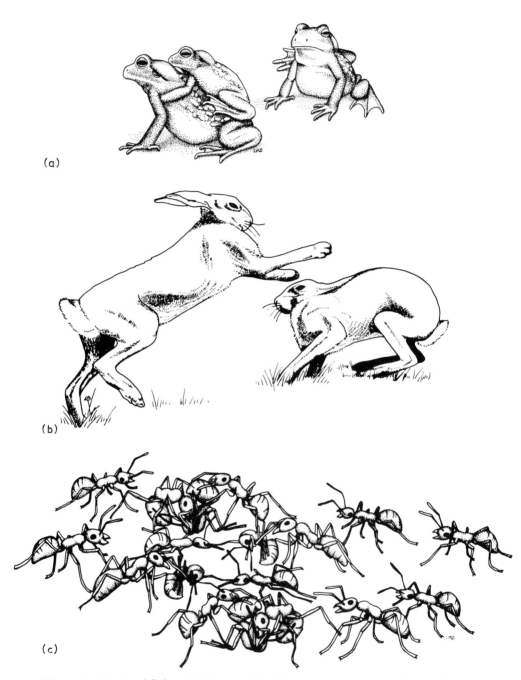

Figure 1.1 Animal fights. (a) Competition for mates: male toads; (b) conflict between the sexes: hares; (c) warfare: a territorial battle between two ant colonies.

might have chosen, because aggression is very common in both human and non-human animals. However, as we have seen, aggression can take a variety of forms from a brief exchange of signals, through intense, injurious encounters between two individuals to large-scale fighting and mass killing. Actual fighting is just one of a number of ways in which animals attempt to come out on top where their interests conflict. One aim of this book is to discuss the relationship between aggression and other aspects of an animal's life. Since aggression is so widespread and so variable we also address the question of why it occurs at all and why it takes a particular form in a given species.

1.2 CONFLICTS OF INTEREST

Throughout their lives, animals frequently interact with other individuals. The results of these interactions may be mutually beneficial: Cape hunting dogs are able to run down and overpower large prey because they hunt communally. Anemonefish are able to live among the tentacles of sea anemones, safely protected from predators by the stinging tentacles of their hosts, which in turn get snippets of food dropped by the fish. In contrast, fish of most other species are killed and eaten by anemones. This last kind of interaction, in which one participant (the anemone) benefits at the expense of the other (the fish) is very common. Although organisms may have mutually beneficial effects on each other, most encounters between animals are riddled with conflicts of interest.

We can loosely divide the causes of such conflicts of interest between animals into two broad and overlapping categories, conflict over *resources* and conflict over other *outcomes* (Table 1.1). Thus, conflicts of interest commonly arise when two or more individuals are competing for something which they both need but which is in short supply: rival ant colonies need the same feeding area or two male toads are both ready to mate with the same female. In a slightly different context, while a hungry young bird may require the worm that its parent has brought to the nest, it may be in the interest of the parent to give the worm to a healthier sibling which has a better chance of surviving.

In contrast, a conflict of interest may arise over outcomes rather than over particular resources. The disputed outcome may be what happens to an animal's body; a hyaena gains a meal if it catches and kills a wildebeest and the wildebeest benefits if it escapes. Less obviously, when a male and female of the same species meet, the male will usually benefit if he mates with the female but the female may have a higher reproductive success if she rejects him in favour of a

Table 1.1   Categories in conflict between animals.

1. Over resources
   Food
   Shelter
   Mates
2. Over outcomes
   Killing or survival of prey
   Occurrence of mating
   Occurrence of parental care
   Distribution of care between parents

different suitor. If mating does take place, it may be in the female's interest to remate but it is almost always in the male's interest to prevent her doing so. In both these cases, the disputed outcome is whose sperm fertilizes the female's eggs. As a final example, if young animals are not self-sufficient, a conflict may arise between the parents over which of them provides the necessary care rather than leaving to look for new mates.

1.3 ADAPTIVE RESPONSES TO CONFLICTS OF INTEREST

The *fitness* of both parties (that is their chances of surviving and reproducing) will depend critically on how these conflicts are resolved. So we should expect to find that animals have evolved ways of increasing their chances of coming out on top. Where a conflict takes the form of competition for limited resources, the participants may effectively appropriate whatever is going by becoming increasingly efficient at locating and exploiting it; this is called *non-aggressive scramble competition*. For example, straight after their final moult, male mites set out to look for females, which are in short supply; males which hatch early and are good at finding females breed more successfully (Potter, 1981). Alternatively, competition may be side-stepped by *mutual avoidance*, either in space or in time. If two troops of green monkeys meet, they both change directions and move away from each other (page 38). Small species of hummingbird are able to exploit good food patches by feeding at dawn and dusk when larger competitors are not foraging.

    A more active response to a conflict of interest is to *meddle* with the activities of rivals. In plants and sessile animals this may involve simply growing on top of competitors; when suitable settlement sites are in short supply, some species of barnacles grow over and destroy their rivals. During courtship in salamanders, males produce packets of sperm (spermatophores) which the females then pick up. When two or more males of the species *Amblystoma maculatum* court the same female, they disrupt each others courtship and spermatophore deposition by nudging and pushing (Arnold, 1976).

In some cases, a favourable outcome is gained by *manipulating* the behaviour of other animals. Males of a different species (*A. tigrinum*) sometimes nudge the opponent's cloaca; this causes him to deposit a spermatophore which the first male then covers with one of his own (Arnold, 1976). A male smooth newt retains the flagging attention of a female by waving the worm-like end of his tail in front of her (T. Halliday, personal communication). Females of species ranging from crabs to tree frogs and seals incite potential mates to fight and then mate with the winner (Cox, 1981; Kluge, 1981; Zucker, 1983). When a male hangingfly (*Bittacus apicalis*) mates he injects into the female a substance which makes her less receptive and promotes ovulation (Thornhill, 1981).

One potential response to a conflict of interest is to make active attempts to *physically coerce* other animals into giving up a disputed resource or into acquiescing to a particular outcome. Conversely, the victims of such attempts resist strenuously. For example, males of a third salamander species (*Plethodon jordani*) bite and chase away rivals for any females thay happen to be courting (Arnold, 1976). In ecological terminology, responses to conflicting interests other than scramble competition are often referred to as *interference competition*.

1.4 THE PROBLEM OF DEFINITION

1.4.1 Does aggression deserve a special name?

The use of physical coercion in response to a conflict of interest is often described as *aggression* and the literature on animal fighting is full of argument about exactly what this word means. In everyday language, aggression has a wide range of meanings; for example, we talk of aggressive cells in a cancerous growth, of aggressive marketing policies and of aggressive driving. If the term is to be useful, it must refer to something which is recognizable and distinct both from other responses to conflict and from other kinds of social interaction.

Aggression has been defined in terms of the *form and nature of the movements used*; thus it is 'a massive display of physical strength', 'stepping or moving forward against' or 'collective or individual fighting behaviour'. Other definitions refer additionally to *antecedent conditions* ('unprovoked attack') or the *internal state* that accompanies the behaviour ('all the emotional states and characteristic physiologi-cal changes that accompany attack'). Perhaps the most commonly used diagnostic criterion for aggression is the *immediate consequence* of the behaviour ('the delivery of noxious or potentially harmful stimuli to another animal'). These noxious stimuli may take the form of direct physical damage, but may include the delivery of aggressive threat displays (page 40).

On the other hand, some scientists, and particularly those interested in ecological questions, see the *longer term function* of aggression as its critical characteristic; thus aggression is 'behaviour which results in withdrawal of a rival or opponent', 'enhancement of status' or 'spac-ing out of conspecifics by means of the repelling principle'. Finally, it has proved necessary to include a statement about the performer's *intentions*; unsuccessful attacks would count as aggressive but accidental injuries would not. Reference to the intentions of the individuals concerned is particularly important in the case of human aggression, and this is discussed in more detail in Chapter 12.

Aggression can be recognized by a collection of features: by the forceful (and deliberate) attempt to inflict harm (either physical dam-age or exposure to an aggressive display) on another (reluctant) individual; this may be accompanied by strong physiological and emotional arousal and often functions to space animals out or deter-mine status.

Aggression only deserves its special name if it really is distinct from other responses to conflicting interests, such as mutual avoidance. Both aggression and what we have called meddling (page 6) may cause injury (overgrowth kills) and both may effect the removal of a competitor (the overgrowing individual gains a place to live). However, meddling does not necessarily involve specific, directed behavioural responses (barnacles just grow). On the other hand, although mutual avoidance and manipulation do not inflict physical injury, both are specific behavioural responses which are sometimes directed at particular individuals and which may result in animals being spaced out. So aggression, meddling, avoidance and manipu-lation are closely related in terms of their beneficial consequences to animals that do them successfully.

The dividing line between aggression and manipulation may be blurred in other respects. Scent marks make rivals avoid a particular area (page 73), but cause no injury. In the case of toads competing for mates (page 1), a small rival male suffers no physical injury, just hears a deep croak which causes him to withdraw. However, if he does not retreat, a potentially injurious fight ensues; the croak is, as it were, a declaration of intent to fight and a signal about fighting ability (page 283). There is a continuum from the exchange of long-term and long-distance signals (such as scents), through threat displays to injurious fights. The two extremes of the continuum are clearly distinct, but it is impossible to draw a clear line dividing manipulation from aggression. Perhaps it is best to think of aggression as a special case of manipulation in which the desired outcome is brought about by intense displays, which can if required lead to direct physical conflict and injury.

As we have seen, apparently co-operative ventures such as the interaction between potential mates may involve strong conflicts of interest. Courtship often involves elements of aggression and in some salamanders, males lacerate the females' skin with their sharp teeth (Arnold, 1977). On the other hand, there may be an element of co-operation in aggressive interactions just as there is in other social encounters. Where groups of animals are fighting on the same side (ants belonging to the same colony, for example) this co-operation is fairly obvious. Even fights between two rivals may involve an element of information exchange and negotiation. For example, it may be mutually beneficial for any obvious differences in fighting ability to be detected at an early stage so that the inevitable end result can be achieved with little cost to either participant (page 283). However, at best the loser just cuts its losses; no mutually beneficial outcome is possible and this distinguishes the exchanges which constitute a fight from those seen in other social contexts.

1.4.2 Is one name enough?

Accepting that there is a kind of behaviour which is special enough to deserve the name aggression, does this term cover just one kind of thing or does it include several different types of behaviour?

It is generally accepted that in both human and animal conflict the initiation of an attack (*offence*) is not the same as protection against such an attack (*defence*). For example, the initiator of a fight often uses one particular set of behaviour patterns (moving forward, threatening and attacking). The responder often behaves differently, retreating, showing submissive behaviour (page 41), keeping a low profile and trying to escape. It is hard to draw a clear dividing line at any point of the continuum from offence or attack through offensive and defensive threat and submission to escape, yet it seems an abuse of language to include escape under the heading of aggression. Therefore this word is often replaced by the term *agonistic behaviour* (Scott and Fredericson, 1951), which refers to 'a system of behaviour patterns having the common function of adaptation to situations involving physical conflict'. (This is not to be confused with the neurochemical and pharmacological use of the term agonist, which refers to substances that enhance the action of a certain chemical in the brain.)

Behaviour that inflicts non-accidental injury on other animals, and can therefore be described as aggressive, is shown in a variety of circumstances: predators (including cannibals of the same species) damage prey, which may launch a defensive counter-attack; female mammals with young fight off males that come too close; young may be bitten during weaning; rival males fight over females, and so on. Are these really examples of the same sort of behaviour? If not, we should give them different names. In principle, it is easy to answer this question; we simply identify the properties of the different patterns of injurious behaviour and see if they are the same. In practice, things are not so simple, because the answer is not always a clearcut 'yes' or 'no' and may differ from species to species.

These points are taken up in later sections of this book; for the purposes of illustration we concentrate here on fighting shown in two very different situations, namely between male conspecifics in competition for a female and between predators and prey of different species. We know that herbivores can be extremely aggressive when fighting for mates (for example, male mountain sheep) so predation and inter-male fighting cannot be identical. What is the situation in carnivores? Clearly when a male fights another male over a female and a predator attacks its prey, a conflict of interest exists (who gets the female and what happens to the body of the prey animal), and injury to one or both participants is a probable consequence of the encounter.

How similar is the form of attack in these two contexts? There are plenty of examples in which quite different behaviour patterns involving different parts of the body are used during fights between rivals and between predators and prey (page 42); many male antelopes lash viciously at predators with their sharp hooves, but use their horns against rivals. In other cases the movements and postures used in the two contexts are very similar so that a clearcut distinction cannot be made between prey killing and fights between rivals on the basis of the form of the movements used.

Perhaps these two patterns are different in the mechanisms within the animal that control their expression. The hormone testosterone increases fighting between males in many mammals (page 104) but rarely has an effect on predatory behaviour. Similarly, the regions of the brain from which predation and attack on a rival can be elicited by electrical stimulation are distinct in cats and in other species (page 134). Here, the internal control of the two patterns of agonistic behaviour is clearly different. On the other hand, levels of attack towards both conspecifics and predators vary with reproductive condition in some fish (for example, sticklebacks) while in some species (for example, cichlid fish) the brain sites which produce attack to a rival and biting at prey coincide. Here it seems that a degree of similarity exists in the mechanisms that control predator–prey interactions and rival fights.

Over and above the very general fact that both potentially enhance fitness, fighting between males and attack on prey clearly differ in their adaptive consequences. Fitness is enhanced in very different ways; in predator–prey encounters, the winner gains a meal or avoids being eaten but in fights over females, he fertilizes some eggs.

It is clear that the patterns of agonistic behaviour shown in different contexts are not necessarily identical or even similar in form, motivation and function just because they all may cause non-accidental injury. It is also clear that the extent to which they are similar varies from species to species. There are no general rules which specify the relationships between any two patterns of agonistic behaviour; each case has to be considered in its own right.

1.5 THE BIOLOGICAL STUDY OF ANIMAL CONFLICT

The science of biology is concerned with the way living things work on a number of levels, from ecosystems and social groups to individuals and their physiological systems; all of these can help us to understand the way animals behave. For example, the role of a species in the ecosystem is both a cause and a consequence of the way its members behave (page 238), just as the concentration of chemicals in a particular part of the brain can be both a cause and a consequence of what an animal does (page 158).

The aim of this book is to ask biological questions about animal conflict, particularly conflict between members of the same species, at these various levels. Biologists from different disciplines tend to concentrate on different kinds of conflicts of interests; thus the field-oriented ecologists have studied fighting between rivals, differences of opinion between potential mates, disagreements between parents and young and many other patterns of conflict. In contrast, laboratory-based physiological studies have mainly been concerned with the more straightforward agonistic responses of fighting over resources. In either case, there is a huge literature which cannot possibly be covered exhaustively in a single book. In what follows we concentrate on selected examples from different groups of animals, referring to published reviews for a broader comparative perspective.

We are trying to answer many different questions about animal conflict. What external stimuli and internal events cause animals to fight in a particular way? How do differences in form and level of agonistic behaviour develop and what are the roles of genotype and environment in this process? What impact do agonistic behaviour and

other responses to conflict have on ecological and evolutionary processes? What is the evolutionary history of a particular pattern of
agonistic behaviour? How is an animal's fitness influenced by the way
it resolves conflicts of interest? While recognizing that human behaviour is uniquely complex and profoundly susceptible to cultural
influences, a biologist would not exclude our own species from
consideration. This book therefore also explores the relevance and
limitations of a biological approach to agonistic behaviour in humans.

2

A survey of animal conflict

2.1 INTRODUCTION

Plants may kill or otherwise overcome competitors by overgrowing
and shading out smaller plants or by using toxic chemicals to kill other
individuals (Newman, 1983). However, it is in animals that the use of
physical force in response to conflicts of interest is most highly
developed.

   In the following sections we describe examples in each of the major
groups of animals to illustrate special features of their behaviour as
well as the main patterns of fighting observed. This comparative
survey shows that, although different kinds of animals may use
distinct behaviour patterns in response to conflict, the end result may
be quite similar. Thus we see such phenomena as dominance
hierarchies, territoriality and mate guarding in animals as different as
shrimps and monkeys. Such general principles about agonistic behaviour are discussed in Chapter 3.

2.2 ACELLULAR ORGANISMS AND PROTISTS

Bacteria: microscopic living organisms with no cellular organization and no nucleus. Protists: slightly larger, still microscopic organisms, unicellular or acellular, with differentiated internal organization including a membrane-bound nucleus.

Neither the bacteria nor the protists are animals, but many of them are freely moving and have evolved techniques for eliminating competitors which deserve mention. Several types of bacteria produce toxins that kill members of the same or closely related species (Reeves, 1972; Hardy, 1975). In bacteria such as *Escherichia coli*, which lives in the gut of animals, some individuals (called colicinogenic) have a genetically controlled capacity to produce a poison called colicin. The individual bacterium that releases the poison dies, as do any non-colicinogenic bacteria in the vicinity. However, other colicinogenic bacteria are immune. In an environment such as the gut, bacteria can attach themselves to a substratum where they multiply. As a result, clumps (or clones) of related bacteria develop. By the self-sacrificing activity of the individual that releases the poison, a clone of colicinogenic bacteria can form an inhibition zone around itself preventing other bacteria from encroaching on food resources (Chao and Levin, 1981).

Some protists contain forms of bacteria living within their bodies; if these symbiotic bacteria are poison-producing the protists release the poisons into the environment (or inject them into their partner during mating), thus removing potential competitors. Such protists, which belong to the so-called killer strains, are themselves resistant to the effects of the poison (Preer, Preer and Jurand, 1974). *Paramecium aurelia* lives at the mud–water interface of ponds. In this habitat the population has a clumped distribution that is probably related to the distribution of its food. *P. tetraurelia* secretes a poison which kills related species of *Paramecium* and is thus able to monopolize patches of food (Landis, 1981).

2.3 SEA ANEMONES (COELENTERATES)

Radially symmetrical, multicellular animals with primitive tissue organization, including sensory structures, a primitive nerve net and cellular contractile elements. Unique, specialized stinging cells, nematocysts, in the skin.

Coelenterates have a primitive immune response, which enables them to distinguish their own tissues from those of genetically distinct

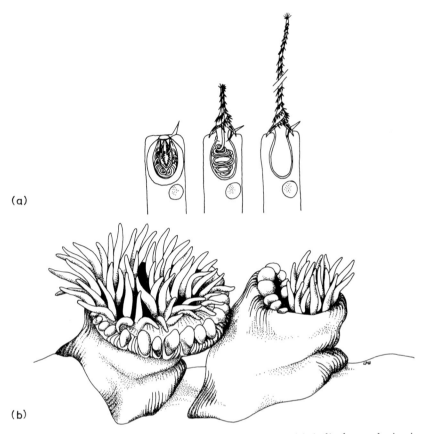

Figure 2.1 Weapons and fighting in sea anemones. (a) A discharged stinging cell or nematocyst; (b) an agonistic encounter between two sea anemones.

individuals (Theodor, 1970; Hildemann *et al.*, 1977,). This means that they are able to direct their responses to specific categories of competitor. Their specialized stinging cells, or nematocysts, explode when touched, releasing a sharp, sticky thread which impales, entangles or injects poison into nearby animals (Fig. 2.1(a)). In other words, they have the capacity for effective, directed aggression.

In sea anemones stinging cells are collected into specialized weapons (known as acrorhagi) around the mouth, and in the tentacles (Francis, 1976; Purcell, 1977). These are used against genetically distinct individuals, derived from a different clone. The beadlet anemone *Actinia equina* can be found spaced out in inter-tidal areas; there are two colour morphs. Red/brown individuals are more aggressive and usually defeat green ones. On contact with another anemone red individuals at first withdraw their tentacles, inflate the acrorhagi around the mouth (Fig. 2.1(b)) and point these towards the opponent. When the acrorhagi touch the opponent they pull patches of ectoderm (skin) off the victim. A red anemone will usually retaliate in like manner; a green one usually withdraws. The eventual loser contracts its upper end, partially concealing its weapons and sometimes moves away. Larger individuals and anemones resident on the site at which an encounter occurs are more likely to emerge victorious (Brace and Pavey, 1978; Brace, Pavey and Quicke, 1979; Brace, 1981; Ayre, 1982).

2.4 PARASITIC WORMS (ACANTHOCEPHALANS)

Small, thin, unsegmented parasitic worms with a muscular proboscis armed with large hooks and specialized for life inside the body of other animals. The life cycle is complex and includes larval forms in different species of host.

Competition for space and food (and mates) is likely to be very severe in the specialized ecological niche of internal parasites (endoparasites). In various kinds of endoparasites (trematodes, for example) the larvae kill and eat competing species (Basch, 1970). The possibility that conflicts are resolved by agonistic interactions can only be investigated indirectly. For example, in the acanthocephalan parasite *Moniliformes moniliformes*, adult females live in the top of the gut and males have to migrate forwards against the current. In the absence of any females, a male's size is unrelated to his position in the gut. However, when a female is present, size and position are correlated, with the larger males being nearer the front and the smaller ones near the back. Males in mixed-sex infections also frequently cover the genital apertures of their rivals, thus preventing release of sperm. These various facts suggest that males compete for females and that larger males may do so more effectively (Miller, 1980).

2.5 RAGWORMS (ANNELIDS)

Segmented worms with a discrete nervous system consisting of anterior brain and ventral nerve cord sending nerves to segmental muscles. Sensory cells scattered over the body surface and specialized sense organs (for light, chemicals and body position) at the front of the body.

Ragworms are mainly omnivorous scavengers, tearing off chunks of food with a pair of hard jaws on an eversible proboscis (or snout). They are solitary, living in tubes made of mucus or in burrows in

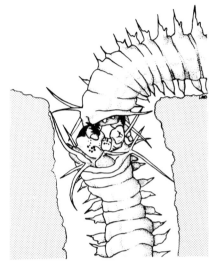

Figure 2.2 Ragworms fighting over a burrow.

mud. Agonistic encounters occur over access to burrows, owners of burrows or tubes attempting to evict intruders. The two worms face each other, touch each other with the sensory tentacles around their mouth regions and finally push at each other with the jaws and proboscis which is everted (Fig. 2.2). The intruder usually retreats or is pushed backwards out of the burrow (Cram and Evans, 1980). Fights over the ownership of burrows occur both within and between species but are more frequent between species (Evans, 1973).

2.6 LIMPETS, OCTOPUSES AND SQUIDS (MOLLUSCS)

Soft-bodied, unsegmented, predominantly marine invertebrates. Mainly protected by a hard shell. Most have eyes and various other sense organs. Nervous system concentrated into several ganglia which form a brain. Nervous system and eyes very well developed in the large, active cephalopods.

Large individuals of the inter-tidal limpet *Patella longicosta* defend patches which contain high quality food (algae), forcing smaller conspecifics onto less desirable plants. The owner of a good patch deals with intruders of any limpet species by first touching them with its tentacles then pushing against them until they retreat (Branch, 1975). The owl limpet responds to rivals in a similar way (Stimpson, 1973). Many individuals, especially small ones, evade contact with a resident limpet. If contact is made the resident usually only needs to pursue the intruder a short distance before the latter retreats, so full-scale fights are rare (Wright, 1982).

Some terrestrial molluscs defend shelters, as protection from dessication and a hiding place for eggs. *Limax maximus* is a particularly aggressive slug, which fights over shelters especially in the summer when these are needed most but are also at their scarcest. When individuals fight the aggressor touches the opponent with tentacles and mouth lobes and then bites, sometimes also rearing up and lunging. The opponent may hit back with its tail or retreat, sometimes exuding mucus to delay the aggressor; the defeated animal is often eaten (Rollo and Wellington, 1979).

The behaviour of cephalopod molluscs such as squids and octopuses contrasts with that of other molluscs because of their highly developed brain and striking visual communication system. During social interactions, including agonistic encounters, cephalopods use rapidly changing visual stimuli based on body pattern and colour. Octopuses are solitary, each individual having its own resting site although feeding ranges can overlap. Fights are rare but when a dispute over a site does occur octopuses use a visual display. The opponents splay their arms in front of their body, suckers extended so that they look larger; they turn red, the aggressor being darker than its opponent (Packard and Sanders, 1971).

Up to now we have dealt with aggression as a means of removing competitors for food, space and shelter; but among cephalopods, females are also a resource to be defended. Squids are social and form large groups when mating and laying eggs so there can be many rivals for a female. Male squid consorting with sexually receptive females warn off rivals with a lateral silver display; if a rival approaches, the defending male changes to a zebra stripe display – the darker he becomes the higher is the level of aggression (Moynihan and Rodaniche, 1982).

2.7 INSECTS

Small, segmented invertebrates with three pairs of jointed legs and two pairs of wings, occupying a variety of habitats. Tough exoskeleton, specialized mouthparts and highly developed nervous and sensory systems.

Most fights between adult insects are concerned with reproduction, conflict arising over mates, over a space for courtship and mating or over sites for laying eggs. Young insects, which often have a different life style to the adults, may defend feeding sites and shelters. An interesting aspect of agonistic behaviour in insects is that in addition to this variability due to age, different individuals of the same species may compete for resources in different ways; in modern jargon, they display alternative competitive strategies (page 312).

With such a large and diverse taxon, a comprehensive review of fighting patterns is impossible. The following section therefore gives only a few examples of the different types of conflict found in insects. Reviews of selected aspects of agonistic behaviour in insects can be found in Blum and Blum (1979), Matthews and Matthews (1980) and Lewis (1983).

2.7.1 Dragonflies and damselflies (odonates)

Dragonflies and damselflies have a short adult life and lay their eggs in water where the fierce, predatory larvae develop. Agonistic behaviour in adults is usually between males over females and, in many cases, takes the form of defence of an area in which breeding takes place (i.e. a breeding territory). Thus, males of the damselfly *Hetaerina vulnerata* defend territories along streams where there are suitable oviposition sites. Intruders are attacked but sometimes they fail to retreat and the two males chase each other in a circle until one of them, usually the intruder, breaks away and leaves the territory. Some males (which are probably young and are called satellites) remain motionless on the territory and so may escape the notice of the owner; if they are seen they are chased out.

Males copulate with females that enter their territories. After mating, females may oviposit there or outside the territory boundaries. In either case the male stays close to the female, perching while she goes underwater to lay her eggs. He can thus prevent another male fertilizing some of her eggs. On returning to his territory the male may have to evict an intruder but usually manages to do so. Territories are held only during the middle of the day; in the early morning and late afternoon males and females concentrate on feeding, which they do by hawking insects from trees along the stream (Alcock, 1983).

2.7.2 Crickets and grasshoppers (orthopterans)

Like many birds, male orthopterans defend mating territories (from which they exclude other males) by singing. Male *Teleogryllus commodus* crickets remain well spaced out when singing by moving away from a nearby calling male and approaching distant ones (Campbell and Shipp, 1979). In *Gryllus integer* some males sing to attract a mate, but the same song attracts other males who remain silent and steal mates from the singers. Since they do not sing themselves, these satellites are not attacked by the singer but they do attack and bite him (Cade, 1979).

Again, as in birds, song is not always an effective deterrent; males are especially aggressive after mating. An intruding male is attacked; if he fails to retreat, a grappling fight ensues. After the fight, the winning cricket shows a conspicuous, bouncing behaviour (Bailey and Stoddart, 1982).

2.7.3 Cockroaches (dictyopterans)

Cockroaches are one of the few invertebrates that look after their offspring, defending them from predators (including conspecifics). Female *Shawella couloniana* fiercely defend their oothecae (egg cases) from other females who would eat the eggs (Gorton *et al.*, 1983).

Cockroaches such as *Nauphoeta cinerea* defend territories, but at high population densities agonistic interactions between males result in a different form of social structure – the dominance hierarchy (page 46). In a dominance system, one individual consistently wins encounters with another, which generally retreats without a fight. In a group, this can produce a ranking system, whereby one animal, the dominant or *alpha*, is regularly victorious over all the others. Below the dominant we find high ranking individuals which are victorious over all but the most dominant animal and so on to the other extreme, where the *omega* or subordinate individual gives way to all other members of the group. Dominant animals usually gain access to favoured resources; the dominant male in a group of *Nauphoeta* gets most matings.

Fights generally occur only when dominance–subordinance relationships are being established. In *Nauphoeta* two rivals charge each other head-on, lock legs and bite each other. One male eventually gives way, by retreating or by folding its legs, tucking in its head and folding back its antennae, in which posture it is no longer attacked. At intermediate population densities, territory and dominance merge into one another. Most males establish territories, but those of the alpha males are larger and more permanent and the lowest ranking males have none (e.g. Breed, Smith and Gall, 1980).

2.7.4 Scorpion flies (mecopterans)

Two relatively unusual types of behaviour occur in scorpion flies: stealing food from others (kleptoparasitism or piracy) and forcible copulation. Male black-tipped hanging flies present arthropod prey to a female (whom they attract initially with a smell or pheromone), and copulate with her while she eats it. If a rival encounters a male carrying prey, he sometimes tries to force him to relinquish it. Pirates also interfere with copulating pairs, usually robbing them of their food and sometimes mating with the female. In the genus *Panorpa* some males with no prey item to offer attempt to copulate forcibly with females, holding the female's wings down and clasping them with the genital forceps to prevent escape; such forcible copulation in animals is sometimes called rape, by analogy with human behaviour (Thornhill, 1980).

2.7.5 Flies (dipterans)

In the Sonoran desert botfly, and several other species of flies, males defend adjacent territories on the top of a ridge, chasing away rivals and mating with any females that arrive. This type of territory, used only for mating and containing no resource required by either the male or female, is known as a lek. The mating territories are often clumped and females come to them only to mate before dispersing to find a woodrat host in which to lay their eggs (Alcock and Schaefer, 1983).

2.7.6 Beetles (coleopterans)

While most insects fight with their mandibles (jaws) and legs, some of the beetles have special, and often spectacular, weapons which may be enlarged mandibles as in stag beetles or specialized outgrowths from the head or centre of the body as in scarabs (Otte and Stayman, 1979; Fig. 2.3).

Figure 2.3 Male stag beetles fighting.

Male *Podischnus agenor* dig burrows in sugar cane stems to which they attract females and over which they fight. The two horns, on the head and body, are used to hold the opponent which is then lifted and dropped free of the burrow. Large individuals, which have relatively larger horns, usually win (Eberhard, 1979). Not all beetles use specialized weapons during fights. Males of *Tetraopes tetraophthalmus* fight by pushing, head butting and locking mandibles. These beetles live on milkweed plants and fight over possession of plants and females and to dislodge males already copulating. Large males usually win (McCauley, 1982).

2.7.7 Ants (social hymenopterans)

In social insects the colony, and sometimes a feeding territory around it, is defended from conspecifics and other intruders. African weaver ants fight fiercely and frequently and produce colony-specific chemicals which help to recruit more ants when intruders are discovered; the intruders are killed and eaten (Hölldobler, 1979).

Several ant species make raids on other ant colonies, stealing and eating their broods. In the so-called slave making species, rather than being eaten, the brood is transported back to the attackers' nest and recruited into the workforce (Stuart and Alloway, 1983). Some slave-making species have developed special weapons such as a strong stinging apparatus or enlarged mandibles; *Raptiformica* uses an alarm pheromone that confuses the workers in the raided nest so that the brood is left unguarded for the conquerors to enslave (Buschinger, *et al.* 1983).

2.8 SHRIMPS AND CRABS (CRUSTACEANS)

Relatively large invertebrates with hard exoskeleton, segmented body, two antennae and various other appendages. Exoskeleton often reinforced dorsally by a carapace and often pigmented. Mostly marine, but some freshwater and a few terrestrial forms. Well developed nervous system and sense organs, especially visual.

Some crustaceans such as barnacles destroy competitors by simple overgrowth, but in the most advanced crustacea (the Malacostraca) directed behavioural responses to conflict are observed. The interest of this group lies partly in their conspicuous visual displays and the fact that many possess potentially lethal weapons, and partly in their commercial significance. Reviews of agonistic behaviour in crustacea are provided by Salmon and Hyatt (1982) and Rebach and Dunham (1983).

Agonistic encounters in advanced crustaceans occur mostly over shelters and mates; males of many species guard the females at the time of mating (for example, amphipods, Ridley and Thompson, 1979), others defend mating burrows (for example, ghost crabs, Brooke, 1981) while yet others defend resources needed by females (some crabs). Males that cannot fight successfully may use alternative methods of competing, for example, by courting females at less favoured times (Brooke, 1981). Encounters often start with conspicuous behaviour patterns in which weapons are displayed, followed by pushing and grappling fights. Although the weapons are used against conspecifics and serious injury does occur, fights often end without physical contact. Large size and prior ownership generally give an advantage in crustacean fights.

2.8.1 Mantis shrimps (stomatopods)

The stomatopods are tropical marine crustacea measuring up to 10 cm in length. They have specialized legs (called raptorial appendages) with knobs and spines which are used both to kill prey and in fights against rivals. Stomatopods (for example *Gonodactylus bredini*) live alone in burrows which are often in short supply and which are defended by both sexes, but especially fiercely by females with eggs (Montgomery and Caldwell, 1984). When a would-be intruder approaches an occupied burrow, the owner spreads its raptorial appendages, lunges at the intruder and grapples with it. If the intruder enters the burrow, a fierce fight ensues and serious injuries may occur. The resident animal usually succeeds in repelling intruders, although if the latter is larger it may take over the shelter. Stomatopods avoid the burrows of individuals by whom they have previously been defeated, recognizing them by smell (Dingle and Caldwell 1978; Caldwell, 1979).

2.8.2 *Leptochelia dubia* (tanaidacean)

The tanaid crustacean *Leptochelia dubia*, a benthic form found at very high densities, is a protogynous hermaphrodite (most males arise by sex reversal in large females). This process is inhibited by the presence of adult males. The females live in burrows in which they tend their eggs. Males, which have long, curved claws on their first legs, fight long and hard over females, making forceful downward swings with the claws, grappling and throwing rivals away. There is no evidence that this injures the victim which, although it may give up, often approaches again. Males found in the burrows of females are usually larger and usually succeed in repelling intruders. While two large males are fighting, smaller males may enter the tube surreptitiously (Highsmith, 1983).

2.8.3 Crabs (decapods)

The crab *Pilumnus sayi* lives in colonies of bryozoa (sessile inverte-brates) from which it gains protection from predators and, in the case of males, access to females. If another male intrudes onto a colony the resident approaches rapidly, spreads its claws (chelae) and swipes at the intruder with them. If the intruder does not retreat, the fight intensifies, and the males may grasp and pinch each other. Large crabs usually win fights and end up with the largest bryozoan colony (Lindberg and Frydenberg, 1980).

2.9 SCORPIONS, MITES AND SPIDERS (ARACHNIDS)

Small, predominantly terrestrial invertebrates with strong exoskeleton, seg-mented body, four pairs of legs and specialized appendages (pedipalps and chelicerae) for seizing and tearing prey. Well developed nervous system and sense organs, especially vision.

Agonistic encounters in arachnids take place over a variety of different resources, including shelters, individual food items, feeding terri-tories and mates. Conflict also occurs during sexual interactions; since most arachnids are predators, males run a real risk of being eaten by their intended mates, and the complex courtship displays of arachnids are a response to this risk. Although many conflicts start with a harmless display of weapons, fierce damaging fights are not at all uncommon. Reviews (for spiders) can be found in Riechert (1982) and Burgess and Uetz (1982).

2.9.1 Scorpions (scorpiones)

'When two scorpions are found beneath the same stone, they are either mating or devouring each other' (Fabre, 1907, cited in Cloudsley-Thompson, 1958); this comment gives an accurate picture of the social behaviour of scorpions. For example, about 10% of the diet of adult *Paruoctonus mesaensis* consists of younger conspecifics. The frequency of cannibalism increases with population density and this behaviour thus plays a major role in the dynamics of scorpion populations. The victim of a cannibalistic attack defends itself by flicking its sting towards the attacker and grasping its claws. Courtship is also a very aggressive affair in this species, with the female repeatedly clubbing the male with the back of her body, grasping him by the claws and often eating him (Polis and Farley, 1979).

2.9.2 Mites (acarids)

Although mites are barely visible to the naked eye and are cryptic in their habits, agonistic encounters between males have been minutely documented in studies aimed at controlling pest species. The water mite can only survive inside freshwater mussels and most mussels contain just one male and up to several dozen females. If a male comes across another adult male in his mussel he attacks immediately, grasping the opponent's legs, grappling with him and sometimes piercing his cuticle, with fatal results. Thus a male defends exclusive access to the mantle cavity of his host and, presumably, to the females that live there (Dinnock, 1983).

2.9.3 Spiders (araneids)

Some spiders live and feed communally but even here conflict is in evidence. Some social spiders defend small areas around themselves while others establish dominance hierarchies (Burgess and Uetz, 1982). However, most spiders are solitary with mating and fighting (over food or mates) being the commonest form of social interaction.

The bowl-and-doily spider builds a large, complex web with a bowl-shaped cup containing a maze of threads (hence the name) below which the owner hangs. The spider runs out to capture and kill any moving prey that become entangled in the threads. When they are mature the males leave their webs and search for mates. The mature females, much the larger sex, are almost continuously receptive and mate with several males. Females are relatively uncommon so encounters with potential mates are rare and males compete fiercely. When a male arrives at a female's web that already has another male in residence, the two rivals approach each other and with very little by way of preliminaries start to grapple furiously with legs and jaws interlocked. These vicious struggles may last for several minutes; although a losing male can release his hold and drop out of the web at any stage, fights quite frequently continue until one contestant is killed (Austad, 1983). Most fights are won by the larger male and where opponents are evenly matched, the resident male usually comes off best (Suter and Keiley, 1984).

2.10 SEA URCHINS (ECHINODERMS)

Relatively inactive marine invertebrates with pentaradiate symmetry and a simple nerve ring. Sea urchins have a very complex, rasping jaw.

Some sea urchins (*Echinometra* and *Strongylocentrotus* spp.) live in burrows which they excavate from rocks at some cost in time and energy; these burrows provide protection against predators and against wave action. Even at high densities, sea urchins usually avoid each other, but intrusions do sometimes occur. The burrow owner then moves out of its shelter, approaches its opponent and the two animals rotate their spines until these point at the rival. The resident pushes the intruder away and if the latter does not retreat, it may be bitten by the resident's complex jaws (Grünbaum *et al.*, 1978; Maier and Roe, 1983).

2.11 FISH

Three distinct living lineages; agnatha; primitive jawless vertebrates with no true fins (lampreys and hagfish); elasmobranchs: large, predatory fish, predominantly marine. Cartilaginous skeleton and heavy, rather unwieldly body (sharks, skates and rays). Osteichthyes and especially the teleosts are a much more diverse group of smaller fish with bony skeleton, flexible fins and light, manoeuverable body. Fertilization, in teleosts, usually external. Occupy range of freshwater and marine habitats. Possess well developed nervous system and sense organs.

It is hard to specify any typical form of agonistic behaviour for such a large and diverse group as the fish. However, there are a number of commonly occurring features which are worth emphasizing here. In the first place, specialized weapons are rare. Most species have a range of agonistic responses in addition to overt attack. Many of these involve raising the fins or gill covers, with a resulting increase in apparent size, and tail beating or mouth fighting, during which rivals may determine their relative strength. Fish are unusual among vertebrates in that males commonly care for the broods of young; this means that competition is often for suitable breeding space rather than for females. Territorial behaviour is common and ranges from year-round, all-purpose territories (as in damselfish, page 27), to small breeding territories defended for several weeks (as in sticklebacks, Wootton, 1976) and lek territories defended for just a few hours a day for the purpose of mating (*Halichoeres melanochir*, Moyer and Yogo,

1980). Not all fish are territorial; guppies and mollies, for example, live in permanent schools with a dominance structure. Although a few observations have been made on lampreys and sharks, most of the available information concerns the teleost fish. Reviews of various aspects of conflict behaviour in fish are provided by Miller (1978) and Turner (1986).

2.11.1 Lampreys and sharks (agnathans and elasmobranchs)

Breeding male brook lampreys build nests in which the females spawn. They fight other males by attaching the sucker-like mouth to their rival's side and pushing it away. The most active and alert males are most likely to achieve copulation while unsuccessful males spiral round copulating pairs possibly releasing their own sperm into the water (Malmquist, 1983).

Grey reef sharks appear to defend territories. They respond to intruders with exaggerated sideways swinging movements of the head and body with the back arched, the snout raised and the lateral fins lowered. This is followed by slashing, biting and extremely damaging attacks (Johnson and Nelson, 1973; Barlow 1974; Nelson, 1981).

2.11.2 Damselfish (pomacentrids)

In the three-spotted damselfish, a small, alga-eating reef fish, each individual defends a territory of about 4 m across usually with a shelter near the centre. Should another fish enter this territory, the resident raises its fins, swims rapidly towards the intruder, thrusting at it with an open mouth and often biting the intruder's fins, If the intruder does not retreat, the two fish engage in parallel display and circling (Fig. 2.4), until one breaks off the encounter and flees. Rival conspecifics are attacked when they are some distance

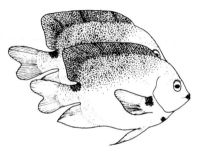

Figure 2.4 Parallel display in three-spotted damselfish.

from the shelter; alga-eating fish are attacked when they get slightly closer and intruders which do not eat algae are only attacked when they are close to the shelter. The centre of the territory is a core area which is defended from all intruders regardless of their species. Damselfish are able to recognize each other, probably by a combination of colour pattern and geographic location, and attack familiar neighbours less vigorously than unfamiliar fish (Thresher, 1976, 1979).

The anemone fish (*Amphiprion akallopisos*) is usually monogamous, each pair living in one anemone, mostly defended by the female, in which the male cares for the eggs. However, in areas of the Red Sea where carpets of anemones occur, this same species lives in groups comprising one female and a number of adult or subadult males. The cluster of anemones in which each group lives is defended by all its members. Within the group a dominance hierarchy exists (similar to that described above for cockroaches, page 20), with the female, which is always largest, at the top and the smallest male at the bottom. Subordinate fish show a characteristic head-down posture which inhibits attack by dominants (as does the submissive posture of subordinate cockroaches, page 20). Only the dominant male mates with the female and all the others have smaller testes. Should the female disappear or be removed from the group, the largest male changes sex within a few weeks and becomes a fully functional female; his role as dominant, breeding male is taken over by the next largest male (Fricke, 1979).

2.12 SALAMANDERS AND FROGS (AMPHIBIANS)

Small, four limbed vertebrates with moist skin and aquatic larval stage. Well developed nervous system and senses, especially vision and hearing. Two main groups, the tailed urodeles (newts and salamanders) and tailless anura (frogs and toads). Fertilization external.

The small size and retiring habits of the amphibians, together with their lack of any obvious weapons, have fostered the impression that they do not fight much. However, this impression is false and amphibians (both urodeles and anurans) fight over shelter, food and feeding territories and, especially, over mates (Wells, 1977). Males of species with a short, explosive breeding season (for example, the common toad, Davies and Halliday, 1979) fight directly for access to females. Where the breeding season is longer, males compete for resources needed by the females (bullfrogs defend suitable sites for egg laying, Howard, 1978) or for sites needed for display. Breeding male tree frogs (such as *Hyla cinerea*) fight for calling perches. Other males remain quietly nearby and intercept the approaching females (Perrill, Gerhardt and Daniel, 1978). The existence in the same breeding population of males which use different ways of competing for females is common in amphibians.

2.12.1 Salamanders (urodeles)

Male and female red-backed salamanders defend territories around moist patches on the forest floor. These areas, marked with scents from glands in the chin and cloaca, are usually avoided by other animals (Jaeger and Gergits, 1979). If an intrusion does occur, the resident adopts an upright position, raising its body off the ground. The intruder frequently drops its whole body flat on the ground by extending its legs sideways. An intruder in this position is rarely bitten but if it takes up the same upright posture as the resident, it is attacked. The resident animal usually suceeds in driving off an intruder. Over 11% of adults caught in the wild have scars on their snouts, where most bites are directed, suggesting that fierce fighting is quite common (Jaeger, Kalvarsky and Shimizu, 1982).

2.12.2 Frogs and toads (anurans)

Male gladiator frogs (tropical tree frogs) build and defend small pools in which the females lay their eggs; the nests provide the vulnerable eggs with some protection against predation. Though slightly smaller than the females, the males have sharp, curved scythe-like spines on their thumbs. Territories are defended by a combination of complex

Figure 2.5 Fighting gladiator frogs.

calls and very fierce fights. After an exchange of calls, a resident male rushes, hissing, at an intruder forcibly knocking him away from the nest or grasping him in a bear hug, growling and cutting and jabbing at his eyes and ears with the thumb spines (Fig. 2.5). Males in possession of a nest usually succeed in repelling intruders. These fights can be very damaging; the majority of males, but none of the females, captured in the field had wounds or scars on their heads and necks, and deaths during fights are not uncommon. This dangerous fighting brings its reward, however, since males that fight get more matings than those that do not (Kluge, 1981).

2.13 LIZARDS AND SNAKES (REPTILES)

Large, diurnal vertebrates with scaly, waterproof skin. Fertilization internal and young hatch from waterproof eggs. Occupy a wide ecological and geographical range. Well developed nervous system and sense organs. Three main groups: squamata (lizards and snakes), chelonia (turtles and tortoises) and crocodilia.

Because of their relatively large size, diurnal habits, use of visual cues and slow movements, terrestrial reptiles have proved very suitable subjects for behavioural studies. They have a rich repertoire of agonistic behaviour; fights may be over shelter and food or in defence of young but access to females is the cause of most conflict in this group. Territoriality is quite common but some species have a system of dominance. Reviews of agonistic behaviour in this group can be found in Gans and Tinkle (1977) and Burghart and Rand (1982).

2.13.1 Lizards (squamates)

Many male lizards (especially iguanids) defend territories which contain mature females but others, such as the nocturnal gekkos, defend exclusive shelters within overlapping home ranges or (as in the case of skinks) are not territorial. Agonistic behaviour may vary within a single species but can be quite fierce (Fig. 2.6); at low population densities, breeding male *Uta stansburiana* maintain sharp territorial boundaries but where the population density is high, they establish hierarchies within a communal mating area, dominant males mating with most of the females (Stamps, 1977).

Adult male and female green iguanas (large herbivorous lizards) are strikingly different in appearance (i.e. they show sexual dimorphism). The males are bigger than females with relatively larger heads, well developed muscular protruberances below the jaw, a larger dewlap

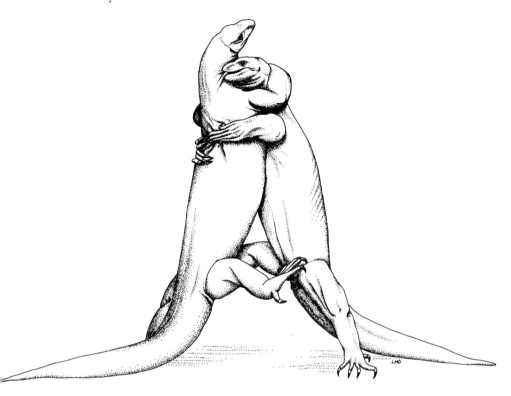

Figure 2.6 Fighting lizards.

(the flap of skin below the throat), sharp scales on the neck and longer spines on the neck and back.

Non-breeding iguanas live in loose aggregations, with overlapping home ranges, but as the breeding season approaches, larger males establish territories. These are defended by displays consisting of dewlap extension and head bobbing and, if the intruder does not retreat, by fights in which opponents circle each other with bodies parallel, hissing and lashing their tails. Eventually a pushing match ensues until one male bites the other on the neck at which the bitten male usually flees (Distel and Veazey, 1982). Several females are usually found in the territory of a large male, with whom they mate. Medium sized males remain at the periphery of these territories, courting any passing females but rarely copulating with them. Small males are tolerated by the large territory holders unless they try to copulate with the females (Stamps, 1977; Dugan and Wiewandt, 1977).

Figure 2.7 Fighting pythons.

2.13.2 Snakes (squamates)

Although fights between snakes are mostly settled without biting, striking and biting at conspecifics have occasionally been reported (Carpenter, Gillingham and Murphy, 1976). Among non-poisonous snakes, male pythons fight by coiling their bodies together, jostling for a superior position (Fig. 2.7), hissing and secreting musk from the cloaca. Both animals raise their sharp pelvic spines and attempt to force these into the delicate skin between their opponent's scales. Intermittent attempts to escape are encountered by bites, but the loser (the lower of the two snakes) is eventually allowed to flee (Barker, Murphy and Smith, 1979).

2.14 BIRDS (AVES)

Generally small vertebrates with light, streamlined bodies, a smooth covering of feathers and paired wings. Well developed brain and particularly sensitive eyes and ears. Fertilization internal and eggs require incubation. Two main groups: primitive birds such as the tinamous and the flightless ratites (such as the ostrich) and the more advanced birds of which the perching birds make up the largest group.

Although birds are mostly similar anatomically, their behaviour is often very diverse, even among closely related species. Territoriality is common in birds; some birds defend territories for roosting, feeding or breeding. The territories may be retained year-round or abandoned after a short time and range in size from a few centimetres around a nest to an area several miles in diameter.

Prolonged parental care, often by both male and female, is usual among birds and direct male competition for females is less intense than in the mammals for example; the emphasis is instead on acquiring a good site for attracting females and for breeding. Because rearing young is so demanding, most birds are monogamous (see Chapter 11, where references are given), often pairing for life. Paired males sometimes mate promiscuously, however, and in some cases males regularly mate with several females (pied flycatchers) or a female may mate with several males (sandpipers).

Some birds such as swallow-tailed manakins, in which males are not necessary to help rear the young, have lekking territories, to which females come only to mate. It is not uncommon among birds for several adults to defend a single territory and rear a common brood, but even in this apparently co-operative system conflict is common (page 314). Reviews of some aspects of conflict behaviour can be found in Oring (1982) and Pulliam and Millikan (1982).

2.14.1 Perching birds (passerines)

Male great tits are territorial, although they will join flocks in winter. There is a dominance hierarchy within the flock with young birds, arriving late after dispersing from their natal site, being subordinates. Males sometimes pair while in a flock and will then guard and roost with their mates. Towards the end of winter, they become more aggressive and set up territories, using song to keep intruders away (Krebs, 1977). Subordinates settle in less preferred habitats to breed. When the population size is large some birds fail to establish a breeding territory (the 'floater population') and attempt to breed within another pair's territory. These 'floater' pairs produce only half the number of fledglings of a territorial pair and also reduce the reproductive rate of the owners of the territory in which they settle (Dhondt and Schille-mans, 1983 and Drent, 1978, cited therein).

In marked contrast are those species in which males aggregate at leks and compete for the attention of females during the breeding season. In the swallow-tailed manakin of South America four to six males communally use a lek area. Within the group of males a dominance hierarchy is established by supplantings, chases, vocal duels and ritualized displays. The hierarchy is stable even between years, positions changing when a member dies. The alpha male is the sentinel of the group, occupying a higher perch than subordinates and calling to attract a female. Males perform co-operative displays when a female arrives but it is the dominant bird that eventually mates (Foster, 1981).

2.14.2 Wading birds (charadriforms)

Flock feeding, both intra- and interspecific, is common amongst birds. Although there are advantages to flock feeding there are also opportunities for conflict including food stealing or kleptoparasitism. This is common in oystercatchers feeding on mussel beds in winter. Aggressive attacks, usually over mussels, consist of a lunge with the bill held horizontally, occasionally with wings flapping. Sometimes a bird stabs its bill at another or even grabs at a wing or tail. Attackers manage to steal the mussel only 21% of the time, however. Some individuals are more aggressive than others, feed in the best areas and are more successful at stealing. Immature birds are less aggressive, less successful and are displaced from the best mussel beds (Vines, 1980; Goss-Custard *et al.*, 1981, 1982, 1984).

2.14.3 Raptors (falconiforms)

It is common for the large raptors to lay two eggs but invariably the older chick kills the younger one (in a so-called Cain and Abel struggle). Gargett (1978) describes this behaviour in the black eagle. The older chick grips the sibling with its feet and repeatedly pecks at it, picking it up, dropping and shaking it, and sometimes pulling out pieces of down (Fig. 2.8). Sometimes the younger chick dies directly from physical attacks, but if it survives a few days starvation also contributes since it is prevented from feeding. The parents make no attempt to stop the aggressive behaviour. This is an extreme case of brood reduction (page 43).

Figure 2.8 Fratricide in eagles.

2.15 MAMMALS

Long-lived tetrapods with a hairy skin. Brain and sense organs (vision, hearing and especially smell) well developed. Capacity for learning well developed. Fertilization internal, young suckled by female. Occupy a very wide range of habitats.

With a few notable exceptions, male mammals show little in the way of parental care. Because of this difference in parental investment, the two sexes generally come into conflict over different things. Females fight mainly over food and feeding territories and in defence of their young, whereas males, which tend to be polygamous, compete intensely for mates. Many mammals live in permanent groups whose members can recognize each other. Individuals form long-term relationships, often along kinship lines, and the resulting social structure can be extremely complex. Mammals fill a wide range of ecological niches and their behaviour is very diverse. A review of social behaviour in mammals can be found in Eisenberg (1981).

2.15.1 Springbok (ungulates)

Mating systems of ungulates vary widely and include the defence of harems (red deer), males sharing a territory with one (roe deer), two (klipspringer) or several (duijker) females, and a lek system (kob). The larger species often have impressive weapons and use striking body patterns as signals (Fig. 2.9). The small species communicate more by olfaction.

Figure 2.9 Weapons and body markings as visual displays in Grant's gazelle (reproduced with permission from Geist, 1978).

In the springbok, large adult males defend territories throughout the year. They do not have scent glands but form conspicuous dungsites on the territory. Before defaecating and urinating territorial males paw the ground and may also stab at the ground and nearby bushes with their horns. Non-territorial males live in bachelor herds and sometimes challenge the territorial males. Fights start by the two rivals facing each other and champing before engaging horns, pushing and wrestling. The tips of the horns are turned inwards and cannot stab the opponent. Females wander through a home range (the area within which an animal concentrates its activity), foraging with their young, sometimes in association with groups of non-territorial males. As they pass through a male's territory he herds and defends them, mating with any that are on heat. Thus a springbok population consists of a number of different social units, adult females and young, territory holders and harems and bachelor herds (David, 1978).

2.15.2 Ground squirrels and mole-rats (rodents)

In certain circumstances, males of a number of rodent species are able to induce abortion in pregnant females that they themselves have not fertilized and may also kill dependent young. These two patterns of behaviour reduce the time the males need wait before mating with the female and siring their own offspring (for example Vom Saal and Howard, 1982). In Belding's ground squirrels (*Spermophilus beldingi*), yearling males disperse over a wide area and kill and eat juveniles that they find unattended, probably benefitting from the source of protein-rich food. Adult females kill entire litters, usually following loss of their own young and subsequent emigration from the former nest site to a safer area. Having removed the local competitors, they then settle near to the victims' burrow and raise a litter of their own the following season (Sherman, 1981).

Naked mole-rats are small, hairless and effectively blind rodents which live in East Africa. They inhabit extensive burrow systems, in colonies averaging 75 individuals (Brett, 1986); of these, only one large female, and 1–3 large males breed. The queen can produce more than 60 pups a year (Jarvis, in press); these are cared for primarily by the breeding males and the queen. The queen also actively patrols the colony, behaving aggressively towards her companions. Colony members change the work they do as they grow larger and older, although some individuals may also specialize in what they do (Lacey and Sherman, in press). The youngest and smallest animals carry food and nest material to the nest and tidy up the burrows. Larger, older individuals dig burrows and defend the colony from snakes (their main predator) and from intruders from other colonies, which are fiercely attacked and injured severely. During an intrusion, the queen nudges workers toward the threatened burrows to face the enemy. If the queen dies or is removed, several of the largest remaining females come into breeding condition; suppression of breeding may depend on urine-borne scents in mole-rats as in other rodents (Jarvis, in press). These females fight amongst themselves, sometimes to the death, until one emerges as the new queen; in contrast, males do not fight for breeding opportunities (Lacey and Sherman, in press). Thus the social organization of mole-rats is comparable in many ways to that of the social insects (page 229), involving reproductive division of labour, communal foraging and colony defence and a high level of within- and between-colony conflict (Jarvis 1981; Lacey and Sherman, in press).

2.15.3 Weasels, stoats and lions (carnivores)

In most mustelids (such as weasels and stoats) males expel rivals from a territory which overlaps the home range of several females. The territory of a male stoat contains discrete defaecation and urination sites. An anal secretion is used to mark both these sites and novel objects in the territory and also to mark over the scent of conspecifics. Other scents are used during direct encounters with rivals. Intruders to a territory usually avoid the owner; should an encounter take place, the owner increases his rate of scent-marking and the intruder generally retreats. Direct encounters are rare and physical aggression rarer still (Erlinge, Sandell and Brinke, 1982).

Several species of carnivores live in groups, hunt communally and defend a joint territory. At low densities, lions have a territory within a larger home range, at high densities a territory only. The pride consists of several related females with their young and one or a few males. As subadult males mature they are driven from the pride by the adults and eventually try to takeover another one from the resident males. If successful, the new males kill any dependent young, bringing the females into oestrus (Bertram, 1975; Elliot and Cowan, 1978).

2.15.4 Whales (cetaceans)

Most whales and dolphins live in small, family groups, sometimes combining into a large feeding herd; sperm whales are polygynous and, in captivity at least, dolphins form dominance hierarchies. Humpback whales live in small groups usually with a central animal, probably a female, a principal escort and secondary escorts, probably males. The secondary escorts sometimes fight and occasionally chase off the principal escort. During such an encounter the opponents charge and lunge at each other, sometimes with the mouth open. The tail is thrashed and slammed into the opponent and escorts often have scarred fins. Singing whales are lone whales; when they join a group they stop singing and when an escort leaves a group it starts singing but the full significance of this behaviour is not clear (Tyack and Whitehead, 1983).

2.15.5 Monkeys and apes (primates)

Primate social organization varies widely from a solitary lifestyle (for example, orangutan) or monogamous groups (for example, gibbons) to permanent, complex, multi-male groups. Green monkeys in Senegal live in groups of females with their young and several males, ranging over an area that overlaps with those of other groups and that has flexible boundaries. Groups sometimes meet each other and this usually results in the adult males of the groups displaying briefly by means of stereotyped and noisy leaping and crashing through the canopy or by showing off their white chests, canines and penises. One or both groups may then retreat or a resource such as water may be shared. Where the resource in question is critical, encounters between groups are longer and fiercer, but with no physical contact. The coveted resource may eventually be divided between the groups. The nature of the interaction between groups is variable; in some cases, the home ranges of neighbouring groups overlap considerably, in others rigid territorial boundaries are defended (Harrison, 1983).

In Old World monkeys and apes, the majority of agonistic encounters take place within the group and commonly result in the establishment of a dominance hierarchy. In a troop of rhesus monkeys, both males and females have a dominance structure, with some high status animals being deferred to by all the group and, at the other extreme, low status animals deferring to all others. The nature of this hierarchy is complex, however, since two or more animals (relatives or animals of adjacent rank) often co-operate to defeat an opponent of higher status (Packer, 1977; Datta, 1983; Bernstein and Ehardt, 1985). These complications make the outcome of aggressive encounters between two particular individuals less than predictable, since even the top ranking male may be forced to retreat in certain circumstances.

Since humans are primates, studies of non-human primates can shed light on the evolution of our own behaviour patterns, but this is discussed in the final chapter of this book.

3

Issues and concepts in the study of animal conflict

The main impression given by the survey of animal conflict in the preceding chapter is one of bewildering variety. In the present chapter we make some general points using examples from throughout the animal kingdom. The literature on this topic uses a number of special terms and concepts: threat and submission, ritualization and escalation, dominance and territoriality. This chapter has the additional aim of explaining what these mean.

3.1 HOW DO ANIMALS FIGHT?

3.1.1 Techniques

Animals employ a range of techniques to overpower other individuals whose interests conflict with their own. At one extreme we have the indiscriminate overgrowth and undercutting by which sessile invertebrates destroy or dislodge rivals. In contrast, fights between more specialized invertebrates and between vertebrates involve an exchange of discrete, directed behaviour patterns. These agonistic actions vary enormously from species to species. Can we identify any rules which determine the form they take?

At a very obvious level, whether and how animals fight depends on the physical equipment they have at their disposal. Complex, directed behaviour of any sort requires well developed sense organs and the ability to use sensory information to find out about an opponent. It also requires an advanced system of nerves and muscles to produce rapid, co-ordinated and finely controlled movements. This combination of features is relatively uncommon in the lower invertebrates. Even among more complex animals, physical characteristics restrict the range of possible agonistic responses. Fish, which lack any jointed appendages, can only fight with variations on the theme of undulating body movements, front-on postures or bites; snakes, almost without any appendages at all, have to use folds of their bodies to overpower opponents.

What happens during fights also depends on an animal's life style. Animals which live in dense forests or murky water are likely to use different kinds of agonistic display from those which live in open country; visual signals are likely to be replaced by auditory or even electrical cues (page 67). The fact that air is not a dense medium places a number of constraints on the behaviour of birds. Their ability to develop specialized weapons (page 253) is limited by their need to remain light and streamlined. For energetic reasons, most social interactions take place when the bird is not airborne; this means that agonistic encounters between birds are often punctuated by short flights between perches. This is in contrast to the continuous nature of fights between fish, for example, which live in a much denser medium.

3.1.2 Overt fighting and displays

Although some animals (such as spiders and mites) may launch straight into a vigorous attack when they meet a rival, most agonistic encounters start with actions that do not involve physical contact. Stomatopods spread their claws and coil before striking and lizards head bob and circle before biting. One animal may retreat before any overt attack occurs; an iguana may retreat after a pushing match or a bird may leave an area in response to the song of a rival. When an agonistic encounter is resolved without overt fighting, it is said to be *ritualized*. This same word is also used in a different (but related) sense to refer to the constellation of changes that take place during the evolution of displays making them more effective (page 262).

These low-key agonistic actions often increase the apparent size of the animals concerned (gill cover raising in fish; body flattening and dewlap extension in lizards) and/or expose their weapons (claw spreading in crustacea; open mouth postures in iguanas and many mammals). When physical contact does occur in the non-injurious

Figure 3.1 Body postures in dominant and submissive wolves.

part of a fight, this often involves pushing, pulling or tail beating contests. Such interactions potentially allow opponents to assess their relative strength and one often gives up at this point, as if it has discovered that it is the weaker animal. The behaviour of the animal once it withdraws from a fight tends to reduce its apparent size (salamanders drop flat onto the ground) and conceal its weapons (crabs fold their claws away).

Behaviour patterns which, without any direct physical coercion, alter what another animal does are called *displays*. The collection of displays which accentuate size and weapons and elicit withdrawal are called *threat displays*. Those which reduce apparent size, conceal weapons and inhibit attack are called *submissive displays*. The contrast between the behaviour of winner and loser is seen particularly clearly in wolves (Fig. 3.1), in which the winner of a fight stands tall with its fur fluffed up, raises its tail and bares its teeth; the loser crouches down, with flattened fur, retracted ears, lowered tail and covered teeth.

It is a very common feature of animal fights that they start off with low key, energetically cheap movements (such as head bobbing lateral posturing or claw displays) that are performed at a distance and are unlikely to lead to overt attack. Then more energetic behaviour, which involves physical contact but does not inflict injury, occurs (for example, beating with the tail, mouth wrestling or pushing). Finally, a stage may be reached in which energetic and dangerous actions (such as biting, kicking or striking) occur. This progressive increase in the intensity of a fight is described as *escalation*. Conflicts can be resolved, by one participant retreating, at any point in the escalation sequence, so that full-scale physical attack is not a necessary component of animal fights.

3.1.3 Weapons

Many animals have at their disposal strong, hard structures used to exert force against their environment (page 253). These may be used in locomotion (claws for example), in digging (the proboscis of the ragworm) or for egg laying (the ovipositor of insects). However, many such structures are used to tear or break living tissue, perhaps during feeding (chelicerae in spiders; stylets in mites; claws (chelae) in crabs; mandibles in insects; teeth in reptiles and mammals and beaks in birds). Often, these strong, hard structures are additionally modified for use as weapons in fights against predators or rivals, witness the long sharp teeth of baboons, enlarged claws of male crabs, the huge modified mouthparts of soldier termites, the sharp claws of lizards and the stings of social hymenoptera. The weapons are often made more formidable by the development of irritant or toxic chemicals; the sting of the honey bee provides an example. In other cases, weapons are specialized structures used only in agonistic contexts; thus we find the horns of beetles and ungulates, and the blade-like thumb of gladiator frogs. Weapons are used in fights both to maintain contact between the opponents and to push, batter or gore an opponent.

A variety of objects is used during animal conflicts: ants (*Conomyrma bicolor*) drop stones into the nests of their rivals (Moglich and Alpert, 1979); some crabs (*Ilyoplax pusillus*) defending burrows on sand flats build barricades to deter rivals (Wada, 1984), others fix anemones to their claws and extend these towards their opponent during threats (Duerden, 1905); chimpanzees wield sticks, shake branches and throw stones – one particular male even learned to intimidate rivals by banging petrol cans together (Van Lawick-Goodall, 1971).

3.1.4 Ritualized fighting and killing conspecifics

These weapons are often used in different ways, depending on whether the enemy is a predator or a rival; red kangaroos kick at predators with their strong, clawed hind legs but use these in intra-specific fights only in passive defence, as shields (Croft, 1981). This fact, together with the rich repertoire of threat and submissive displays seen in many species, suggests that fights between animals of the same species are settled without injury, taking the form of harmless, ritualized trials of strength (Eibl-Eibesfeldt, 1961; Lorenz, 1966). However, fights between conspecifics do often escalate, (Fig. 3.2) weapons are used and used effectively with the result that injury and even death are not at all uncommon (Table 3.1). In what circumstances does this happen?

Figure 3.2 (a) Threat and (b) escalated fighting in robins.

1. *Fights over vitally important resources.* Much intraspecific fighting occurs over limiting resources such as food or mates (Chapter 11). As one might expect, fierce, damaging fights are most likely to occur when the resource in question is very important for the fitness of the animals concerned. All-out attacks with 'tooth and claw' are relatively common when animals are fighting over mates, as when a bowl-and-doily spider launches immediately into a grappling fight with an intruder. Since rapid growth at an early age can give an animal a life-long advantage over its contemporaries (see Chapter 8), fights between young animals over food are often fierce and can be fatal (page 193).

2. *Infanticide.* Parents sometime kill or fatally neglect their own offspring if, for example, the whole brood or litter cannot be reared successfully (Elwood, 1980; Sherman, 1981). The young of other individuals may be killed to eliminate potential competition. For example, dominant Cape hunting dogs kill the pups of their subordinates, increasing the food supply for their own pups (Van Lawick, 1974). Young animals may also be killed to bring their parents into breeding condition. For example, male lions, langurs and ground

Table 3.1 Examples of injury and killing in animal fights.

Species	Observation	Reference
Bacteria	Release poison, killing some members of their own species.	Chao and Levin (1981); see page 14
Slugs	Conspecifics are injured (and eaten) when fighting for shelter.	Rollo and Wellington (1979); see page 18
Ants	Large numbers of ants are killed in battles over feeding territories.	Hölldobler (1979); see page 22
	New queens kill queens and workers during slave raids by biting and stinging.	Buschinge, Winter and Faber (1983)
Polistine wasps	Attempted nest take-overs are common and involve biting, grappling fights and stinging. Corpses often found under nests.	Little (1979)
Fig wasps	78% of males in a single fig were injured, many seriously.	Hamilton (1979)
Spider crabs	Legs get broken off in fights over symbiotic anemones. In the wild, crabs of both sexes have missing legs and claws.	Wirtz and Diesel (1983)
Fiddler crabs	25% of males have signs of injury probably inflicted by conspecifics.	Jones (1980)
Mantis shrimps	Species nesting high on shoreline where shelters are rare and essential fight fiercely and often fatally.	Dingle (1983)
Mites	Males can injure each other seriously by piercing cuticle during fights for females.	Potter (1981); Dinnock (1983); see page 25
Spiders	Long fights, sometimes to the death, between males over females.	Austad 1983; see page 25
Blind gobies	Fierce, injurious attacks on conspecifics approaching shelter.	MacGintie (1939)
Red-backed salamanders	11% of animals in the wild have signs of injury to olfactory apparatus, probably sustained in fights over shelter.	Jaeger et al. (1982); see page 29

Table 3.1 (*contd.*)

Species	Observation	Reference
Gladiator frogs	Extremely fierce, injurious fights over females; males often wounded or killed.	Kluge (1981); see page 29
Skinks	Males engage in violent, injurious fights.	Duvall *et al.* (1980)
Grey heron	Immature male killed by adult during prolonged territorial encounter.	Richner (1985)
Meadow voles	82% of males and 57% of females in the wild have scars.	Rose (1979)
Musk oxen	5–10% of males are killed per year during the rut.	Wilkinson and Shank (1976)
Mountain sheep	Many males have badly damaged nose and head, including broken bones. Rams have been found dead or dying from wounds gained during the rut.	Geist (1971)
Cervids	84% of hides of male cervids of more than 1.5 years have scars from fights. Up to 225 scars per individual and 20% longer than 100 mm.	Geist (1986)
	23% of stags from five years old upwards show signs of injury (blinding, leg and antler breakage); in 6% this is permanent.	Clutton-Brock *et al.* (1982)
Cheetah	Males often injured and sometimes killed in territorial encounters.	Caro and Collins (1986)
Chimpanzees	Males and females of a small group chased and killed by groups of adult males from a larger group following troop fission.	Goodall *et al.* (1979)
Seals	Males engage in long, damaging fights over females. Approximately 65% of seals identified as male in a single haul-out site have fresh neck wounds (some severe) probably inflicted during fights in the water.	Le Boeuf (1977); Paul Thompson (unpublished data)

squirrels and female jacanas destroy dependent young when they enter a new group or territory (Hrdy, 1979; Stephens, 1982).

3. *Cannibalism*. Eating conspecifics is a common but diverse form of behaviour occurring in many herbivores and granivores as well as in carnivores (page 309). While some cannibalism is just scavenging, there are a number of species which regularly kill and eat members of their own species, especially when food requirements are high. Adult social wasps will eat pupae (a safe and nutrient-rich food supply) from their own colony in times of food shortage (Wilson, 1971). The heavier a female mantid is, the more eggs she produces, so by allowing himself to be eaten the male increases his own reproductive output as well as that of the female (Buskirk, Frolich and Ross, 1984). In the toad *Scaphiopus* large tadpoles specialize in cannibalism on their smaller companions, thus gaining the food they need to mature before their pool disappears (Polis, 1981).

4. *Accidental killing*. Sometimes during the course of fights between competitors, individuals other than the participants get injured or killed (page 306). Thus female toads, dungflies and fleas may be killed in the middle of a mass of males which are fighting over them, or as a result of incessant copulations; young elephant seals may be crushed by large bulls fighting over females.

Thus in spite of the widely accepted picture of animal aggression as a harmless exchange of signals, fierce fighting, injury and killing are quite common features of conflict among members of the same species.

3.2 CONTEXT-SPECIFIC AGONISTIC BEHAVIOUR

One reason why overt fighting is not as common as it might be is that animals often adapt the form and intensity of their agonistic responses to the context in which an encounter occurs. Where the modification depends on the identity or properties of the opponent the result is a system of dominance; where the critical factor is the place in which an encounter occurs, territoriality is observed.

3.2.1 Dominance and subordination

Some fights are resolved quickly, by one animal withdrawing at an early stage in response to nothing more than an approach by the eventual winner. When such behavioural imbalance is a consistent feature of the relationship between two individuals, this is often summarized by the terms *dominance* and *subordination*. Although the behaviour of the persistent loser (the subordinate animal) is as impor-

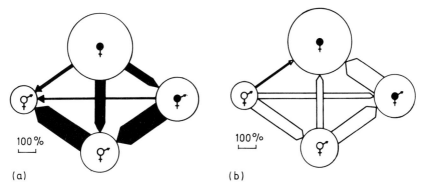

Figure 3.3 Dominance structure in a group of anemonefish. (a) direction of aggressive acts (b) direction of submissive acts. ♀ functional breeding female; ♂ functional breeding male; ♂ non-breeding male. ○ Width of circle represents body size. Width of arrow represents percentage of responses directed at each individual. (Reproduced with permission from Fricke, 1979.)

tant as that of the winner (the dominant), the phenomenon is usually simply called dominance. In group-living animals, pairwise encounters occur within a nexus of social interactions. In some cases the members of a group can be arranged in terms of who bosses and who defers to whom, to produce a linear ranking system which summarizes the pattern of agonistic relationships within the group (Fig. 3.3(a)).

This kind of situation, with predictable dominance–subordinance relationships in pairs and linear hierarchies in groups, is likely to occur when the outcome of an encounter between two individuals depends only on their relative fighting ability. If other factors in addition to fighting ability are important, simple dominance relationships are less common. For example, in crustacea, a moulting animal may give way to one which it normally dominates, a female primate in oestrus may dominate an individual to which she usually defers and a primate holding a baby may gain in status. In primate groups, alliances may be formed, especially between relatives, and these have a major influence on the outcome of interactions between two group members (Datta, 1983; Bernstein and Ehardt 1985). Such complicating factors mean that dyadic relationships are not directly predictable from fighting ability and, in a group context, that the pattern of agonistic interactions in a group cannot be summarized in a single, linear rank order.

The concept of dominance takes on additional value if other behavioural differences can be predicted reliably from a knowledge of the dominance status of two animals. For example, in anemone fish the direction of submissive postures is the converse of the ranking based on attack direction (Figure 3.3(b)). In primate social inter-actions, the direction of attention and greeting between two animals is predominantly from the subordinate to the dominant. The dominant member of a pair often gets priority of access to favoured objects or limiting resources; dominant crabs get shelters, dominant primates may get more food and water, at least when these are in short supply, while dominant seals and mice (but not chimpanzees) get copulations (page 287).

The favoured resources may not be in dispute at the time dominance relations are first established, status being competed for as it were in anticipation of conflict over resources. Dominant female primates only get more food on the relatively infrequent occasions when this is scarce. Tom cats compete for status in the absence of receptive females but dominants get most matings when a female does come into heat. Dominance does not always give priority of access to all valued resources and, consequently, there may be different ranking systems for different resources. The order in which male mice drink is not the same as the ranking for access to females and female primates may dominate males in arguments over sleeping sites but not over food. Dominance is not a permanent, immutable feature of an animal (like the size of its feet) but just one aspect of its relations with particular companions and in particular circumstances.

3.2.2 The behavioural mechanisms underlying dominance interactions

For a dominance–subordinance relationship to exist, rivals must assess their relative fighting ability prior to or early on in a fight, and submit straight away if this assessment is not in their favour. Table 3.2 shows some mechanisms by which this might be achieved. Mice and crickets which have previously lost several fights tend to give way to all rivals; this effectively provides a mechanism for an animal to *assess its own fighting ability relative to the population average*.

Decisions about whether to engage in a fight may be based on an *assessment of the opponent's fighting ability*. The most obvious cues an animal could use to assess an opponent's fighting ability would be physical attributes such as size or the possession of weapons. Where opponents are of different size, the larger is usually consistently vic-torious and therefore dominant over the other. Dominant cockerels and mollies are larger than subordinate ones. The regular dominance of one species, sex or age category by another probably results from this sort of system. If frequent defeats modify behaviour, assessment of an opponent's fighting ability could be on the basis of the way it behaves.

Table 3.2 Mechanisms giving rise to dominance–subordinance relationships.

---

1. Assessment of own fighting ability in relation to population average

2. Assessment of opponent's relative fighting ability from its
    Physical attributes, e.g. size
    Behavioural attributes, e.g. 'winner's' posture
    Individual identity

Crabs, crickets and mice which have won fights all get fiercer, walk tall and approach a rival more readily. If would-be opponents can detect this they submit without a fight and so a dominance–subordinance relationship is formed.

Finally, animals that have the ability to recognize each other individually may use *experience in encounters with specific individuals* to gain a direct measure of their relative fighting abilities. For example, in Indian pythons, the usual sequence of events when two unfamiliar males encounter each other is an exchange of agonistic acts, by the end of which one animal has clearly overpowered its opponent, which retreats and shows submissive postures. In subsequent encounters between the same two males, the same animal wins but the fights become progressively shorter until mere approach by the previous winner makes the other snake retreat; a dominance–subordinance relationship has thus been established (Barker *et al.*, 1979).

Thus there are a number of different behavioural mechanisms which can produce a dominance–subordinance relationship between two animals or a dominance ranking in a group. So even if the patterns of agonistic interactions within groups of cockroaches and among the members of a rhesus monkey troop, for example, can both be summarized in terms of dominance status, this does not necessarily mean that the hierarchy arises in the same way.

3.2.3 Territoriality

Another aspect of context which determines whether or not an animal will engage in a fight is the place in which an encounter occurs, and a common manifestation of this site-dependent agonistic behaviour is territoriality. Breeding male sticklebacks restrict their activity to particular areas and chase any intruding males up to the edge of the area, using a combination of display and overt attack (Wootton, 1976). This is the classic territory, a fixed area from which other animals are expelled by means of some sort of agonistic behaviour.

Territorial behaviour has four separate components:

1. *Site attachment.* Restriction of aggression to a specific area, the geographical component, differentiates territorial behaviour from the defence of personal space or of a moving resource, as when a red deer stag drives rivals away from a harem of females or a gull defends a group of foraging eiders from rival kleptoparasites.

2. *Exclusive use of the area.* The situation for sticklebacks is quite distinct from that of female stoats, for example, which show site attachment but whose home ranges overlap.

3. *Agonistic behaviour.* An animal may have exclusive use of a particular area for a number of reasons, the most trivial being simply low population densities. Groups of coatis occupy exclusive home ranges by mutual avoidance; all the groups benefit from having an undisturbed feeding area and when they encounter each other they exchange a few friendly signals and then separate (Kaufman, 1962). This is quite different from the aggressive response of a resident stickleback to the presence of another fish and deployment of agonistic behaviour is a diagnostic feature of territoriality. Scent marks are often used to discourage other animals from entering a particular area; in stoats, for example, a stranger almost always withdraws from another's territory so that rivals rarely encounter each other. Here territoriality grades into a system of mutual avoidance.

4. *Attack changes to retreat at the territory boundary.* An important aspect of territoriality is that the aggressive behaviour of a territorial animal is restricted to the confines of its own patch. Male sticklebacks chase and bite intruders inside their territorial boundary but leave them alone or even flee from them once the boundary has been crossed (Fig. 3.4). Within the territorial boundary, the owner's behaviour is not uniform, however. There is commonly a focal point or core area of a territory containing a key resource such as a nest, a shelter or a clumped food supply which is defended intensely. The male dragonfly *Leucorrhina intacta* defends his perch very fiercely and sticklebacks persistently attack any intruder that comes near their nest (Wootton, 1976; Wolf and Watts 1984).

3.2.4 Who gets excluded?

Not all intruders pose a threat; those that are likely to be the fiercest competitors are those most likely to be attacked by the territory owner. Among potential conspecific intruders, an established territory owner represents less of a threat than an animal without a territory which is looking for a take-over. Thus in damselfish and seals, to give just two examples, familiar neighbours are attacked less

Figure 3.4 The dependence of agonistic behaviour on territory in stickle-backs (after Tinbergen, 1953).

fiercely than are strange rivals. In some damselflies other breeding males are rivals for the spawning site and are attacked. In many species, this differential aggression may lead to a large territory containing within it the smaller home ranges or territories of one or more members of the opposite sex. For example, jacana females defend territories within which a number of males have smaller territories, while the converse situation exists in stoats, with the large territory of a male overlapping with the smaller ranges of a number of females.

3.2.5 Types of territory

Some sort of territorial behaviour is clearly a very common phenomenon but, equally clearly, the form that it takes is extremely variable. People have tried to classify territories into types. In the first place, territories can be distinguished on the basis of the resource to which an owner gets access, which may be food (hummingbirds), shelter (stomatopods), mates (ruffs, iguanas, bowl and doily spiders), or a nest site or something else needed to rear young. Of course, many territories provide several or even all of these resources (owl limpets, damselfish).

A number of other features derive from this difference in the critical defended resource, including the size of the territory. Thus damselfish defend their whole home range as a multipurpose territory but territorial slugs defend just their shelter, feeding in overlapping home ranges. Equally, territories may differ in the length of time for which they are maintained, from lek territories defended for a few hours per day just during mating, to territories defended for a particular season (the breeding season in great tits, feeding territories during migration in hummingbirds) and those defended all the year round (damselfish). Territories may also be defended by a single individual (a stoat, a dragonfly, a hummingbird), a mated pair (anemonefish, gibbons) or a group of animals; in the latter case we may find a family territory (gibbons) or a territory defended by a larger group which may represent an extended family (Cape hunting dogs, jackals and ants) but may contain unrelated animals (anemonefish).

3.2.6 Are territory and dominance different?

So far dominance hierarchy and territory seem quite distinct. In one, the direction of agonistic interactions between freely mixing individuals is determined by permanent (or semi-permanent) properties of the animals concerned. In the other, rivals are aggressively excluded from a specific area. However, it is not easy to make a clear distinction between these two phenomena when the full range of agonistic behaviour is considered. A number of intermediate cases will be discussed here, to illustrate the complexity of the situation.

Within a single species some animals may show dominance while others display territoriality. Territoriality can grade into a dominance structure as population density increases and maintaining an exclusive area becomes too costly (Chapter 11); the good fighters who are able to defend territories become the dominant animals in a hierarchy. Rather than being either territorial or having a dominance system, animals such as cockroaches switch between the two states as a response to population density.

The same individuals may be simultaneously part of a dominance and a territorial system. In territories defended by groups of animals, a hierarchy may exist within the group (anemonefish in dense areas). Where a number of small, contiguous territories exist, and particularly in lek systems, territory size and quality may vary. Older or larger dominant animals may occupy central territories, for example.

Social structures can be found which have some, but not all, features of territoriality together with some but not all the features of a dominance system. Thus, it is possible to have site-related dominance without exclusive use of an area. Steller's jays feed as mobile groups with a dominance structure, but the status of each animal decreases with distance from its nest site; each animal has a series of zones of decreasing dominance centred on its nest site (Brown, 1963).

Because of all these complicating factors, deciding where dominance ends and territoriality begins is very much a matter of personal preference. This is why there are so many different definitions of these two concepts, and why we have tried to describe the behavioural phenomena rather than produce precise and clearcut definitions. A whole range of mechanisms exists by which good fighters get priority of access to important resources. Which system is found in any particular circumstances depends on the costs and benefits of the behaviour concerned; these are discussed in more detail in Chapter 11.

3.3 ALTERNATIVE WAYS OF WINNING

We have seen that in species such as iguanas, some strong, fierce individuals aggressively deny others access to limiting resources. Under these circumstances the excluded animals may use various tactics to gain a share of the sequestered resources. Animals without a territory may remain inconspicuously on the territory of a more successful one as a tolerated or even a valued satellite. Damselflies ignore rivals in their territory so long as they remain still and inconspicuous. Juvenile iguanas are allowed on the territory of breeding males so long as they do not copulate with the females. Satellite ruffs help to attract females and satellite pied wagtails help to defend the territory on which they live. Alternatively, non-territorial animals may spend most of their time on the edge of a territory, exploiting its resources unobserved. Subordinate hummingbird species feed in the territories of dominant species by sneaking in at dawn and dusk; some non-territorial male frogs and toads intercept females as they move into the territory of a calling male.

The fact that individuals of the same species do not necessarily compete in the same way does not mean that we cannot make any generalizations about their behaviour. What is typical of the species is the set of rules that determines how the animal deploys its repertoire of agonistic responses. The more complicated the rules, the less useful it is to think of a given species as having a typical pattern of agonistic behaviour. This situation applies particularly to longer lived, social animals with large brains and the capacity for learning, such as primates in general and humans in particular.

PART TWO
The causes of agonistic behaviour

From observations of animal conflict it is clear that some circumstances are more conductive to fighting than others, that animals do not always fight when given the opportunity to do so and that individuals vary in the readiness with which they resort to fighting. Such differences in agonistic behaviour often reflect an animal's chances of winning or getting injured and the importance of the issue at stake. This section looks at the mechanisms that underlie these behavioural phenomena and so allow animals to make strategic decisions during agonistic encounters. Because the mechanisms involved are likely to be different in different kinds of animals, we consider studies of the causes of conflict in a number of species and then use the comparative method to see if we can draw any general conclusions. Research into causal factors has concentrated on one particular way of resolving conflicts of interest, namely the use of physical force, or aggressive behaviour (see above), a bias which is reflected in the examples discussed in this section.

The ultimate basis of controlled movement is contraction of muscles stimulated by nerves, so it seems obvious that explanations of behaviour should take the form of statements about the activity of nerve cells. However, the extreme complexity of the nervous system makes it difficult to relate aggressive behaviour to underlying neural activity. Some biologists have therefore looked elsewhere for causal explanations of fighting, seeking, for example, to understand the occurrence of aggressive behaviour in terms of the action of hormones.

In such studies of nerve activity and hormone levels, direct attempts are made to discover the physical bases of the mechanisms that control aggression. It is possible to approach the problem of what causes animals to fight indirectly by trying to work out the properties of these mechanisms from the overt behaviour they generate. If the behaviour of intact animals is studied carefully, logical deductions can be made about the systems that control it, without any reference to underlying physiology. Of course, the mechanisms involved must ultimately have physiological reality, but this may be highly complex and its exact nature need not be known in order to understand the resulting behaviour.

These three approaches are not mutually exclusive; instead they seek to explain at different levels and all have essential and complementary roles to play in our attempts to understand agonistic behaviour. Therefore in this section on the causes of agonistic behaviour we consider all of them; Chapter 4 looks at studies which rely on behavioural observation and experiment in the absence of any physiological manipulation. The hormonal and neurophysiological bases of fighting behaviour are discussed in Chapters 5 and 6.

4

Behavioural mechanisms

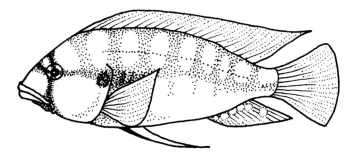

4.1 INTRODUCTION

As it goes about its business, an animal is exposed to many stimuli
from its environment, some of which may advertise the presence and
nature of a potential rival. These stimuli influence the way the animal
behaves, causing it to initiate an agonistic exchange when appropriate
and modifying what it does once a fight has started. However, an
animal does not always respond identically even when it encounters
the same opponent in the same circumstances; clearly factors within
the animal itself also influence its behaviour. These two sources of
control, the stimuli an animal receives from its environment and
factors within its own body, together determine the course of events
during an agonistic encounter. This chapter considers behavioural
studies of how this dual system works.

1. What systems within the animal control the expression of agonistic
   behaviour?
2. How do environmental stimuli influence the state of these systems?
3. How do internal and external factors interact to produce the
   behavioural shifts that occur during a fight?
4. Is it possible to produce coherent explanations (or models) of when
   and how animals fight?

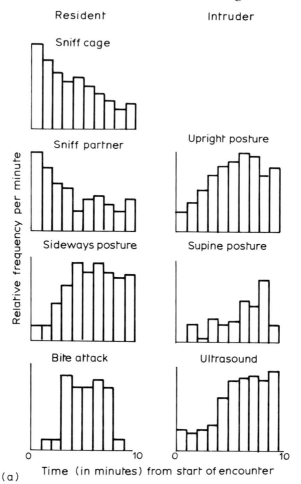

Figure 4.1 The organization of agonistic behaviour in male laboratory rats (simplified from Lehman and Adams, 1977). (a) Changes in frequency of seven behaviour patterns performed by resident and intruder in the first ten minutes after the start of the encounter. Note none of these acts occupies much time. (b) A summary of the systems involved.

4.2 THE SYSTEMS THAT CONTROL AGONISTIC BEHAVIOUR

When a male rat meets the male owner of a territory in which he is intruding, the two animals launch into a complex social interaction involving many different behaviour patterns. Figure 4.1(a) shows the frequency of some of these actions at various stages of the encounter.

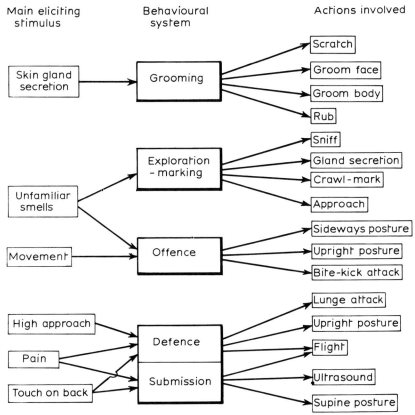

| Main eliciting stimulus | Behavioural system | Actions involved |

(b)

During the first few minutes, sniffing by the resident rat (at both the cage and the opponent) decreases in frequency, while biting, kicking and offensive sideways attack increase. Over the same period, three behaviour patterns increase in frequency in the intruding animal, namely upright posture, ultrasound production and lying in a supine posture.

The simplest explanation for this pattern of changes is that three separate systems exist, controlling exploration and marking, overtly aggressive acts and defensive responses. More detailed analysis of frequencies and sequences of actions during the encounter support this distinction and identify an additional group of causally related actions, namely grooming (Lehman and Adams, 1977). At a finer level of analysis, different agonistic actions are performed in different cir-cumstances and come into play at different stages in the fight, so these systems must be quite complex.

Putting this together (Fig. 4.1b), four independent causal systems underlie the behaviour shown during an encounter between two male rats: an *exploration and marking* system (activated particularly by unfamiliar scents), a *grooming* system (activated by secretions on the skin surface), an *offensive* system (initiated by the sight, sound but especially the scent of an intruder) and finally a system controlling *defence and submission* (elicited by tactile stimulation of the snout and back). Defence is shown in response to an upright approach by an opponent; submission (freezing flat on the ground, accompanied by ultrasonic vocalizations) occurs when flight is impossible or inappropriate.

Thus by looking at the behaviour of undisturbed rats, without any reference to their physiology, this study has produced a scheme which can be used to explain what happens during fights. In particular, it has identified independent systems controlling the behaviour patterns used during overt aggression or offence and those used in defence or submission. During the early stages of a fight, aggression on the part of the resident and submission/defence in the intruder both increase.

To show that this kind of organization is not special to rats or to animals in confined conditions, selected examples of similar studies on a range of other animals, carried out in the laboratory and in the field, are summarized in Table 4.1. The animals concerned live in different social systems and have their own special repertoire of agonistic responses. All the same, a number of similarities in the way agonistic behaviour is organized emerge from this comparison. In each case a system has been identified which controls agonistic behaviour and which is distinct from the mechanisms underlying other things animals do in a social encounter. This agonistic system consists of two separate entities, aggression (or offence) and fear (which features as flight, defence or submission). In social mammals at least, additional processes controlling exploration, grooming and affiliation come into play during agonistic encounters. In order to understand the behaviour observed when members of these species come into conflict, we need to consider their general affiliative tendencies as well as their aggression and fearfulness.

From the data presented here, it is not clear whether the behavioural regularities used to identify these systems are caused by factors inside the animal or whether they are the result of environmental influences. Broadly similar groups of behaviour emerge when a rat responds to a constant stimulus, such as an unresponsive opponent, so internal factors must be at least partially responsible. Analysis of interactions between male territorial cichlid fish (*Aequidens portalegrensis*) and female intruders suggests that the associations between the agonistic behaviour patterns are not the result of their all being influenced in a similar manner by the same cues from the intruder. Performance of a particular category of action depends primarily on the internal state of

Table 4.1 Studies of the organization of agonistic behaviour in animals.

Species	Circumstances of study	Behavioural systems	Representative behaviour patterns	Reference
Wolf spiders	Males competing for dominance in laboratory colonies	Location	Jerky tapping Extend legs Approach	Apsey (1977)
		Aggression	Chase Follow Vibrate-thrust	
		Escape	Run Retreat Vertical extend	
Poma-centrid fish	Males freely interacting with territorial intruders	Aggression	Parallel swim Mouth push Chase	De Boer (1980)
		Escape	Fast swimming Flight	
		Courtship	Dipping Leading	
		Nesting	Nipping Skimming	
Arctic skuas	Males defending breeding territories in the field	Aggression	Oblique Bend Attack Long call	Andersson (1976)
		Escape	Neck backwards Escape Long call	
		Mating	Staccato call Neck forward Copulation	
		Nesting	Squeaking Nest activity	
Stumptail macaques	Pairs of males in neutral cage in the laboratory	Offence	Hit Bite the back Restrain	Adams (1981)
		Defence/ submission	Crouch Run away Scream	
		Grooming	Groom self Groom other	
		Display	Bouncing Walk towards Patrol	

the fish, although choice of specific actions from within that category may depend on external stimuli (Baerends, 1984).

4.3 THE EFFECTS OF CUES FROM AN OPPONENT

On meeting a potential opponent, an animal needs to know its species, gender, social or reproductive status, fighting ability and motivational state. In a stable social system, individual identity may also be important. Together, these various kinds of information indicate whether another individual is an appropriate target for attack and, if so, how hard and how effectively it is likely to retaliate. All this information is potentially available in the complex collection of cues (sights, sounds, smells and physical contact) that animals send out.

As the following sections show, all of these different kinds of cues (together with less usual ones such as the electrical fields generated by some fish) are used during agonistic encounters by animals of one sort or another. The emphasis placed on the different modalities during agonistic encounters depends on the taxonomic position of the species concerned (most birds have a poorly developed sense of smell), the stage of the fight (encounters between territorial birds switch from auditory to visual cues as the rivals come closer) and on the physical environment in which the animals live (electric communication is used by fish living in murky water).

In any particular species, several modalities are often used during an agonistic encounter. Although some components of the whole con-stellation of stimuli provided by an opponent have stronger effects than others, different cues interact with each other and with the internal state of the receiving animal, to produce both immediate responses and longer-term motivational changes. In addition, the effect of a particular stimulus also depends on the context in which it is experienced. Far from being mechanical responses to a single, simple aspect of a potential opponent, as the older literature suggests (Tin-bergen, 1951), agonistic behaviour is the result of a complex decision-making process. In this way animals respond to the probability that a fight will be won and to the costs and benefits which are likely to accrue in the process (page 277).

The role of external stimuli in the control of agonistic behaviour can be studied by tampering with the animal that receives them. For example, blocking the nasal passages of mice drastically reduces levels of aggression, suggesting that olfactory cues normally cause attack. An alternative approach, which provides more scope for fine-scale analy-sis, is to manipulate the signal itself, either by tampering with a live animal or by using artificial mimics of the normal stimulus in which different aspects can be systematically altered (dummies). Encounters between firemouth cichlid fish escalate quickly when the red opercular spots are obscured, suggesting that these normally inhibit aggression (Radesater and Ferno, 1979).

Dummy sounds, sights, electric fields, smells and tactile cues can all be used to tease apart the role played in a fight by different cues. In a classic experiment, breeding male three-spined sticklebacks responded with attack to dummies bearing a red spot on their undersides but ignored dummies without this feature. This suggests that the colour red acts as a trigger for attack in these fish (Tinbergen, 1951). To keep the record straight, more recent experiments have shown that a red stimulus does not always elicit attack in sticklebacks and that other factors (including size and shape) influence the response (Wootton, 1976).

4.3.1 Auditory cues

As Table 4.2 shows, many different mechanisms are used to generate sounds during fights and an effect of these sounds on agonistic behaviour has been demonstrated in a range of species. In the text we consider in more detail a few examples in which the role of sound is particularly clearly demonstrated.

Adult cichlids (*Cichlasoma centrarchus*, a cryptically coloured fish living in reedy, turbid water) growl when they attack each other. Short sounds with long gaps are associated with low intensity agonistic displays while long, rapidly repeated pulses frequently accompany overt attack. When one fish makes a sound its next act is likely to be aggressive, regardless of what its partner does. Silent actions are frequently matched or imitated by the other fish; the same actions accompanied by sounds usually elicit retreat and fewer encounters escalate when the fish are exposed to recorded growls. Growling apparently indicates an aggressive state in the producer and inhibits aggression in the receiving fish (Schwartz, 1974a,b).

Various lines of evidence indicate that sound is important in agonistic encounters in song birds. In great tits, the peak of song production occurs after pair formation, ruling out mate attraction as a primary function and suggesting that sounds play a role in territorial defence. Birds produce different songs when they patrol their territories or confront rivals (Smith *et al.*, 1978) and as agonistic encounters intensify (Kramer and Lemon, 1983).

To the owner of a territory, the sound of a rival singing elicits various components of the agonistic repertoire, exactly which depending on properties of the song itself, its position and on how familiar it is. White-throated sparrows (Falls, 1969) and blackbirds (Dabelsteen, 1982) reply to recorded song played on their own

Table 4.2   Examples of the role of sounds in agonistic interactions.

Species	Observation	Reference
Spiders	Forelegs and palps drummed on ground during fights, Sound elicits threat.	Rovner, 1967
Crickets	Serrated wing edges rubbed together producing calling song which elicits calling or approach and aggressive song during fights, especially by winner. Inhibits attack and stimulates retreat.	Heiligenberg, 1976; Alexander, 1961; Phillips and Konishi, 1973
Frogs	Territorial males call by vibrating vocal sac. Elicits approach by rivals.	Brzoska, 1982
Gekkos	Territorial males produce vocalizations, including churrs in high intensity fights. Elicit avoidance.	Marcellini, 1978
Rats	Loser of a fight produces ultrasonic vocalizations which inhibit attack.	Lore *et al.*, 1976
Rhesus monkeys	Short pulsed screams given by young when threatened by dominant. Elicit aid from mother.	Gouzoules *et al.*, 1984

territory by singing, approaching, displaying and sometimes by attacking the microphone. In chaffinches, recorded song played inside a male's territory elicits approach and aggressive display, whereas the same song on a neighbouring territory triggers approach and singing (Slater, 1981). In indigo buntings, tape recorded songs of familiar neighbours elicit a mild approach, songs recorded from unfamiliar individuals produce immediate approach, threatening and attack (Emlen, 1971).

Potential intruders are inhibited by the sound of another bird singing. Recorded great tit song broadcast over uncolonized woodland discourages settlement by pairs looking for territories; artificial sounds have no such deterrent effect (Krebs, 1977, see also Fig. 4.2). Territorial male red-winged blackbirds muted before finding a mate generally fail to maintain their territories, suggesting that song is important in this context. However, it cannot be all-important, since males muted later on successfully defend their territories (Peeke, 1972).

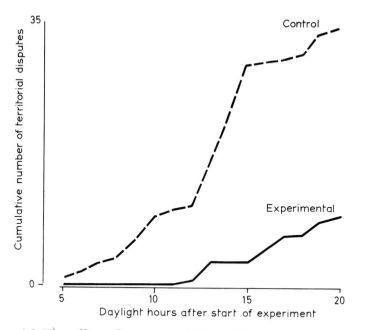

Figure 4.2 The effect of tape recorded sounds on the establishment of territories in great tits. Broadcast sound was great tit song in the experimental group and white noise in the controls (after Krebs, 1977).

Males which have been operated on so that they can only produce unstructured sounds are no less successful in territorial defence than normal birds (Smith, 1976). Song is a first line of defence but can be backed up by other methods, especially visual cues.

Rutting red deer stags are notorious for the roaring sounds they exchange during fights over females. Many encounters do not proceed beyond the roaring stage, so stags may use these sounds to assess the fighting ability of a potential rival (page 284). Recorded roars, especially those of larger, older stags, elicit roaring in response. Stags respond most strongly to roars repeated at a rate of five per minute (the upper end of the natural range); roars repeated at a higher rate intimidate them (Clutton-Brock and Albon, 1979).

4.3.2 Electrical cues

Electrical stimuli serve as distance cues in agonistic encounters in the turbid fresh-water environment in which the South American gymnotid and the African mormyrid fish live. These fish generate rapid, repeated species- and sex-specific electrical discharges by modified muscles or nerves situated near the tail. The pulses, which are detected

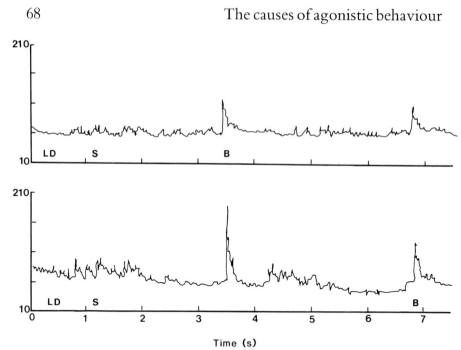

Figure 4.3 Electrical discharge rates during an agonistic exchange between two freely interacting *Gymnotus carapo*, indicated by upper and lower traces. LD = antiparallel lateral display. S = serpenting display. B = bite (after Black–Cleworth, 1970).

by specialized low-resistance sense organs embedded in the skin near the front end of the body, are used to locate objects in the environment, but also play a role in social communication.

Artificially generated electric pulses mimicking the sustained discharge rate of a male gymnotid (*Gymnotus carapo*) elicit attack in male fish; pulses typical of other species tend to be ignored. In the same species, defeated fish stop producing pulses for several minutes and during this silent period they are rarely attacked. A sharp increase in pulse rate occurs before or during biting attacks in both participants (Fig. 4.3) and the fish usually retreat in response to an artificial increase in pulse frequency (Black–Cleworth, 1970). So attack is stimulated by the sustained discharge which signals the presence of a rival and is inhibited by the sharp increase in pulse rate indicative of an aggressive animal.

4.3.3 Visual cues

An aggressive opponent presents a formidable array of visual cues. Threat displays often make weapons more conspicuous and increase the apparent size of the animal concerned; submissive displays have the opposite effect (page 41). In many cases, fighting animals are brightly coloured, and colouration may change during fights. Finally, agonistic movements and postures impose distinctive shapes on the performer. An influence of shape, size, colour and behaviour on agonistic behaviour has been demonstrated in a range of species (Table 4.3). As in the previous section, only a few well documented examples are discussed in the text.

(a) Shape

In damselfish (*Eupomacentrus altus*) an intruder's shape is critically important in determining whether it will be attacked. Dummies which are flat and circular elicit much more attack than do standard fish-shaped dummies, regardless of their colour or size; slender, elongated dummies are hardly attacked at all. These fish compete with

Table 4.3 Examples of the role of vision in agonistic encounters.

Species	Observation	Cue	Reference
Spiders	Smaller dummies elicit attack, larger dummies elicit flight.	Size	Crane, 1949
Mantis shrimps	Models with coloured ventral spot elicit most attack.	Colour	Hazlett, 1979
Sunfish	Dummies with raised gill covers are attacked most.	Shape	Colgan and Gross, 1977
Lizards	Smaller opponents elicit no response, same sized elicit display, attack and some flight, larger ones elicit flight.	Size	Stamps, 1977
Red-winged blackbirds	Males with red shoulder patches obscured are less successful in territory defence.	Colour	Peeke, 1972
Hamsters	Males with enlarged black chest patch elicit more flight.	Colour	Grant *et al.*, 1970

many other species for the algae growing within their territories. In the reef environment, most herbivorous fish are flat and more-or-less circular in outline, while carnivores tend to be elongated and stream-lined. Thus a simple rule ('attack intruders whose body depth to length ratio exceeds a critical value') generally ensures that rivals are attacked but that energy is not wasted on fish which pose no threat. There are a few thin herbivores which avoid attack and some fat carnivores which, though not rivals, are driven out of the territory; but on the whole, mistakes are rare (Kohda, 1981).

(b) Size

Males of the territorial cichlid fish *Pseudotropheus zebra* respond dif-ferently to dummies of different sizes (Fig. 4.4). The details are complex but, in general, as dummy size increases overt attack (but-ting) gives way to display. Display to a medium sized dummy is predominantly from a facing position (frontal display), but when fish

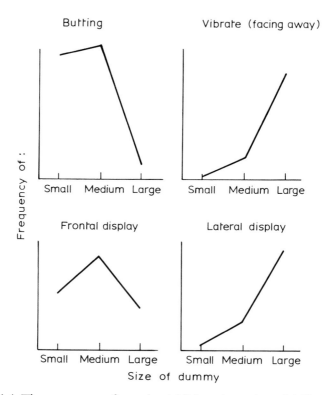

Figure 4.4 The responses of a male cichlid to dummies of different sizes (after Vodegal, 1978).

confront larger dummies this is replaced by displays performed from a sideways position (lateral display) or while facing away (vibrate). These results can be explained by suggesting that both attack and flight tendencies increase with size of stimulus, but that the effect on flight is stronger. In the presence of a small dummy, attack predominates but in the presence of a large one, flight predominates and the fish turn away as if about to flee. The frontal and lateral displays of fish to a medium sized dummy reflect moderate attack and flight tendencies (Vodegal, 1978).

(c) Colour

Colour patterns and colour changes are spectacularly developed in fish and play a rather special role in the control of agonistic behaviour. Some colour patterns are permanent or semi-permanent, persisting over whole phases of the animal's life and thus potentially allowing recognition of species, gender and reproductive condition. Others change drastically, over periods ranging from seconds to days. These short-term colour changes can often be related to motivational state and so potentially allow fighting animals to predict how an opponent will behave. In the firemouth cichlid some patterns (spot-bar or bar) are strongly associated with the performance of agonistic acts while non-aggressive fish generally show different colouration (spots or stripes; Neil, 1984). In guppies and several other species, eye colour reflects internal state, with aggressive fish having darker eyes (Kingston, 1980; Martin and Hengstebeck, 1981).

Male pumpkinseed sunfish have a blue-green body, a black and red opercular flap (with a white border) and a red iris, colour patterns that are particularly clearly marked in breeding, territorial fish. In contrast, male longear sunfish have reddish brown sides, a yellow throat and belly, black pelvic fins and large black opercular flaps with red and white borders. Breeding males of these two species attack preferentially dummies painted to resemble conspecific males, their most serious rivals; colour patterns typical of a conspecific female inhibit attack (Keenleyside, 1971; Stacey and Chiszar, 1978).

Single components of a colour pattern can alter agonistic behaviour, but the overall effect of a rival depends on its complete colour pattern. Territorial male cichlids (*Haplochromis burtoni*) living in semi-natural conditions have a cluster of orange spots on the operculum. In the laboratory a fair proportion of males have a black bar below the eye although this colour pattern is rare in the wild (see Fig. 4.5, Fernald, 1977). A dummy with a black eyebar causes a rise in attack rate in breeding males; this response has two components: a large incremental effect which decays rapidly and a smaller effect, also

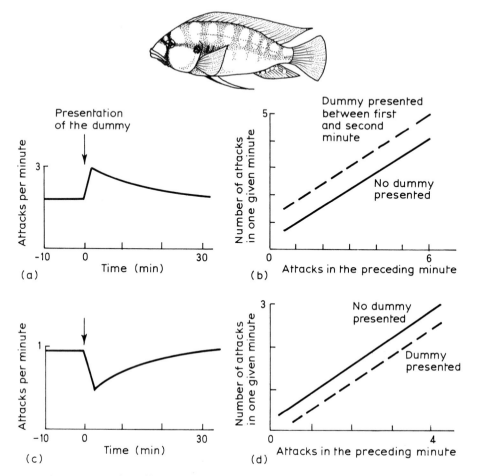

Figure 4.5 The effects of colour patterns on attack rate in male cichlids (*Haplochromis burtoni*). (a) The effect of brief presentation of a dummy with a black eyebar; (b) attack rates in minutes before and after *either* no stimulus *or* presentation of a dummy with a black eyebar; (c) the effect of brief presentation of a dummy with orange spots; (d) the same as b for presentation of a dummy with orange spots; (Heiligenberg, 1976, adapted by Huntingford, 1984).

incremental, which gradually disappears over a period of hours (Fig. 4.5). Dummies with an orange spot have the converse effect, producing a sharp decrease in attack rate, with the fish returning to its base level in a matter of minutes (Fig. 4.5(b)).

In both these cases, the effect of the external stimulus is added onto the base level attack rate, the eyebar causing a fixed increase and the spots a fixed decrease in bite rate, regardless of previous attack levels

(Fig. 4.5(c) and (d)). A dummy bearing both eyebars and spots produces a small increase in attack rate, suggesting that the effects of the two stimuli combine additively to determine how the fish behaves (Heiligenberg, 1976).

(d) Posture and behaviour

Animals often respond to the sight of an aggressive opponent with flight but attack a fearful opponent. For example, male blue crabs tend to retreat from a dummy approaching in the frontal position or with chelipeds fully extended. A dummy approaching in the lateral position or with the cheliped partially folded elicits attack (Jachowski, 1974). The predominant response of territorial glaucous-winged gulls to an intruder (a radio-controlled wooden gull) in a high neck posture (indicative of submission) is attack and to a bird in a horizontal position (indicative of an aggressive bird) is avoidance (Galusha and Stout, 1977).

Decisions during a fight may depend on very complex features of the behaviour of an opponent. For example, the same behaviour can produce a different response depending on what it indicates about changing agonistic motivation. When a (dummy) gull intruder changes from an upright to the more aggressive oblique posture, the resident bird retreats. When an oblique posture is reached from a horizontal body position, which reflects a decrease in aggressiveness, the resident attacks (Galusha and Stout, 1977).

4.3.4 Olfactory cues

Scent marks play two broadly distinct roles in agonistic interactions: they are used for long-term territorial marking but may also provide close-range stimuli, controlling both the occurrence and the course of a fight. Table 4.4 gives some examples in which these two broad functions of olfactory cues have been demonstrated. The special role of olfactory cues in territory marking lies in the fact that they may persist for long periods. Animals of many different groups deposit scent marks which, in their absence, discourage certain other animals from entering the marked area.

The most carefully controlled studies of behavioural responses to territory marking in mammals have been carried out on laboratory rodents (Fass and Stevens 1977; Ropartz, 1977). Male mice (particularly dominants) deposit urine marks at the boundaries of their territories. Given a choice, other males avoid areas which have been marked in this way, especially when the urine comes from a dominant male, one who has just won a fight or a male who has been housed in

Table 4.4 Examples of the role of olfactory cues in agonistic encounters.

Species	Observation	Role of scent	Reference
Cockroaches	Females fight more when exposed to scent of males.	Promotes attack	Raisbeck, 1976
Ants (*Myrmica rubra*)	Deposit Dufour's gland secretion on unfamiliar ground. Ants of other colonies avoid marked areas.	Territory marking	Cammaerts *et al.*, 1978
Blennies	Nest site marked with anal gland secretion; deters intruders and reduces their attack rates.	Territory marking; inhibits attack	Liley, 1982
Red-backed salamanders	Secretions from dermal glands deposited on ground and avoided by intruders.	Territory marking	Tristram, 1977
Tortoise	Model tortoise vigorously attacked but only if daubed with secretion from chin gland.	Promotes attack	Rose, 1970
Bank voles	Preputial gland extracts make dominant males urine-mark more but make subordinates mark less.	Territory marking; badge of status	Brinck and Hoffmeyer, 1984

Figure 4.6 Tail flicking in the ring-tailed lemur. The tail is marked with scent from a gland on the arm (the arrow indicates the most intensively marked region) and then flicked towards a companion (in the sequence indicated 1–3). This produces a conspicuous visual signal as well as dispersing scent. (Original drawing by D. L. Cramer, reproduced with permission from Evans and Goy, 1968.)

isolation. Mouse urine must contain a substance which determines whether another mouse will enter an area and which identifies the gender, status and past history of the animal concerned. The scent is produced by a pair of glands associated with the penis (the coagulating glands, Albone and Shirley, 1984).

Besides its role in territory marking, scent also plays a part in actual fights. For example, it can be used directly as a weapon; male rabbits squirt smelly urine at each other. The behaviour patterns by which scents are produced and broadcast can also act as signals in their own right (Fig. 4.6). The preorbital gland in red deer is a conspicuous pocket of skin below the eye which is used in scent marking but also plays a direct behavioural role during fights. It is closed in subordinate males, but opened during roaring bouts in dominants; the gland is more in evidence as agonistic encounters become more intense and this may reflect aggressive intent (Bartos, 1983).

Chemical cues elicit or inhibit attack in species ranging from cockroaches to fish, lizards and primates. When the olfactory system of owl monkeys is inactivated, fights become much less frequent and those which do occur involve less contact aggression. This suggests that olfactory cues are important both in the initiation and in the maintenance of aggression (Hunter and Dixson, 1983). The montane vole is a highly aggressive species which, in the wild, often has scars in the region of the scent glands, which are situated on each hip. If one scent gland is removed, attacks are directed towards the other side of the body, suggesting that chemicals produced by the hip glands help to orient attacks towards territorial intruders (Jannett, 1981).

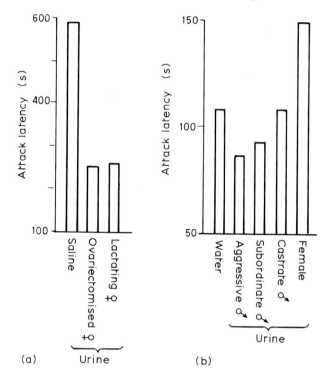

Figure 4.7 The role of scents in agonistic encounters between mice. (a) Attack latency of castrated males to female intruders marked with urine from ovariectomized and from lactating females (after Haug and Brain, 1979); (b) attack latency of intact males to castrates marked with urine from different sources (after Mugford and Nowell, 1970).

Once again, the most detailed studies of the role of olfactory cues during actual agonistic encounters (rather than in territory marking) have used mice as subjects. Mice (and other mammals) pick up chemical cues by their nose (which leads into the main olfactory pathway) and by the vomeronasal organ (in the mouth and leading into the accessory olfactory system). Interfering with either of these two sense organs reduces attack in male mice, and when both are inactivated, fighting is completely eliminated (Bean, 1982; Clancey *et al.*, 1984).

When housed in groups, both female mice and castrated males are particularly aggressive towards lactating females. Intruders painted with urine from either an ovariectomized or a lactating female receive the full force of this aggression (Fig. 4.7(a)). So the urine of lactating female mice seems to contain a substance that provokes attack (Haug and Brain, 1978; Brain, 1980). On the other hand, intact territorial male mice direct rather fewer attacks towards intruders painted with the urine of an intact female. This is not accompanied by any marked increase in sexual activity, suggesting that the urine of female mice contains a substance which directly inhibits attack by males (Mugford and Nowell, 1970; see also Lee and Ingersoll, 1983, and Leshner, 1978 for reviews).

Male mice are not normally very aggressive towards castrates, but a castrated mouse daubed with the urine of an intact male mouse (particularly if the donor is a dominant or has been isolated) is vigorously attacked (Fig. 4.7(b); Mugford and Nowell, 1970). The source of this attack-eliciting scent seems to be a pair of glands (the preputials) associated with the penis (Brain, 1980). These regress following castration and are particularly large in dominant males and in isolates. Castrated males painted with an extract of the preputial gland of an intact male are vigorously attacked (Mugford and Nowell, 1971; Albone and Shirley, 1984).

Therefore, in addition to an aversive, territory marking scent produced by the coagulating glands, the urine of male mice contains a substance produced by the preputial gland which elicits attack. Production of this substance depends on status, housing conditions and on the presence of an intact testis. The smell of urine from intact females suppresses attack in intact males but lactating and ovariectomized female mice produce a urine-bound substance which triggers aggression in group-living castrated males and in females.

4.3.5 Tactile cues

In some species, substrate-borne vibrations act as distance cues during social communication; these are clearly similar to the airborne vibrations which are usually described as sounds. Social spiders (*Metabus gravidus*) bounce up and down on the hub of their web when they detect another spider, setting up vibrations which inhibit intrusion. If the intruding spider persists, larger animals are attacked at a greater distance, presumably because they cause larger web vibrations (Buskirk, 1975). Larval caddis flies (*Hydropsychidae*) hide in retreats in front of which they spin feeding nets. Larvae fight over retreats and may stridulate as they do so; this sets up vibrations in the web which are detected by specialized hairs on the forelegs. A defending larva that stridulates has a greater chance of fending off an attack by a larger opponent than one that fails to do so. So stridulation-produced vibrations inhibit aggression in intruding larvae (Jansson and Vuoristo, 1979).

However, tactile stimuli play their most spectacular role during direct physical contact. In sessile invertebrates, the overgrowth by

Table 4.5 Examples of the role of tactile cues in agonistic encounters.

Species	Observation	Interpretation	Reference
Nereid worms	Removal of sensory palps abolishes differential aggression to heterospecifics.	Worms use tactile cues to decide whether to attack.	Evans, 1973
Fiddler crabs	Territorial males drum with their claws on the substrate in a species–specific pattern. Receivers respond with upright postures and drumming.	Possible role as distance signal.	Salmon and Horsch, 1972
Snapping shrimps	Males respond to artificial water currents with threat postures.	Water currents elicit agonistic responses.	Hazlett and Winn, 1962
Hermit crabs	Tapping shell or body elicits chela flicking.	Tactile cues elicit shell defence.	Hazlett, 1970
Crickets	Lashing with a flexible bristle results in submissive behaviour.	Tactile cues from antennal lashing inhibit attack.	Alexander, 1961

which competitors are destroyed is triggered by physical contact and in mobile animals tactile cues are exchanged during escalated fights, whether these involve butting, wrestling, kicking or biting. The importance of painful stimuli in eliciting and maintaining agonistic behaviour has long been recognized (page 89). Exchanges of contact tactile stimuli are clearly of great importance in determining the course a fight takes and who eventually wins and loses, but for the most part little systematic evidence exists as to the exact nature of their effects. Table 4.5 gives some examples.

Cockroaches start fights by lashing each other with their antennae and this may escalate to direct biting and kicking attacks. The predominant response of *Periplaneta americana* to artificial antennal stimulation is to back away. In contrast, although subordinate *Nauphoeta cinerea* retreat from such stimulation, dominant animals turn towards it and attack. In response to simulated kicks, male *Periplaneta* retreat, jerk their bodies away or kick out with the stimulated leg (Bell, 1978). During agonistic encounters between two male rats, accidental vibrissal contact produces immediate defensive upright postures in both animals (Lehman and Adams, 1977). Removing the whiskers of a resident male has no effect on his response to an intruder, but if an intruding male has no whiskers he shows fewer defensive upright postures (Blanchard *et al.*, 1977).

4.4 THE EFFECTS OF OTHER STIMULI FROM THE ENVIRONMENT

Animals do not respond to the agonistic stimuli provided by an opponent in an absolutely mechanical way; whether two animals fight and if so what form this takes is influenced by various aspects of the environment in which the encounter occurs. The influential factors tend to be those which indicate the costs of fighting, the probability of victory and the rewards gained by the winner (page 282).

4.4.1 The physical environment

Juvenile brook char respond to intrusion into their feeding territory with a vigorous charge at low current speeds. Charges would use up too much energy at high current speeds and so the fish then use displays instead (McNicol and Noakes, 1984). Spiders (*Agelenopsis aperta*), and a number of other animals, fight particularly fiercely over good quality feeding territories (Riechert, 1984).

4.4.2 Familiarity

Quite apart from any intrinsic features of the environment, how familiar it is to the animals concerned is also important. In many different species (including anemones, spiders, fish, birds and mammals, pages 16, 36 and 50), an animal on its home ground has an advantage in a fight over one to which the venue is unfamiliar. In some cases at least, this behavioural difference arises from the reduced levels of fear and exploration that familiarity brings. The prior residence advantage can be eliminated in convict cichlids and in mice and rats by moving a resident to a new cage or tank and the more different the new home from the old one, the greater the reduction in aggression it produces (Figler and Einhorn, 1983; Jones and Nowell 1973).

4.4.3 The social environment

Communal spiders living at high densities allow intruders to approach more closely than those living at low density (Buskirk, 1975), while in hens (Banks and Allee, 1957), mice (Calhoun, 1962) and mongooses (Rasa, 1979) the converse is the case. The situation in primates is more complex; acute crowding in confined spaces increases aggressiveness in pigtail macaques (Alexander and Roth, 1970), but animals experiencing gradually increasing densities in large enclosures get involved in fewer fights (Eaton, Modahl and Johnson, 1981).

The composition of the social environment and the behaviour of its members is also important, especially those aspects which determine the probability and rewards of victory. A rhesus monkey will either attack or submit to the same rival, depending on the presence of potential allies (Datta, 1983; see also page 215). The presence of a receptive female enhances levels of aggression between males, for example in mice and spider mites (Taylor, 1980; Potter 1981). Male red-spotted newts and bowl-and-doily spiders fight longer and more fiercely over large females (which produce more eggs) than over small ones (Austadt, 1983; Verrell, 1986).

4.5 THE DYNAMICS OF UNDISTURBED FIGHTS

The previous sections described internal systems which control agonistic behaviour and the effects (both activating and inhibiting) that environmental stimuli can have on these. To find out how these internal and external factors interact to determine the course and outcome of an agonistic encounter, we need to look at the signals exchanged between opponents.

This can be done by constructing what is called a transition matrix, showing just how often each item of behaviour is followed in agonistic sequences by every other action. If we look at the matrix based on transitions in behaviour of one particular animal, we can see how *performing* each category of agonistic behaviour influences subsequent behaviour. The matrix based on cases where an action by one participant is followed by a change in the behaviour of its opponent identifies behavioural shifts caused by *experiencing* (seeing, hearing or feeling) an agonistic action.

This sort of interpretation relies on a number of assumptions, for example, that the sequential relationships between action patterns remain the same and that a particular action by one animal depends only on what happened just before it. The very nature of fights mean that both these assumptions are almost certainly violated, but there are ways around this problem (Losey, 1978; Van der Bercken and Cools, 1980).

Although complex, such data can tell us a lot about the changes animals undergo as they struggle for victory during a fight and about what causes these changes. For example, from a study of agonistic exchanges between lizards, Cooper (1977) concluded that '. . . either lizard is more likely to approach immediately after its own act than after an action by its opponent, and . . . it withdraws or flees more often after activity by the opponent'. An intimidating effect of experiencing an aggressive action is also evident in pomacentrids. Small, temporary increases in aggression (indicated by darkening eye colour) in one opponent are matched by small, temporary decreases in aggressiveness in the other. Fear levels (indicated by raised fins) fluctuate reversibly, independently of aggression levels but with the same reciprocal relationship. At a critical point in the encounter one fish, the winner, shows a larger, longer term increase in aggression, accompanied by a decrease in fear; the defeated fish shows the converse. At the behavioural level, opponents match each other when performing moderately intense agonistic displays; but once the dispute is resolved the winner chases the intruder out of its territory (Fig. 4.8(a), Rasa, 1969).

Fights between mantis shrimps (Fig. 4.8(b)) involve a relatively straightforward escalation of both participants, but there is still a tendency (represented in the figure by slightly more arrows going to the left in the inter-animal case) for a high-level act in one animal to be followed by a less aggressive behaviour pattern in its opponent. In cichlids (Fig. 4.8(c)) fights are much more complex, with the fish escalating and de-escalating in parallel. The eventual winners and losers behave very similarly right up to the end of the fight. Then the loser becomes less likely to retaliate when bitten and more likely to de-escalate (there are more leftwards arrows in the winner–loser matrix of the figure).

Clearly the internal state of a fighting animal is continually changing as a result both of performing and of observing agonistic behaviour patterns. The nature of these effects depends on the species and behaviour concerned, on the stage of the fight and on whether the animal is winning.

Fish's behaviour   Opponent's behaviour

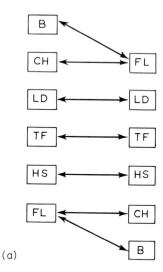

(a)

Figure 4.8 Some studies of fight sequences. (a) Summary of behavioural exchanges during territorial fights between pomacentrid fish (*Pelmatochromis jenkinsi*); B = bite, Ch = chase, LD = lateral display, TF = tail fight, HS = head scrape, FL = flight (after Rasa, 1969). (b) Summary of main intra- and inter-animal transitions in the first ten minutes of fights between mantis shrimps; A = approach, MS = meral spread, IA = intention attack, OA = overt attack. Width of arrow indicates percentage of times each act was followed by every other act (from figures in Dingle, 1972). (c) Summary of main inter-animal transitions in the final section of fights between male cichlids (*Nannacara anomala*) for winners and losers separately; FD = frontal display, LD = lateral display, TB = tail beat, B = bite. Width of arrows as in part (b) (from figures in Jakobsson *et al.*, 1979).

4.6 MODELS OF THE CONTROL OF AGONISTIC BEHAVIOUR

The studies discussed so far all produce isolated pieces of information about the factors (both inside the animal and in its environment) that control agonistic behaviour. If we are to explain what happens in an agonistic encounter, these must all be put together into some sort of coherent explanatory scheme, or model (Toates, 1983; Huntingford, 1984). Two rather different kinds of model can be found in the literature on agonistic behaviour; some attempt to explain detailed results for one particular species while others seek global explanations which can be applied to a whole range of different species.

Mantis shrimps

Intra-animal transitions

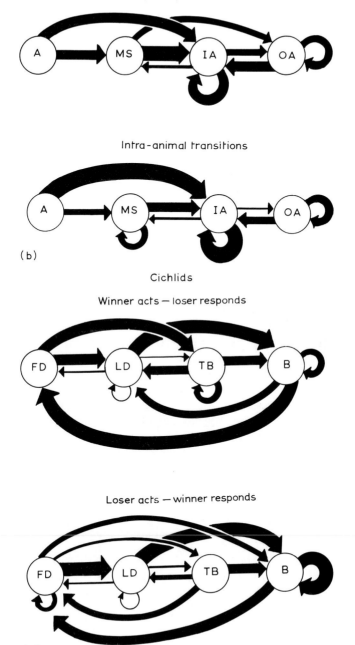

Intra-animal transitions

(b)

Cichlids

Winner acts — loser responds

Loser acts — winner responds

(c)

4.6.1 A specific model: the causes of territorial fights in spiders

The previous sections described models for agonistic behaviour in rats (page 61) and fish (page 81), but these do not really accommodate the complex interactions that occur once a fight is underway. For a model which does do this we have to turn to aggressive interactions between funnel web spiders (*Agelenopsis aperta*). These animals defend feeding territories and spiders from desert habitats fight more fiercely than those from slightly wetter (riparian) habitats (Maynard Smith and Riechert, 1984).

The spiders show four main categories of agonistic behaviour, each incorporating a number of different actions: locate, signal, threaten and contact, representing successive levels of escalation. Resident spiders have an advantage over intruders, and fights are fiercer if the resident is large and the territory a good one. According to the model (which is presented here in a simplified form), choice of particular agonistic actions is controlled by the balance between two independent tendencies, aggression and fear (Fig. 4.9). If fear tendency is greater than aggression tendency, the spider retreats; the absolute level of aggression determines whether the spiders show locate, signal, threaten or contact.

The intruder starts the encounter with both aggression and fear levels set at zero. The resident animal starts with a fixed advantage in aggression (the size of which depends on the environment the spiders come from) and with fear reduced by half the value of the territory over which the spiders fight (the intruder is not supposed to know how valuable the territory is).

Fights take the form of a series of acts performed by one animal and observed by the other, with the participants taking turn about in these two roles. Aggression and fear of the observing spider shift after each agonistic action, depending on the precise behaviour of the opponent and according to a set of rules, the main features of which are outlined in the equations below. In the original model, aggression and fear were divided into temporary and permanent components to allow for the bout structure of spider fights. In this model, performance of an agonistic act has no effect on internal state. The new aggression and fear levels determine what behaviour is shown when the observer becomes the performer of the next act. Whenever the aggression of one animal falls below its fear level, it retreats (or flees) leaving its opponent victorious.

By assigning appropriate values to the parameters of this model, and running computer simulations, predictions can be made about the course a fight would take if the model correctly depicts what is going on. These predictions can be compared to the course of real fights. As

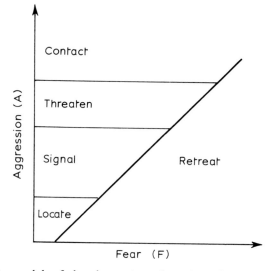

Figure 4.9 A model of the dynamics of territorial encounters between spiders (simplified from Maynard Smith and Riechert, 1984), showing the rules that relate overt behaviour to levels of aggression and fear.

$$A_{after} = A_{before} + X + (G \times D) + E$$
$$F_{after} = F_{before} + (G \times B)$$

where A = aggression before and after observing opponent
     F = fear before and after observing opponent
     D = weight difference between the spiders
     E = a random variable from −1 to +1
X and B = constants representing rate of change of aggression and fear; both greater in grassland spiders
     G = 1 if opponent locates
         2 if opponent signals
         3 if opponent threatens
         4 if opponent makes contact

Table 4.6 shows, the length, structure and outcome of simulated and real fights are comparable. Overall therefore the model does accurately represent the mechanisms which control agonistic sequences in these spiders. However it turns out that intruding spiders as well as residents adapt their behaviour to the quality of the disputed territory.

4.6.2 Global theories

The model for spider fights described in the previous section

Table 4.6   Some observed and predicted parameters of fights between spiders (Maynard Smith and Riechart, 1984).

a) The proportion of encounters won by the owner in within-population fights (taken from a figure, using values for very high quality territory).

	Owner heavier		Equal weights		Owner lighter	
	Predicted	Observed	Predicted	Observed	Predicted	Observed
Grassland spiders	100	95	100	91	8	28
Riparian spiders	100	100	66	70	0	14

b) Outcomes and estimated costs incurred by each participant in encounters between grassland and riparian spiders of equal weight on a neutral territory

	Grassland spiders		Riparian spiders	
	Predicted	Observed	Predicted	Observed
Number of wins	9	9	4	4
Cost	86	88	30	76

postulates separate aggression and fear systems, fluctuating during an encounter as a result of the behaviour of an opponent. This sort of effect has been identified for a number of species (page 80), so the model potentially provides a starting point for similar explanations of the behaviour of other animals. However, it fails to take into account the fact that performing an aggressive action often influences subsequent behaviour or that observing such an act sometimes depresses aggression. The model could be expanded to allow for these facts, but the changes would make it more cumbersome. Turning an explanation which works for a particular species into one which is applicable to others often reduces both its elegance and its accuracy.

In spite of this drawback, general theories that seek to explain agonistic behaviour in a range of animals are attractive because they appear to offer an immediate understanding of what is going on in any new species we might encounter. Some influential theories of this kind are considered in this section as they relate to fighting in non-human animals; their relevance to our own species will be discussed in Chapter 12.

(a)   The endogenous drive-catharsis hypothesis

According to this theory, aggression is an internally generated drive

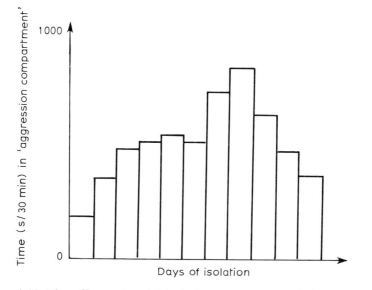

Figure 4.10 The effects of social isolation on aggressive behaviour (time spent in a compartment from which an opponent could be threatened) in juvenile damselfish (*Microspathodon chrysurus*) (Rasa, 1971).

that gradually accumulates inside an animal, motivating it to fight. Aggressive drive builds up in a peaceful animal which therefore becomes progressively more inclined to attack. The accumulated drive is dissipated by the act of fighting and by the end of an aggressive encounter aggressive motivation is at a very low level (Lorenz, 1966).

Isolation sometimes increases aggression in crickets (Alexander, 1961), crabs (Vannini, 1981), fish (Rasa, 1971, see also Fig. 4.10; Chauvin-Muckensturm, 1978; Franck, Marek and Winter, 1978), birds (Eiserer, Emerling and Scardina, 1976) and mice (O'Donnell, Blanchard and Blanchard, 1981). However, it fails to do so in several species of fish (Chizar *et al.*, 1976; Rohrs, 1977; Ferno, 1978), in female mice (Buhot-Averseng and Goyens, 1981) and, under certain circumstances, in rats (Barr, 1981).

Even where isolated animals are more aggressive this does not necessarily support the endogenous drive theory, because there are alternative explanations. Isolated animals are often more exploratory, more active and more responsive to pain and other tactile stimuli and this may explain why they are more aggressive (Poshivalov, 1981). High levels of aggression in isolated animals could equally well reflect a process of forgetting. For example, a subordinate animal may gradually forget the experience of being dominated and become more

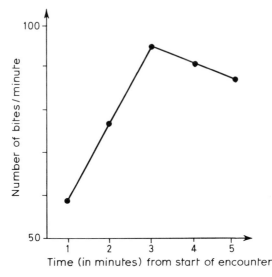

Figure 4.11 Aggression often increases during the course of a fight. Attack rate by territorial male sticklebacks towards a male intruder (confined in a glass tube; after Ukegbu, 1986).

aggressive (swordtail fish, Rohrs, 1977; mice, Parmigiani and Brain, 1983).

Contrary to the predictions of the model, aggressiveness often increases in the early stages of a fight. This has been observed for cockroaches (Breed, 1980), spiders (Riechert, 1984), fish, (see Fig. 4.11; Dow, Ewing and Sutherland, 1976; Peeke, 1982; Barlow, Rogers and Bond, 1984), lizards (Rand and Rand, 1978) and rodents (Parmigiani and Brain, 1983; Potegal and ten Brink, 1984). Performance of an aggressive act clearly has complex effects on subsequent attack probability, some positive and some negative.

If the model is right, and fighting does bring about a reduction in aggressive drive, then we might expect animals to work hard for the rewarding (reinforcing) drive reduction they experience during a fight. A clear distinction needs to be made: fight-induced reinforcement could either (1) arise from gaining access to a particular resource or from removal of an unpleasant stimulus, or (2) stem from the actual performance of an aggressive action regardless of its consequences. Only the latter is specifically predicted by the drive–catharsis hypothesis.

Various species of fish (Rasa, 1969; Sevenster, 1973; Bols, 1977), birds and mammals (Potegal, 1978; Taylor 1979) can be trained to perform some particular task if by doing so they gain access to a member of their own species whom they can attack. The more aggressive the subjects, the more quickly they learn the task. For an aroused animal, aggression may be reinforcing because it reduces the physiological consequences of stress; rats exposed to electric shocks get fewer ulcers if they have an opportunity to fight (Berkowitz, 1982). So it looks as if the opportunity to engage in a fight represents a sufficiently rewarding experience to bring about and maintain learning.

However, in Siamese fighting fish at least, engaging in a fight may simply increase general activity rather than promoting learning (Bronstein, 1981). In addition in many of these examples, the animals were not offered an alternative stimulus, so the opportunity to fight is confounded with the chance to observe or interact socially with a conspecific. In fact, given a choice, aggressive rats and mice prefer to associate peacefully with conspecifics than to fight them (Taylor, 1977). These observations suggest that general stimulation or social contact rather than aggression may be what provides the reinforcement that supports learning.

Overall, the drive–catharsis theory, attractively simple and influential as it is, at best presents an oversimplified picture of the causes of aggression and, at the worst, profoundly misrepresents them. This has important implications for the control of aggression. The theory tells us that it is impossible to prevent animals (including humans) from fighting; all that can be done is to find harmless ways for them to work off their inevitably accumulating aggressive drive. In contrast, theories of habituation and social learning suggest that a permanently low level of aggression can be maintained, given appropriate manipulation of stimulus and experience.

(b) The pain–aggression hypothesis

In complete contrast to the drive–catharsis theory, with its emphasis on factors within the animal, is the suggestion that fighting is an almost reflexive response to the experience of pain. At its most extreme, this theory suggests that all fighting is caused by pain and pain always causes fighting. It has been clearly demonstrated that when an animal, be it a pigeon, mouse, rat or monkey, is subjected to a painful stimulus, this can cause immediate attack even in socially naive subjects (Berkowitz, 1982, 1983). However, the form of the response suggests that it is defensive rather than offensive in nature. In mice, for example, electric shock-induced fighting often takes the form of upright postures, with bites being directed at the opponent's snout, which is typical of a defensive animal (Blanchard and Blanchard, 1981; Hutchinson, 1983).

Electric shock can provoke other responses, such as freezing, escape or sexual behaviour (Bandura, 1983). Whether or not shocked animals fight depends on the nature of the target (familiar mice are not attacked) and on the animal's past experience. Rats reared in a non-competitive environment fiercely attack a conspecific when shocked; those previously exposed to unavoidable shock or punished for aggressive behaviour are less likely to do so. Clearly attack is not the only or even the predominant response to pain (see Berkowitz, 1983 and references therein).

It is also impossible to maintain that all agonistic behaviour is the result of painful stimuli. In the early stages of an agonistic encounter combatants exchange actions involving no direct contact, so physical pain cannot be the cause of the behaviour. One might argue that other aversive events, including the sight of a rival, cause agonistic responses by means of psychological pain, but this is stretching the concept of pain so far that it ceases to have any value in predicting when attacks will occur.

(c) The frustration–aggression hypothesis

Another general theory about the causes of aggression, and one which has intuitive appeal, is that frustration is a necessary and sufficient cause of attack. A pigeon which has been trained to expect a reward (food or drink) when it presses a key will attack another bird if the reward ceases to be forthcoming. With feathers erect and calling aggressively, it pecks fiercely at the target's head and throat, a spectrum of behaviour which is reminiscent of natural fighting in this species; merely being hungry does not cause attack. Similarly, rats respond to withdrawal of reinforcement with species-typical agonistic sequences and both pigeons and rats trained to expect a reward at fixed intervals attack a target animal during non-reward periods (see Looney and Cohen, 1982 and references therein).

However, frustrated animals do not inevitably attack a target; instead they may escape or, when the appropriate stimuli are available, show quite different responses (Yoburn et al., 1981). In addition, it is clear that frustration is not the only circumstance in which animals fight any more than is the experience of pain.

(d) Control theory models

Disparate as all the circumstances which elicit attack may be, many of them do still have one particular feature in common, namely the existence of a discrepancy between what the animal expects (no pain, no rivals within a specified area or the appearance of food when a lever is pressed) and what it experiences (electric shock, the sight, sound and

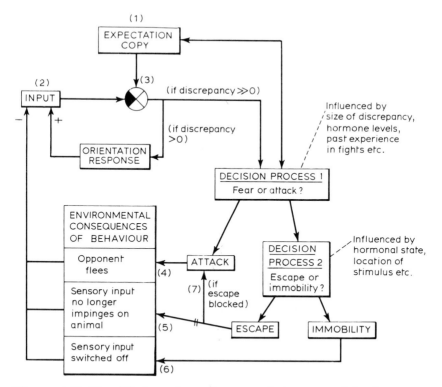

Figure 4.12 Simplified version of a control theory model of agonistic behaviour in vertebrates (from Huntingford, 1984, adapted from Archer, 1976).

smell of an intruder or an empty tray where a food pellet is usually delivered). Perhaps this discrepancy triggers agonistic behaviour which, if successful, removes the cause of the discrepancy (Wiepkema *et al.*, 1980). This sort of explanation is based on what is called control theory (Toates, 1983).

Figure 4.12 shows in simplified form a control model of aggressive behaviour, in which the presence of a rival generates a discrepancy between an animal's (continuously updated) mental picture of the world and its observed external environment. This activates attack, flight or freezing, depending on a variety of internal and external factors. If the animal attacks successfully its rival leaves and, if it flees itself, it can no longer see the opponent. If it freezes, it cuts itself off from most environmental stimuli. In each case, the discrepancy is removed and the behaviour stops (Archer, 1976). Although the concept of a discrepancy is a vague one, this sort of model is useful in stressing the self-regulatory aspects of agonistic behaviour.

4.7 OVERVIEW

4.7.1 Have our questions been answered?

The behavioural observations and experiments described in this chapter go some way towards answering the questions at issue in this section, about what causes animals to fight, when, and how they do.

1. *Controlling systems.* In a wide variety of species, agonistic behaviour patterns are controlled by two independent causal systems which can be described as aggression and fear (page 63). Exploratory and, in social species at least, affiliative tendencies interact with aggression and fear to determine how animals behave in a situation of conflict.

2. *External stimuli.* Cues in various modalities emitted by a potential opponent (indicative of its species, gender, status, reproductive condition, individual identity and fighting ability) influence offensive and/or defensive behaviour over various time scales and in animals of many different species. Some promote aggression (song in blackbirds, page 66; black eyebars in cichlids, page 71; base level pulse rate in electric fish, page 68; preputial gland secretion in mice, page 77). Others have the reverse effect (sounds in cichlids, page 75); sight of an aggressive gull, page 73; increased pulse rate in electric fish; urine of intact female mice, page 77).

Together, these influences on behaviour mean that an animal will only initiate a fight when this is appropriate or advisable. Members of non-competing species and potential mates are not readily attacked (e.g. pumpkinseed sunfish, page 61), nor are much larger opponents, which are likely to be better fighters (cichlids, lizards, stomatopods, pages 69 and 70).

Signs of an opponent's motivational state can influence behaviour shown during fights. In particular, indications of high or increasing aggressiveness inhibit attack while signs of fear promote it (in crabs and gulls, page 73). The observation that fighting animals often wear their hearts on their sleeves, displaying signs of their present motivational state and probable future behaviour, runs counter to the predictions of functional analyses using game theory (page 279). From the latter standpoint, animals are expected to give no hint of how likely they are to attack or flee. Game theory does predict, however, that if these cues are available, for whatever reason, animals will make use of them.

Some cues have stronger effects on agonistic behaviour than others, but different stimuli interact to determine the response it produces. Sights, sounds and feelings combine with smells to determine the course of fights between male rats (page 61) while opercular spots and eyebars combine additively to determine attack rate in cichlids (page 73). Agonistic behaviour is not triggered by just one simple, all-important cue (the sign stimulus of classical ethology) but depends on a complex interaction between the various features of a potential opponent.

The context in which an encounter occurs is also important. The response to a given stimulus depends on various features of the physical and social environment, especially those indicative of the probable outcome of a fight, its costs and benefits (page 79).

3. *Interactions between internal and external factors during a fight.* Once a fight is underway, the behavioural tendencies of both participants are constantly changing as a result both of performing and of observing agonistic behaviour patterns. The details of these motivational shifts depend on the behaviour patterns concerned, the stage of the fight and on the species of animal involved. In general, performing an aggressive action has a positive effect on aggressive motivation; observing an aggressive action, particularly a high intensity one, may have the opposite effect (page 81). Responses of this sort allow animals to escalate when they are winning (or have a good chance of doing so) and to de-escalate when they are losing, thus avoiding serious injury.

4. *Putting things together.* For some of the species discussed (rats, page 61; spiders, page 84) sufficient information is available on the causes of agonistic behaviour for explanatory schemes or models to be constructed depicting the mechanisms involved. These allow us to explain, sometimes in considerable detail, when and how particular animals fight.

4.7.2 General principles

Clearly, similar general processes are involved in the control of agonistic behaviour in a range of species. These include independent systems controlling aggression and fear, complex interactions between inhibitory and excitatory effects of external stimuli and feedback from performing and observing agonistic responses. However, the details involved vary greatly and attempts to find globally applicable models of the mechanisms controlling agonistic behaviour (that it is an endogenously-generated drive, that it is inevitably and uniquely a response to pain or frustration) have not been successful.

Each of these global theories contains a germ of truth and the processes they depict probably contribute to the shifting levels of aggressiveness seen in nature. For example, animals which live solitary lives must often experience long periods without fighting and this may make them very sensitive to aggression-inducing cues, even if this reflects forgetting rather than drive accumulation. Frustration is a common feature of social life and escalated fights often inflict pain, and either may contribute to ongoing agonistic behaviour. However, none provides an all-embracing, generally applicable explanation for the occurrence and form of animal fights.

5
The role of hormones

5.1 INTRODUCTION

Behavioural explanations of the causes of agonistic behaviour (Chapter 4) are perfectly valid in their own right. However, the systems they incorporate must have physiological bases, even if these are very complicated. We now turn to studies which try to identify what these might be, considering first the role of hormones and then, in Chapter 6, the role of the nervous system.

In both vertebrates and invertebrates, the endocrine system consists of a number of glands which secrete chemicals (hormones) into the circulatory system. These are then carried via the blood to their sites of action, the target organs. Hormones can be produced by the axons of nerve cells (neurosecretion) as well as by ductless (endocrine) glands. In both cases, there are hormones whose actions control secretion of other hormones and intimate links exist between the endocrine system and the central nervous system. Together these make up a set of feedback loops which allow precise control of endocrine state.

This provides a powerful mechanism for adapting an animal's physiological processes to environmental conditions. For example, moulting in crustacea involves a complex sequence of morphological, physiological and biochemical changes which must occur in the right order if moulting is to be successful. These changes are brought about as and when required by the action of a special moulting hormone. Behavioural changes are also critical; lobsters and crayfish become less aggressive during a moult, when they are vulnerable to attack from predators and conspecifics. In addition to its physiological effects, moulting hormone also mediates this shift in agonistic behaviour (for example Lipcius and Herrnkind, 1982; see also page 124). It is this last sort of effect, the role of the endocrine system in the control of agonistic responses, that forms the subject matter of the present chapter.

1. How important are hormones in the control of agonistic behaviour?
2. Do different hormone levels cause the differences in agonistic behaviour that exist between species, between the sexes and between individuals?
3. Do changes in agonistic behaviour during development depend on altered patterns of hormone secretion?
4. Can reversible shifts in agonistic behaviour (over the breeding season, following a period of isolation or during the course of a fight) be explained in terms of endocrine activity?

Scientists have tried to answer these questions by relating changing rates of production of a particular hormone to fluctuating levels of agonistic behaviour, by removing the gland which produces the hormone and observing the behavioural consequences, or by artificially introducing the hormone and seeing what happens. For example, aggressive, dominant female wasps (*Polistes gallicus*) produce more of a particular neurosecretory hormone than do subordinates. The glands involved (the *corpora allata*) lie in the brain and their product (a complex secretion usually referred to as juvenile hormone) promotes ovarian development in some adult insects. Removing the *corpora allata* reduces aggression and artificially increasing juvenile hormone levels makes wasps particularly fierce. This all suggests that the aggressiveness which characterizes dominant female wasps may be caused by high circulating levels of juvenile hormone (Roseler *et al.*, 1980; Breed and Bell, 1983).

However, a number of complications, all of which will come up in later sections of this chapter, make the task of unravelling the relationships between hormones and agonistic behaviour a daunting one.

● Because of the feedback loops in the endocrine system, changing the levels of one hormone often influences production of others, and it is not always clear exactly which is responsible for any behavioural effect (page 118).

- The action of one hormone may be influenced by the levels of others (page 122).
- Hormones may undergo metabolic changes after they are released and different metabolites have different physiological and behavioural effects (page 109).
- A single hormone can have many, complex effects on behaviour (page 106).
- The effect of a given hormone often depends on the social circumstances in which it is administered (page 113).
- Hormones influence agonistic behaviour, but fighting experience can also influence endocrine activity so that there is a continual two-way interaction between hormonal state and agonistic behaviour (page 116).

5.2 MECHANISMS

There are many different routes by which hormones can exert an influence on agonistic behaviour. These are all exemplified in the following sections and need to be distinguished if we are to understand what is going on. As far as the behavioural nature of the effect is concerned (Table 5.1(a)), this may be *indirect*, with hormones producing changes in the social or non-social environment which then influence aggression. Castrated mice do not fight much, partly because production of attack-eliciting scents is reduced, so they get involved in fewer fights. Hormones may also have *direct* effects on behaviour, but these can be of a very *general* nature. Thyroxine increases aggression in cichlids, but this is part of a general increase in activity. Vasopressin interferes with a rat's ability to avoid circumstances in which it has previously been hurt (avoidance learning) and this is likely to have an effect on its response to potential opponents. Other effects of hormones are *specific*, producing either an alteration in the perception of agonistic cues or a change in agonistic responsiveness.

A number of different mechanisms produce these various behavioural effects (Table 5.1(b)). Some differences in aggressiveness in adult animals are the result of different hormonal environments experienced during early *development*. In adult animals, hormones may influence *peripheral structures* used during conflicts, such as body size (in squirrel monkeys), weapons (in soldier termites), colour patterns (in lizards) and scent production (in mice). Hormones may also influence the sense organs which detect agonistic cues (in toads giving release calls) or the muscles (in stags) and neuromuscular junctions (in lobsters) required to respond to them. On the other hand, in many species hormones act *directly on the brain* (in lizards and in mice).

Table 5.1   The routes from hormones to behaviour.

(a) *Behavioural effects*
    1. Indirect   (i) via a change in the non-social environment
                 (ii) via a change in the social environment
    2. Direct   (i) via a general effect (on activity or learning processes)
                (ii) via a specific effect on agonistic responsiveness

(b) *Mechanisms*
    1. In the developing animal   – producing more-or-less permanent
                                    changes

    2. In the adult animal   – producing reversible changes
       (i) Peripheral effects   – on sense organs
                             – on body size
                             – on agonistic cues (colours, sounds,
                                smells)
                             – on muscles
                             – on neuromuscular junctions
      (ii) Central effects   – on the activity of the CNS

5.3 HORMONES AND THE PATTERNS OF ANIMAL CONFLICT

As discussed in Chapter 1, animals encounter conflicts of interest over many different resources and outcomes, and physical aggression is just one of the many ways in which these conflicts can be resolved.

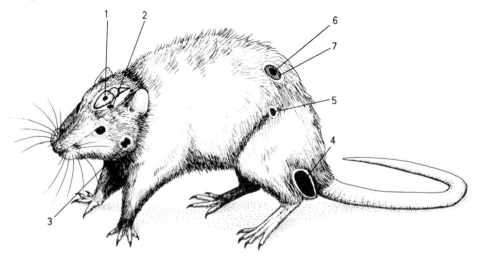

Figure 5.1 Simplified picture of the endocrine system of a rat, showing the glands and hormones mentioned in this chapter.

Gland	Product	Some effects
1. Pineal	Melatonin	Regulates gonadotropin production
2. Pituitary	Vasopressin	Stimulates water resorption in kidneys
	Adrenocorticotropic hormone (ACTH)	Stimulates production of hormones by adrenal cortex Production inhibited by glucocorticoids
	Thyroid stimulating hormone (TSH)	Stimulates production of thyroid hormones Production inhibited by TH
	Melanocyte stimulating hormone (MSH)	Increases synthesis of melanin and disperses melanocytes
Gonado-tropic hormones (GTH)	Follicle stimulating hormone (FSH)	Stimulates gamete production
	Luteinising hormone (LH)	Stimulates production of gonadal hormones Production influenced by gonadal steroids
	Prolactin	Stimulates milk secretion
3. Thyroid	Thyroid hormones (TH)	Stimulates metabolic activity
4. Testes	Testosterone (and other steroids)	Stimulates development of male secondary sexual characteristics and promotes sperm production
5. Ovaries	Oestrogen	Stimulates development of female secondary sexual characteristics
	Progesterone	Prepares uterus for implantation and maintains pregnancy
6. Adrenal medulla	Adrenaline and noradrenaline	Mobilises energy reserves, stimulates heart beat and 'Flight or Fight' response
7. Adrenal cortex	Glucocorticoids	Promotes energy mobilisation and controls electrolyte balance

Some information is available on the endocrine control of non-aggressive responses to conflict. For example, frogs use a release call to deter unwelcome mating attempts. These calls are only produced when levels of vasopressin (which promotes water retention) are low, perhaps because body distension interferes with detection of a would-be mate or with production of the call (Diakow and Raimondi, 1981).

However, research has concentrated on actual fighting. The various patterns of conflict (battles over mates or food, in defence of young, infanticide) play different roles in the lives of the animals concerned and are almost certain to have different hormonal bases. Most studies concern the role of hormones in the control of fighting between adult males and so we discuss this in most detail. Agonistic behaviour shown in other contexts is considered wherever possible.

The following sections provide a (necessarily selective) review of current understanding of the role of hormones in the control of agonistic behaviour. The invertebrates have a section of their own, because their hormone system is so different. For the vertebrates (whose endocrine system is shown in Fig. 5.1), we concentrate on the reproductive hormones and adrenal hormones (involved in an animal's response to stress) since it is here that most information is available. In both cases, we start with data for mammals and then see what happens in other vertebrates. Finally, we look back over the examples to give an overview of the role of hormones in the control of agonistic behaviour and to see if there are any generalizations to be made.

5.4 REPRODUCTIVE HORMONES IN VERTEBRATES

For a number of reasons, stemming from the fact that eggs are more expensive to produce than are sperm (page 303), although male animals fight over various resources and outcomes, they do so particularly fiercely over access to mates. Female animals also fight in a sexual context and in some case they compete for mates. However, a more common observation is the aggressive rejection of unwanted mates and, anyway, most fighting by females takes place over other resources and outcomes, such as food and shelter or in defence of young.

This suggests that, as a very broad generalization, agonistic behaviour in males should depend on the same hormones that control mating physiology and behaviour. It is of no benefit to a male to be ready to mate if he cannot compete effectively for females, so the two should be switched on together. Since competition over mates is usually less intense for females, and since an egg is useless unless a male can get close enough to fertilize it, it is more likely that the hormones controlling receptivity will inhibit aggression.

On the other hand, reproduction and aggression have close functional links when females are defending young, the presence of conspecifics often putting the young at risk. In cases where a special hormonal state is associated with care of the young, for example in lactating mammals, the chemicals involved may be responsible for the fierce aggression such females commonly show. In the following sections, we investigate the extent to which these expectations are confirmed.

5.4.1 Reproductive hormones in female mammals

The hormones most closely concerned with reproductive physiology in female mammals are the gonadotropins, the ovarian hormones (progesterone and oestrogen, which fluctuate in a complex way over the oestrus cycle, Fig. 5.2(a)) and the pituitary hormone prolactin which is involved in lactation (Fig. 5.1). In general, as predicted, gonadal hormones are of limited importance in the control of agonistic behaviour in female mammals.

Aggressiveness changes with the oestrus cycle in female mammals of a number of species. In female rats (Hood, 1981) and hamsters (Fig. 5.2(b) and (c)), aggression is at its lowest at oestrus, when fertile eggs are released and the female is sexually receptive, although there are some conflicting reports (see Siegel, 1985 for a review). In contrast, female deermice are reported to be least aggressive when not in oestrus (Gleason, Michael and Christian, 1979). Initiation of mild aggression peaks at ovulation in female rhesus monkeys (Mallow, 1981) and female ruffed lemurs are fiercest just before their eggs are released (Shideler, Lindberg and Lasley, 1983). Both these changes may be an indirect effect of the increased social contact that accompanies oestrus (Walker, Wilson and Gordon, 1983).

The lack of a strong causal relationship between ovarian hormones and aggression in female mammals is evident in the complex and conflicting effects of ovariectomy and hormone replacement. Aggression in female lemmings is increased by ovariectomy and reduced by injections of progesterone and oestrogen (Huck *et al.*, 1979). Depending on housing conditions and on the gender of the opponent, the same effects are sometimes found in hamsters, with progesterone decreasing aggression on oestrogen-primed ovariectomized females (see Siegel, 1985 for a review). Attack on a female intruder by female rats housed in small groups persists after ovariectomy; however, injections of oestrogen or of oestrogen and progesterone enhance their aggression to female (but not male) intruders (De Bold and Miczek, 1981).

The behaviour of female mammals can be modified by injections of androgens. This is not as unnatural as it might seem, since they do

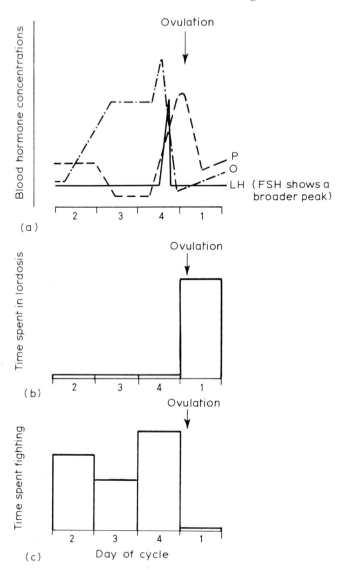

Figure 5.2 The oestrus cycle and behavioural changes in hamsters. (a) Hormone changes during the oestrus cycle (after Siegel, 1985), P, progesterone; O, oestrogen, LH, luteinising hormone, FSH, follicle stimulating hormone. (b) sexual behaviour (lordosis) during the oestrus cycle (after Floody and Pfaff, 1977). (c) aggressive behaviour (rolling fights) during the oestrus cycle (after Floody and Pfaff, 1977).

naturally produce these hormones (in the adrenal cortex). For example, testosterone injections alter agonistic behaviour (usually increasing aggressiveness) in females of various species including deer, cattle (see Bouissou, 1983, and references therein), guineapigs (Goldfoot, 1979), voles (Taitt and Krebs, 1982) and mice (Barkley and Goldman, 1978). Prolonged exposure to testosterone increases attack rates of ovariectomized, group housed female rats when responding to a male but not to a female intruder (De Bold and Miczek, 1981).

The role of prolactin in the control of maternal aggression in mammals is also unclear. Lactating female rodents and ungulates are extremely fierce (mice, Svare, 1981; gerbils, Leshner, 1978; ibex, Nievergelt, 1974; deermice, Gleason, Michael and Christian, 1980; hamsters, Wise and Prior, 1977). Lactating ground squirrels are both aggressive and inclined to show infanticide (Betts, 1976; Hoogland, 1985). In hamsters, maternal aggression is reduced by prolactin-suppressing drugs and this is reversed by prolactin injections (see Siegel, 1985).

In contrast, in laboratory mice, on a day-to-day level, aggressiveness does not correlate with prolactin levels. Prolactin-depleting drugs do not impair maternal aggression, nor do prolactin injections enhance it (Broida, Michael and Svare, 1981; Svare, 1983). Non-lactating female mice which have been induced to suckle experimentally show maternal aggression, so any effect of prolactin in this species appears to be via the very indirect route of promoting suckling (Svare et al., 1980).

5.4.2 Reproductive hormones in other female vertebrates

The endocrine system of non-mammalian vertebrates is broadly comparable to that of mammals, in that the same or similar hormones are produced by the same or similar glands. The tiny amount of information available for other vertebrates suggests that the effects of female gonadal steroids on agonistic behaviour are variable. In electric fish the rate at which pulses are produced influences agonistic interactions (page 67). In *Sternopygus dariensis* females generate pulses at a higher frequency than do males. This sex difference seems to depend on levels of gonadal steroids, since discharge frequency is increased by oestrogen injections and reduced by androgens (Meyer, 1983). Ovariectomy increases and systemic oestrogen injections reduce aggression in mature females of several fish species including sticklebacks and blue acara (cichlids, Liley, 1969; see also Munro and Pitcher, 1983 and references therein). However, neither ovariectomy nor oestrogen injection influences aggression in female green anole lizards (Greenberg and Crews 1983 and references therein).

Some androgens (dihydrotestosterone but not testosterone propionate) induce aggressive behaviour as well as male-like courtship in ovariectomized female green anole lizards (Greenberg and Crews, 1983 and references therein). Some species of whiptale lizards (*Cnemidophorus uniparens*) consist only of females, which produce young from unfertilized eggs (parthenogenesis). Females have colour patterns and scent glands similar to those of males in related species. They also fight and show pseudocopulatory behaviour, both as a dominance display and to stimulate egg production. This male-typical display can be induced by androgen injections but in unmanipulated females its occurrence correlates with a surge in progesterone production at ovulation. Progesterone may induce pseudocopulatory behaviour by acting on androgen receptors in the brain, which in related species induce male behaviour (Crews and Moore, 1986). Female phalaropes are particularly aggressive, competing with each other for mates. They are reported to have unusually high androgen levels (Hohn, 1970) which might explain their aggressiveness. However, recent studies on spotted sandpipers (another polyandrous bird) failed to confirm this finding (Rissman and Wingfield, 1984). Testosterone levels peak in female white crowned sparrows at the time when they participate in territorial defence (Wingfield and Farner, 1978).

5.4.3 Reproductive hormones in male mammals

As predicted, androgens are indeed involved in the control of fighting between adult male mammals of many species, but this relationship is variable. The more strongly immediate access to females depends on fighting, the closer the link between androgens and aggression. Where males compete for mates as they appear, fighting with rivals is often controlled, in part at least, by the gonadal steroids. Where competition for females is indirect (via competition for status, as in primates) or in advance (when territories need to be won, boundaries marked and nests prepared), the link is looser.

(a) Seasonal changes

Increased testosterone secretion is accompanied by enhanced aggressive behaviour, scent production and muscle development during a seasonal period of mating or rutting in red deer (Fig. 5.3) and various other species of mammals. These include elephants (Poole *et al.*, 1984), camels (Yagil and Etzion, 1980), rams (Lincoln and Davidson, 1971), white tailed deer (Brown *et al.*, 1983), reindeer (Leader-Williams, 1979), roe deer (Sempere and Lacroix, 1982) and ground squirrels (Betts, 1976; Dunford, 1977). All this suggests a causal link between enhanced aggression (and all the trappings of fighting) and increased testosterone production.

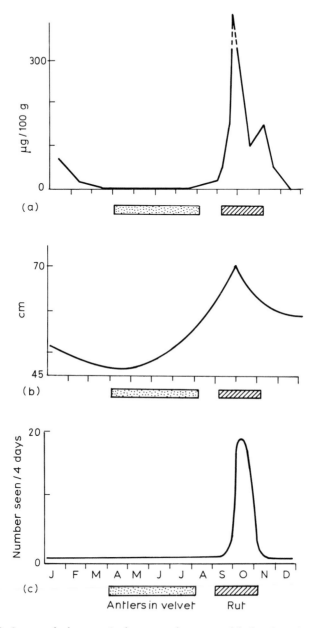

Figure 5.3 Seasonal changes in hormonal state and behaviour in red deer stags. (a) Monthly testosterone levels (after Clutton-Brock *et al.*, 1982); (b) monthly neck girth (after Lincoln, 1971); (c) monthly occurrence of fierce fights between adult males. Outside the rut, both males and females engage in occasional low-intensity threats and fights (after Clutton-Brock *et al.*, 1982).

In contrast, male hamsters become less aggressive as the photo-period increases (in nature, signalling the approach of the breeding season) and testosterone secretion picks up. This allows females safe access to the territories of these normally solitary and intolerant animals (Garrett and Campbell, 1980). In primates the higher test-osterone levels found in breeding males are not usually accompanied by increased aggressiveness, although, in male squirrel monkeys, testosterone stimulates lipid deposition, thus increasing body size and enhancing status. Male rhesus monkeys show more aggression during the breeding season, but this is an indirect effect of more frequent encounters over receptive females (see Dixson, 1980 and references therein).

(b) Between-individual variation

Relatively long-term differences in aggressiveness among male mam-mals can be related to differences in androgen production. In feral baboons, testosterone levels correlate with readiness to initiate a fight (Sapolsky, 1982). Encounters between male guineapigs are more likely to escalate if the two animals have similar testosterone levels prior to the fight (Sachser and Prove, 1984). Testosterone levels are higher in aggressive, dominant males compared to subordinates in rats (Schuurman, 1980), mice (Lagerspetz et al., 1968; Selmanoff et al., 1977), reindeer (Stokkan, Howe and Carr, 1980) and rhesus monkeys (Rose, Holaday and Bernstein, 1971), although the hormonal varia-tion may well be an effect rather than a cause of the behavioural difference.

(c) Short-term, reversible fluctuations

In rats, temporal changes in aggressiveness correlate with shifts in the concentration of testosterone in the blood (averaged to allow for very rapid fluctuations, Schuurman, 1980) and with the activity of the enzymes which convert testosterone to oestrogen (see below, Dessi-Fulgheri et al., 1976).

(d) Castration and androgen replacement

Castration reduces and testosterone injections reinstate aggression in various male mammals (Table 5.2). The many studies of laboratory mice and rats show that the behavioural changes following castration can be complex.

- Castration severely reduces levels of aggression to adult males of several strains. Testosterone increases aggression in castrates in a dose-dependent manner, but this effect tails off at very high doses

Table 5.2 Examples of the effects of castration and androgen replacement in adult mammals.

Species	Result	Reference
Hamsters	Castration reduces and testosterone (T) reinstates scent marking. Castrates show less aggression in neutral area and are attacked more by females. T increases attack on males (not females) and enhances status. Some conflicting results.	Johnston (1981); Siegal (1985), review
Voles	Castration reduces aggression in free-ranging voles. Testosterone increases aggression.	Turner *et al.* (1980); Gipps (1984)
Gerbils	Castration reduces scent marking and reduces attacks received but has little effect on aggression. T reinstates marking.	Yahr (1983)
Cats	Castrates show a decline in urine marking and fighting.	Hart and Barrett (1973)
Dogs	Urine marking and aggression in home area not reduced; both lower outside home area.	Hopkins *et al.* (1976)
Red deer	Castration abolishes all aspects of rutting, including aggression. T reverses the effect, but depending on the time of year.	Lincoln *et al.* (1972)
Tamarin monkey; Rhesus monkey	Castration has little effect on aggression or scent marking.	Epple (1978, 1981); Wilson and Vessey (1968)
Talapoin monkey	Castration does not reduce aggression; T injections increase attacks to subordinate males but not dominants or females.	Dixson and Herbert (1977)

(Fig. 5.4; Selmanoff *et al.*, 1977; Bermond, 1978; Brain, 1980).
- Castrates have smaller preputial glands, show less urine marking and elicit fewer attacks from intact males (Lee and Ingersoll, 1983; De Bold and Miczek, 1981; Yahr 1983). Testosterone injections reverse these effects.

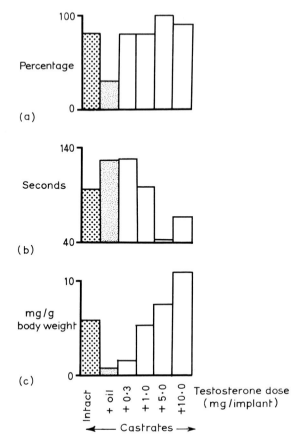

Figure 5.4 The effects of castration and androgen replacement on aggressive behaviour and seminal vesicle weight in adult male mice (after Barkley and Goldman, 1977). (a) Percentage attacking; (b) latency to attack; (c) seminal vesicle weight.

- Castration reduces infanticide (Gandelman, 1983) but increases aggression by group-housed males to lactating females (Brain, 1980); both are reversed by testosterone treatment.
- Castrates no longer discriminate between urine from donors of different status (Sawyer, 1980), but this is reinstated by testosterone.
- Neither castration nor testosterone treatment has an effect on retaliation, submission or on learned avoidance of aggressive opponents (Schuurman, 1980; Leshner, 1983).

Testosterone produces its direct behavioural effects by acting on specific brain regions. Various gonadal steroids are accumulated by the brain and implants of testosterone (for example, in the preoptic area of castrated rats enhance offensive behaviour and partially reinstate dominance (Bean and Conner 1978; Bermond, 1978). Castration raises and systemic testosterone injection lowers the threshold current required to elicit attack from the hypothalamus of rhesus monkeys (page 145; Perachio, 1978).

Testosterone seems to act on the brain after conversion (aromatization) to oestrogen, since its aggression-enhancing effects can be blocked by interfering with this conversion (in some strains of mice at least; Brain, Hang and Kamis, 1983; Simon, 1983). The effect of testosterone on scent production depends on conversion to dihydrotestosterone (DHT; Christie, 1979; Clark and Nowell, 1979; Brain, 1980, 1983).

(e) Complications

One reason why an effect of altered androgens on agonistic behaviour is by no means universal in male mammals (Table 5.2) is that social and experiential factors are also important determinants of agonistic behaviour. Castration has a much less marked effect on aggression when the subject is on its home territory (Christie and Barfield, 1979; Schuurman, 1980, in rats; Edwards, 1969, in mice) or if it had past experience of fights (Schechter and Gandelman, 1981, in rats).

The effects of experience and context which complicate the relationships between testosterone and aggression in rodents are even more important in the primates. Although androgens have peripheral effects on agonistic cues in some cases, they generally have little effect on aggressiveness or status, which are strongly dependent on social factors. Neither scent marking nor the intense aggression shown by male tamarins to strangers are affected by castration in adulthood (Epple, 1978, 1981). Exogenous testosterone has no consistent effects on aggressiveness and no effect at all on status in rhesus monkeys (Gordon et al., 1979). Where other relevant factors are evenly balanced, however, altered concentrations of testosterone can influence dominance interactions, tipping the balance in favour of the animal with the highest levels (Dixson, 1980), as in talapoins (Dixson and Herbert, 1977) and rhesus monkeys (Cochran and Perachio, 1977).

5.4.4 Reproductive hormones in other male vertebrates

More is known about the role of reproductive hormones in male than in female non-mammalian vertebrates. As in mammals, the pituitary–gonadal hormones are involved in the control of agonistic behaviour,

Table 5.3 Effect of castration and androgen replacement in some non-mammalian vertebrates.

Species	Result	Reference
Haplochromis burtoni	Testosterone (T) increases attack rate and enhances colouration.	Fernald (1976)
Paradise fish	Antigonadal agent impairs and T reinstates aggression.	see Villars (1983)
Sunfish	Aggression persists after castration but site-related component is lost.	Smith (1969)
Siamese fighting fish	Castration has no effect on aggression.	Weis and Coughlin (1979)
Skinks	T increases aggression and status in subordinate males.	Done and Heatwole (1977)
Box turtles	T implants increase aggression and promote growth of claws and bright eye colouration.	Evans (1952)
Cockerels	Castration reduces and T reinstates aggression.	Berthold (1849) cited in Harding (1983)
Grouse	T implants increase aggressiveness in males on winter feeding territories and increase comb size.	Watson and Parr (1981); Harding (1983) and references therein
Redwinged blackbirds	T implants enhance escalated attack on subordinates but not success at supplanting rivals.	Searcy and Wingfield (1980)
Pied flycatchers	Males with T implants fight more, have larger territories and more mates, but do not fledge more young.	Silverin (1980)

but this relationship is variable and depends on the context in which fights occur. Table 5.3 summarizes data on the relationship between reproductive hormones and agonistic behaviour in males of selected species of non-mammalian vertebrates. A few examples are discussed in the text for the purposes of illustration and reviews of the various groups can be found in Brain (1980) and Svare (1983).

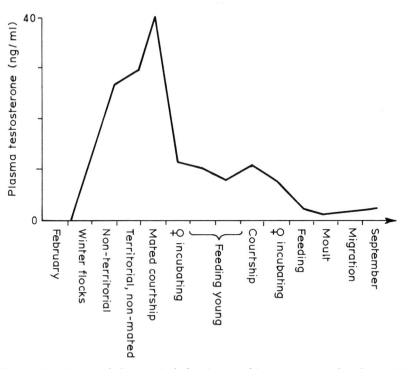

Figure 5.5 Seasonal changes in behaviour and in testosterone levels in white crowned sparrows (after Wingfield and Farmer, 1978).

(a) Seasonal changes

Full establishment of aggressive behaviour during the breeding season in green anole lizards depends closely on increased secretion of gonadal hormones. As androgen production starts up, males of this species develop secondary sexual characteristics and start to compete for females (Greenberg and Crews, 1983). In white crowned sparrows, territorial aggression and song production are most intense when testosterone levels peak in May and June. All three fall during incubation, finally dropping to a low level when the territory is abandoned at the end of the breeding season (Fig. 5.5; Wingfield and Farner, 1978; see Wingfield and Ramenofsky, 1985 for review).

However, in other species fighting is not just about females, so agonistic behaviour persists beyond the breeding season and is not dependent on reproductive hormones. In both sheathbills and in western gulls breeding territories are retained during the winter (for feeding or for safe-keeping). Except for a small increase during the mate guarding period in gulls, testosterone levels are unrelated to agonistic behaviour in these species (Burger and Millar, 1980; see also Wingfield and Ramenofsky, 1985).

In male three-spined sticklebacks, where territories are established and nests built early in the breeding season, aggressiveness is not directly influenced by testosterone levels, but depends instead on gonadotropin (see Fig. 5.1). Seasonal changes in agonistic behaviour parallel gonadotropin production; injections of this hormone increase attack rates and chemical suppression of gonadotropin production has the reverse effect (Wootton, 1976 and references therein). However, recent genetical studies (page 187) have thrown doubt on this conclusion.

(b) Between-individual variation

In young male cichlids (*Tilapia maria*) the length of the genital papilla (a protrusion from the genital aperture) correlates with plasma androgen levels. When two males fight, the fish with the larger papilla, and therefore higher androgen levels, generally wins (Schwank, 1980). Aggression during a fight by two adult male Japanese quail, who have not fought before, is related to their plasma testosterone levels prior to and shortly after the initial encounter; this relationship disappears in subsequent fights as dominance relationships are established and testosterone levels in the winner decline to those of the losing bird (Ramenofsky, 1984, Wingfield and Ramenofsky, 1985).

(c) Reversible fluctuations

Quite rapid shifts in agonistic behaviour may be caused by changes in hormone metabolism. For example, the sight of a female raises androgen levels in male ring doves (O'Connell *et al.*, 1981). This initially causes aggressive courtship, but since the androgens are converted to oestrogen in the pre-optic region of the brain, the activity of the aromatization enzyme is enhanced, leading to accelerating oestrogen production and a rapid switch from aggression to nesting (Hutchinson and Steimer, 1983).

(d) Castration and testosterone replacement

Castration eliminates bright body colouration and scent production, as well as agonistic behaviour, in green anoles. Systemic injections and implants of testosterone into the brain reinstate aggression in castrates (Morgentaler and Crews, 1978; Greenberg and Crews, 1983). In a number of other non-mammalian species, the effects of castration and testosterone replacement are more complex. Castrated gouramis fail to defend their nest sites but still fight in non-reproductive contexts (Johns and Liley, 1970). Following surgical or chemical castration, three-spined sticklebacks lose their bright colouration and their kidneys cease to produce the glue needed to maintain their nests. Since sticklebacks fight more when they have intact nests, castration reduces aggression, but only by a very indirect route (Rouse et al. 1977; Wootton 1976). Castration and testosterone treatment fail to disrupt established dominance relationships in quail or to alter territorial boundaries in grouse, although levels of aggression are modified (Trubec and Oring, 1972; Tsutsui and Ishii, 1981).

(e) Complications

These last examples show that, as in mammals and even in species where androgens are clearly implicated in the control of agonistic behaviour, the relationship is complicated by social experience. Testosterone implants in castrated song sparrows fail to increase aggressiveness in established groups but a transient increase is seen towards an unfamiliar bird (Wingfield and Ramenofsky, 1985). Other aspects of the context of an agonistic encounter are also important; for example, castration has a much less marked effect on aggression in green anoles when encounters take place on the subject's territory rather than in an unfamiliar environment (Greenberg and Crews, 1983).

5.4.5 Reproductive hormones and the development of agonistic behaviour in mammals

A clear temporal correlation exists between the development of sexual maturity, with its consequent increase in androgen levels, and the appearance of adult agonistic behaviour (including infanticide, Gandelman, 1983) in mice (Barkley and Goldman, 1977; Vom Saal, 1983), rats (Meaney and Stewart, 1981) and bank voles (Gipps, 1984). Although morphological changes related to fighting are observed at puberty in primates (skin colouration in patas monkeys and large size in talapoins), neither aggression nor status increase, social factors being more important than hormones in this context (Dixson, 1980; see also page 215).

Exposure to androgens at an early stage in development can have marked effects on the way adult agonistic behaviour is organized in mammals, although it is not at all clear how these effects come about. Adult male mice castrated just after birth take longer to fight when injected with testosterone as adults, and require higher doses before they do so than, for example, males castrated just before puberty (Fig.

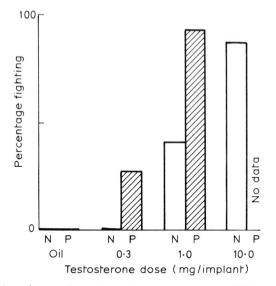

Figure 5.6 Attack rate four days after treatment at different testosterone doses in isolated male mice castrated at birth or castrated before puberty (after Barkley and Goldman, 1977). N, neonatal castrate; P, prepubertal castrate.

5.6). On the other hand, (ovariectomized) females given injections of testosterone soon after birth more readily show aggressive behaviour when treated with testosterone as adults. Treatment of females at a later date is also effective, but much larger doses are required (Leshner, 1978; Vom Saal, 1983; Gandelman, Rosenthal and Howard, 1980; Mann and Svare, 1983). These effects may depend on conversion of the early dose of testosterone to oestrogen (Vom Saal, 1983).

In addition to their obvious implications for the development of gender-specific responses, differences in the early hormone environment may be responsible for some naturally occuring variation in aggressiveness between members of the same sex. Female mice foetuses which lie in the womb (uterus) between two males have unusually high testosterone levels in their blood. As (ovariectomized) adults they show a stronger testosterone-induced aggressive response to males than do females which grow *in utero* next to just one male or to none at all (Gandelman *et al.*, 1977; Vom Saal 1983). Males developing between two females are relatively insensitive as adults to the aggression-enhancing effects of testosterone (Vom Saal, 1983).

Similar effects on agonistic and related behaviour of the hormonal environment in which early development takes place has been reported for rats (Barr, Gibbons and Moyer, 1976), hamsters (Payne, 1977), dogs (Beach, Buchler and Dunbar, 1982) and cattle (Bouissou and Gaudioso, 1982).

Female spotted hyaenas have a false penis and scrotum which they use during social displays; females reproduce normally but are extremely aggressive and compete for dominance with members of both sexes. They have unusually high androgen levels both as adults and as foetuses; these high androgen levels in early life are probably responsible for their morphological masculinization and may cause their aggressiveness (Racey and Skinner, 1979). Female rhesus monkeys exposed to high androgen levels during foetal life have slightly masculinized genitalia but show menstrual cycles and can reproduce. When young, these masculinized females engage in rough and tumble play and fighting more frequently than do normal females. They are also more aggressive as adults, but aggressiveness is not further enhanced by testosterone treatment (Dixson, 1980 and references therein).

5.4.6 Reproductive hormones and the development of agonistic behaviour in other vertebrates

Increased testosterone synthesis at sexual maturation is accompanied by the development of an anal scent gland and the appearance of crowing and strutting (shown by adult males during courtship and aggression) in Japanese quail (Ottinger, 1983). Levels of aggression rise sharply in paradise fish as they reach reproductive maturity (Davis and Kassel, 1975).

Early exposure to gonadal steroids has a profound effect on morphology in non-mammalian vertebrates, often effecting complete functional sex reversal. There is a little, scattered evidence suggesting that this can also influence development of aggressive behaviour. In birds, the female is the heterogametic sex (having the XY rather than the XX sex chromosome complement of female mammals) and both structure and behaviour are influenced by exposure to gonadal steroids during development; for example, turkeys develop male plumage patterns in the absence of oestrogen (Scott and Payne, 1934). Male chicks treated as embryos with testosterone or oestrogen do not show crowing and strutting as adults. Females treated with anti-oestrogens during development respond with male behaviour to testosterone injections as adults (Adkins-Regan, 1983). In contrast, the brain mechanisms controlling song in zebra finches (normally a male prerogative) are masculinized by an oestrogen present in male but not female chicks (Hutchinson, Wingfield and Hutchinson, 1984). Long term post-hatching testosterone treatment reduces chest fighting and feather pecking in male ducklings (Deviche and Balthazart, 1976). Exposure to androgens during development in cichlid fish (*Oreochromis mossambicas*) enhances aggressiveness in both males and females (Billy and Liley, 1985). Morphological and behavioural masculinization by high androgen levels in development may be responsible for the aggressiveness of parthenogenetic female whiptail lizards (see page 104; Crews *et al.,* 1983).

5.4.7 Fighting experience and reproductive hormones in mammals

The fact that social experience can alter endocrine activity was mentioned previously (page 97) as a technical complication and is considered later (Chapter 9) as one of the many consequences of agonistic behaviour. Here it is discussed briefly as one mechanism by which behaviour during conflicts is continually adapted to social circumstances. For example, the presence of potential mates enhances testosterone levels in rats, sheep, white crowned and song sparrows and ring doves, to give just a few examples (Kamel *et al.*, 1975; Illius *et al.*, 1978; Moore, 1984; Wingfield, 1985). As a result, males fight most fiercely when they have something worth defending.

In general, short-term increases in testosterone levels are observed at the start of a fight, probably contributing to its escalation. On a longer time scale, levels stay high in the victor and fall in the loser (Bronson and Stetson and Stiff, 1973; Sachser and Prove, 1984), so that dominant and subordinate animals differ in hormonal state. The difference is reversed when the subordinate animal is artificially rendered dominant (mice: Machida, Yonezawa and Noumara, 1981;

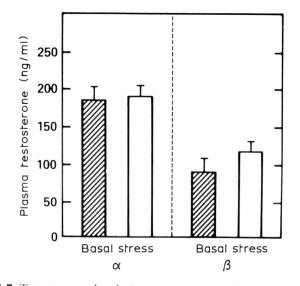

Figure 5.7 Testosterone levels (means ± s.e.) in dominant (α) and subordinate (β) male squirrel monkeys in stressed and unstressed conditions (reproduced with permission from Coe *et al.*, 1979).

Ely and Henry, 1978; rats: Schuurman, 1980; bank voles: Gustafsson, Andersson and Meurling, 1980; red deer: Bartos, 1980). The effects of winning and losing a fight on hormone production perpetuate and exaggerate small status imbalances. They provide a method by which dominant animals incapacitate reproductive competitors (psychological castration) but also allow weak fighters to avoid continual involvement in hopeless fights.

Socially-induced changes in hormone secretion have been particularly well documented in primates. Dominant squirrel monkeys, rhesus monkeys and talapoins have higher testosterone levels than do subordinates and these rise and fall with changes in status (Fig. 5.7; Perachio, 1978; Coe, Mendoza and Levine, 1979; Eberhart, et al., 1980). Resting levels of testosterone are independent of status in male baboons observed in natural conditions, but in response to stress, they rise in dominants and fall in subordinates (Sapolsky, 1982). The hormone system of females is also influenced by dominance interactions; subordinate female talapoins fail to show the oestrogen-induced surge of gonadotropin (luteinizing hormone) which elicits ovulation in natural cycles (see Fig. 5.2; Bowman, Dilley and Keverne, 1978; Keverne, 1979).

5.4.8 Fighting experience and reproductive hormones in other vertebrates

Rapid changes in circulating hormone levels have been observed in birds following a bout of agonistic behaviour. Territorial red-winged blackbirds caught while fighting show more variable testosterone levels than those caught while feeding. Hormone levels are very high in the most aggressive of the fighting birds and very low in those showing more fearful responses (Harding and Follet, 1979; Harding, 1983). In song sparrows, short, simulated territorial intrusions elevate testosterone levels, and males in the process of setting up territories have higher testosterone levels than those on established territories. Testosterone levels therefore rise when the birds need to fight but fall once boundaries are established and costly aggressive behaviour is no longer necessary (Wingfield, 1985).

In male swordtails, both winners and losers of pairwise encounters experience a rapid decrease in testosterone levels. This effect is weaker and shorter in winners, who end up with higher androgen levels than their defeated rivals (Hannes, Franck and Liemann, 1984). Dominant females in groups of cichlids (*Haplochromis burtoni*) take on male colouration, perhaps because their androgen levels rise (Wapler-Leang and Reinboth, 1974). Socially-induced hormonal changes of this sort may well be going on during sex reversal in regularly hermaphrodite fish (page 28).

5.5 ADRENAL HORMONES

Production of adrenaline (under neural control) and of glucocorticoids (stimulated by adrenocorticotrophic hormone or ACTH, see Fig. 5.1) causes a complex of physiological changes (such as increased blood pressure and carbohydrate metabolism) allowing animals to respond vigorously to environmental events and modulating the effect of long-term stress. Fights demand energetic action and their consequences can be stressful, so the pituitary–adrenal hormones are likely to play a role in agonistic behaviour.

5.5.1 Adrenal hormones in mammals

Secretion of catecholamines by the adrenal medulla is stimulated during and after an agonistic encounter in mice. Conversion of noradrenaline to adrenaline is enhanced in subordinates, probably as a result of glucocorticoid action on medullary function, so that dominants have higher noradrenaline:adrenaline ratios than do subordinates (Brain, 1980).

In mice (Brain, 1980) and rats (Schuurman, 1980) both participants in a fight experience an increase in circulating ACTH and glucocorticoid levels, but this is stronger and longer lasting in losers (Fig. 5.8). Even after glucocorticoid levels have returned to normal, losers show enhanced secretion of these hormones when faced with a potential opponent. ACTH levels increase in proportion to the number of attacks received when rhesus monkeys are introduced into a strange captive group (Scallett, Suoni and Bowman, 1981) and male vervet monkeys have high blood concentrations of glucocorticoids when they are ascending in status. In stable hierarchies, dominant male talapoin monkeys and free-living baboons have lower glucocorticoid levels (Bowman et al., 1978; Keverne, 1979; Sapolsky, 1982).

Differences in glucocorticoid levels, socially induced or otherwise, can influence agonistic behaviour. Young rhesus monkeys with low plasma glucocorticoid levels are more likely to win fights than are those with high levels (Golub, Sassenrath and Goo, 1979). There is experimental evidence in mice that both ACTH and glucocorticoid hormones influence agonistic behaviour, but because of the negative feedback relationship between these hormones, the effects are complex. Glucocorticoids have a direct facilitatory effect on submissiveness; they also increase aggression, but indirectly by feed-back suppression of ACTH production. After an initial facilitation, in the long term, ACTH has a direct, inhibitory effect on aggression; it also enhances submissiveness, but indirectly via the glucocorticoid production that it causes (Brain, 1980; Leshner, 1983).

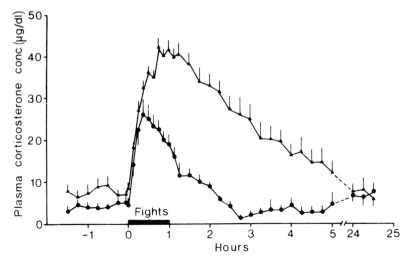

Figure 5.8 The effects of agonistic experience on glucocorticoid (corticosterone) levels (mean ± s.e.) in winning (●) and losing (▲) rats; (reproduced with permission from Schuurman, 1980).

Glucocorticoids can cross the blood–brain barrier and are taken up by neurons in the brain, including those in the limbic system, which is implicated in the control of agonistic behaviour (page 150). ACTH does not cross the blood–brain barrier (Adams, 1983 and references therein), but peptide fractions similar to parts of the ACTH molecule improve memory retention in avoidance learning tasks in mice (Flood et al., 1976). In the context of an agonistic encounter, this effect on memory would consolidate the behavioural effects of defeat.

So experience in fights alters production of pituitary–adrenal hormones and these are known to influence agonistic behaviour. There is some evidence to suggest that this represents one route by which behaviour is modified when an animal is losing a fight. Preventing fight-induced hormonal changes (by removing the pituitary and injecting fixed amounts of ACTH) delays establishment of dominance–subordinance relationships in mice. Accentuating hormone changes following defeat (by injections of small amounts of either ACTH or corticosteroid) accelerates and enhances submissive behaviour (Leshner, 1983).

5.5.2 Adrenal hormones in other vertebrates

Production of catecholamines and glucocorticoids is influenced by fighting experience and social status in vertebrates other than mammals. For example, both participants in a fight between male sword-

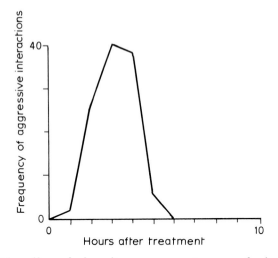

Figure 5.9 The effect of adrenaline on aggressiveness of subordinate male skinks (after Done and Heatwole, 1977).

tails experience a rapid increase in plasma corticosteroids but this effect is weaker and shorter in the winner (Hannes *et al.*, 1984). Experience-mediated changes in the activity of the adrenal gland (or its equivalent, see Fig. 5.1) generally suppress gonadal activity; in rough-tailed newts, for example stress reduces secretion by the brain of the chemicals that promote LH release (Moore and Zoeller, 1985). However, in female paradise fish increased glucocorticoid production caused by attacks from males induces ovulation and stimulates sexual receptivity (Villars, 1983).

In unstable hierarchies, subordinate green anoles have higher gluco-corticoid levels than their dominant counterparts; these depress gonadal function and speed up conversion of noradrenaline to adrena-line. Since chromatophore pigments are dispersed by adrenaline (producing a general darkening of the skin) and aggregated by nor-adrenaline (making the skin bright green), dominant lizards are green and subordinates brown (Crews and Greenberg, 1981; Greenberg, 1983). The relationship between status and adrenal activity can be complex. Dominant Harris' sparrows have lower glucocorticoid levels in snow-free conditions but after heavy snowfall the situation is re-versed because the larger dominants suffer more sever nutritional stress (Rohwer and Wingfield, 1981).

As far as influences on agonistic behaviour are concerned, the adrenal hormones play a special role in reptiles, because of their effects on body temperature and colouration. Secretion of noradrenaline and adrenaline rises sharply when fence lizards exchange agonistic displays. This increases metabolic rate, raises their temperature and so allows these ectothermic (or cold–blooded) animals to fight more energetically (Engbretson and Livzey, 1972). In male skinks, injections of adrenaline increase aggressiveness but fail to influence dominance relationships (Fig. 5.9; Done and Heatwole, 1977).

5.6 ENDOGENOUS OPIATES

It has recently been discovered that the body produces its own opiates. These are peptide molecules that are produced by neurosecretion and that bind to specific sites (opiate receptors) in the brain. One class of these molecules, the endorphins, is found in the pituitary gland and so we discuss them here. These naturally produced opiates seem to be part of a system for suppressing pain. This being the case, a number of studies have looked for an influence of endogenous opiates on agonistic behaviour in animals.

The results are variable but on the whole they suggest that endogenous opiates may sometimes be both a cause of differences in agonistic responsiveness and a consequence of fighting experience. Injections of endorphin suppress defensive, shock–induced fighting in various laboratory rodents; naloxone (a drug which blocks both endorphin and encephalin receptors) enhances this behaviour in rats, although not all opiate antagonists have this effect (Tazi et al., 1983). Thus artificially induced differences in endogenous opiate activity can modulate agonistic behaviour. Spontaneously aggressive mice have low endogenous opiate levels, so similar effects may contribute to natural variation in agonistic responsiveness (Ray, Sharma and Sen, 1983).

Prolonged exposure to intense attack (for example from a lactating female in rats or from an isolated territorial male in mice) makes animals less sensitive to pain. This is partly counteracted by naloxone, implicating the endogenous opiates, but other factors (psychological rather than physiological) are involved (Rodgers and Hendrie, 1984). These authors suggest that animals engaged in a fight experience an initial increase in sensitivity to pain, which facilitates defence and escape. If escape is impossible and the attacked animal is involved in a hopeless encounter, a suppression of sensitivity to pain follows, partly as a result of enhanced production of endogenous opiates.

5.7 OTHER HORMONES

Several other components of the endocrine system are involved to a lesser extent in the control of agonistic behaviour. Vasopressin given immediately after a defeat increases future submissiveness in male mice (Roche and Leshner, 1979), possibly via its known effect of promoting avoidance learning (De Weid, 1980; Siegfried, Frischknecht and Waser, 1984). Vasopressin is released in response to various stressors. If it is also released in response to the stress of defeat, it will help a beaten animal to remember to show submission.

As a few final examples:

- Removal of the pineal gland (see Fig. 5.1) reduces territorial aggression in mice and injections of its product (melatonin) reverse this effect, which depends on functional adrenal glands (Patterson et al. 1980).
- MSH increases production of attack-eliciting smells in intact male mice (Nowell, Thody and Woodley, 1980) and rabbits (Ebling, 1977) and reduces territorial aggression in mice, probably by inhibiting melatonin release (Patterson *et al.*, 1980).
- Isolated male mice have high circulating thyroxine levels and this may contribute to their enhanced activity and aggressiveness (Valenti *et al.*, 1981). Prolonged thyroxine treatment increases territorial aggression in cichlids, probably via increased activity (Spiliotis, 1974).
- Insulin injections enhance aggression in rats, via the hypoglycaemia that they produce (Davis, 1978).

5.8 INVERTEBRATES

Very little is known about the behavioural effects of hormones in invertebrates, and most of the information that we do have concerns the insects and crustacea (Breed and Bell, 1983). Although the life style and endocrine systems of arthropods are very different from those of vertebrates, there are parallels to be drawn between the way hormones and agonistic behaviour interact in these two groups.

5.8.1 Reproductive hormones

In insects and crustacea, as in vertebrates, much aggression by males is against rivals for mates while females fight over other things, including the need to reject unwelcome suitors. The hormones that are involved in the physiology of sexual reproduction might therefore also influence fighting behaviour.

In many insects females are receptive only for limited periods; at other times they respond aggressively to copulation attempts. This behavioural shift is under the control of a hormone produced by the *corpora allata* (neurosecretory glands in the insect's brain). The products of these glands (sometimes called juvenile hormones) promote gonadal development in some female insects. Female grasshoppers (*Gomphocerus rufus*) consistently reject males if the *corpora allata* are removed; injections of gland extract suppress this aggressive response and promote receptivity. In the grasshopper *Diploptera punctata* the developing eggs produce a hormone which inhibits activity of the *corpora allata* so that the females respond to males with aggression rather than with sexual behaviour (Stay *et al.*, 1980).

Secretion of juvenile hormone is reduced in subordinate female wasps (*Polistes gallicus*) and in this case, both gonadal development and aggressiveness are reduced. Therefore, as in vertebrates, dominant animals breed without competition and subordinates avoid repeated defeats (Roseler *et al.*, 1980).

Male crustacea often have large claws, long spines and robust bodies compared to females and these are used in fights. Together with other secondary sexual characters, these weapons develop under the influence of a hormone produced by the androgenic gland, lying near the testis. In the shrimp (*Macrobrachium* spp.) claw size varies from male to male and this can be related to the volume of the androgenic gland. The variation in fighting efficiency which stems from differences in claw size depends on variable androgenic gland secretion (Nagamine and Knight, 1980).

5.8.2 Rapid behavioural effects of neurosecretory hormones

A second parallel to be drawn between invertebrates and vertebrates is that neurosecretory hormones can exert rapid effects on agonistic behaviour. During encounters between lobsters, a victorious animal stands high on the tips of its legs, with open claws held to the front. In contrast, a defeated animal collapses onto the ground with limbs folded and claws closed. Secretory cells in the nervous system of lobsters produce chemicals called serotonin and octopamine. These can remain in the nervous system where they act as neurotransmitters (page 133) but are also released into the body fluids where they act as hormones. Injections of serotonin produce a sustained 'winner's' posture; octopamine has the opposite effect. Serotonin and octopamine act on neuromuscular junctions of the flexor and extensor muscles which generates the two postures (page 138; Kravitz *et al.*, 1983).

5.8.3 Developmental effects

Permanent effects of hormones on behavioural development can be seen in social insects where caste determination depends on juvenile hormone secretion. When soldier termites are in short supply, some nymphs develop into presoldiers and then into soldiers; this involves a constellation of morphological and behavioural changes, including development of weapons and an increase in aggressiveness. Presoldiers have hyperactive *corpora allata* but these get smaller after the final, formative moult. Implanting or injecting juvenile hormones into newly moulted larvae makes them moult into presoldiers and then soldiers. In *Macrotermes*, the juvenile hormone content of newly laid eggs (derived from the queen) is variable and this may bias development towards or away from the soldier path. Existing soldiers may produce an anti-juvenile hormone which suppresses soldier development (Brian, 1979).

5.8.4 Moulting hormones

The need for periodic moults and the increased vulnerability that these entail have no obvious equivalent in vertebrates (except perhaps the vulnerability of parturient mammals, which makes them avoid or actively discourage social contact). The offensive component of agonistic behaviour is suppressed by a moulting hormone (ecdysterone, produced by an endocrine gland in the head), reducing levels of aggression at a time when this behaviour would be inappropriate. In lobsters threat drops to a minimum and avoidance peaks when levels of moulting hormone are high (Fig. 5.10; Tamm and Cobb, 1978). In stomatopods, injections of moulting hormone can reverse the rank of animals which would otherwise remain dominant (Steger and Caldwell, 1983).

5.9 OVERVIEW

5.9.1 Have our questions been answered?

The questions outlined at the start of this chapter can be answered in the affirmative:

1. *Hormones are implicated.* The examples given in the preceding sections show that, while hormones cannot be said to cause agonistic behaviour in any simple sense, in almost every species considered one or several hormones play a part in determining when and how fights occur.

2. *Permanent differences* between animals in their patterns of agonistic responses can sometimes be understood in terms of endocrinological differences. The aggressiveness of female parthenogenetic lizards, phalaropes and spotted hyaenas, when compared with related species, may be linked to high androgen levels both in development and in adulthood. Differences in patterns of agonistic behaviour between the two sexes are caused at least in part by different hormone levels

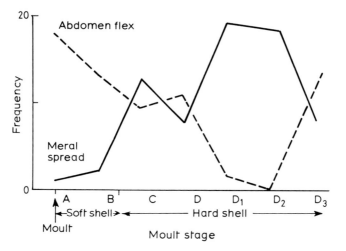

Figure 5.10 Changes in threat (meral spread) and avoidance (abdomen flex) over the moult cycle in lobsters (after Tamm and Cobb, 1978).

both during development and in adulthood in mammals, birds, lizards and fish (page 113).

Permanent differences in aggressiveness between individuals may stem from differences in hormonal environment either during development (page 114) or in adulthood (page 106). For adult animals, cause and effect are confused, but there are cases where hormones cause behavioural differences between individuals. The outcome of agonistic encounters in young rhesus monkeys (page 118) depends on their relative glucocorticoid levels and dominance of male guineapigs (page 106) and quail (page 112) can be predicted from plasma testosterone levels before a fight. Differences in fighting ability between male shrimps depend on hormone-induced differences in claw size (page 123).

3. *Development.* In males of some species (paradise fish, mice, quails but not primates, pages 113 and 115) the appearance of adult patterns of aggression can be related to the endocrine changes of puberty. As mentioned in the previous paragraphs, the early hormonal environment can have more-or-less permanent effects on the development of agonistic behaviour and related structures. Adult patterns of sensitivity to the aggression-promoting effects of androgens in mammals develop in response to foetal hormone levels (page 113). The equipment and behaviour necessary for fighting in soldier termites develop under the influence of high juvenile hormone levels (page 123).

4. *Reversible fluctuations.* Changing patterns of hormone secretion and metabolism are responsible for shifts in agonistic behaviour on a

number of time scales. Changes in agonistic behaviour over the moult cycle in crustacea depend on fluctuating levels of moulting hormone (page 124). In female grasshoppers (page 122) and hamsters (page 101), the endocrine states which accompany egg production bring with them a reduction in aggressiveness; in female rhesus monkeys, the converse occurs (page 101).

Rising gonadotropin levels induce territorial aggression in male sticklebacks at the start of the breeding season (page 112). In male green anoles (page 111), white-crowned sparrows (page 111) and deer (page 104), to name just a few examples, changes during the breeding season in aggression and in the equipment used in fights depend on seasonal increases in testosterone secretion. Again, primates do not show such a relationship between seasonal changes in reproductive hormone levels and aggressive behaviour.

Relatively rapid behavioural shifts can be caused by endocrinological events. For example, changing patterns of secretion of androgens (page 116) and of glucocorticoids (page 118) accompany and probably contribute to the behavioural shifts that occur during an agonistic encounter as fights warm up and then as winners and losers are determined (page 119).

5.9.2 General principles

Although sweeping statements to the effect that a particular hormone always controls aggression are bound to be wrong, generalizations can still be made about the role of hormones in agonistic behaviour:

1. Fighting being a stressful experience, ACTH, catecholamines and glucocorticoids (sometimes called the stress hormones) form part of the hormonal basis of agonistic behaviour in a number of vertebrate species. They can cause behavioural changes; ACTH inhibits aggression and glucocorticoids facilitate submissive responses (page 118). However, these hormones also vary in response to fighting experience; dominant animals often have lower levels of glucocorticoids and higher noradrenaline:adrenaline ratios (pages 118, 120). This two-way interaction contributes to the behavioural changes observed as fights are settled.
2. The relationship between the reproductive hormones and agonistic behaviour is finely and adaptively tuned to the context in which fighting occurs and to the needs of the animals concerned. Thus aggression serving purposes other than securing mates is independent of the gonadal hormones; this is the case for female of many species (page 101) and for male western gulls and sheathbills (page 111). On the other hand, where fighting is over reproductive issues a

causal link is found between reproductive hormones and agonistic behaviour. This is negative in grasshoppers and hamsters (pages 122, 101), where unreceptive females aggressively repel unwanted mates. The relationship is positive in female wasps (who compete over breeding rights and over brood care (page 123) and in males of many species where most fighting is over access to mates.

In males, the link between reproductive hormones and aggression is strong when fights are over immediate access to females and mating success is closely dependent on their outcome (mice, lizards, pages 106 and 112). It is weaker and less direct when males compete for females but when success depends, for example, on previously acquired status (as in primates, page 109) or resources (as in sticklebacks, page 112). The fact that testicular androgens enhance aggression in vertebrates ranging from fish to mammals suggests that a relationship between male reproductive hormones and aggression has a long evolutionary history. The control of agonistic characteristics by androgenic gland secretions in male crustacea suggests that links between reproductive hormones and aggressive behaviour have evolved independently in several different lineages.

3. Where an influence of a particular hormone on agonistic behaviour has been demonstrated, it often turns out to be very complex; this allows subtle and efficient control of the responses of fighting animals. The effect of testosterone on agonistic behaviour depends on other hormones, so a variety of different behavioural outputs is possible for a given level of androgen. Even when acting on a constant hormonal background, testosterone affects a number of target organs (muscles, scent glands, sense organs, the brain) whose threshold of response may vary. So the hormone is able to modify several different aspects of the agonistic response and call them into play in an appropriate sequence. Since testosterone exerts these different effects after conversion to different metabolites (oestrogen and DHT) they are potentially amenable to separate control.

4. In almost all the animal groups discussed, social experience can produce complex and adaptive changes in endocrine state. The increase in testosterone levels induced in male rats and doves (page 116) by the presence of a female means that these animals will fight more fiercely when a valued resource is at stake. The enhanced testosterone production caused by a short agonistic encounter (page 116) promotes and maintains effective fighting while the encounter lasts. The hormones that promote aggression (mainly androgens in vertebrates, page 116) are often depleted in defeated animals and those that promote submission (glucocorticoids, page 118) are enhanced. These endocrine changes protect subordinates living in permanent social groups from repeated involvement in hopeless but damaging

fights (Brain, 1980).

5. In many species (of vertebrate), factors such as experience and the context in which an agonistic encounter occurs modify the behavioural effects of a particular hormone. The more complex the social environment, the weaker the link between hormones and aggression. Particularly in primates, although gonadal steroids and glucocorticoids modulate agonistic behaviour, their influence is probably small compared with the role of other, social factors.

6
Neural mechanisms

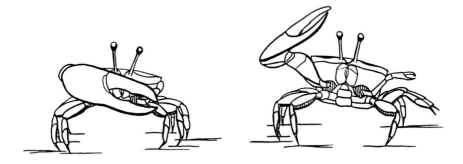

6.1 INTRODUCTION

An animal faced with a potential opponent must assess its species, gender, size and reproductive state. On the basis of this assessment, it must make decisions about whether and how to fight and must put these decisions into effect in the form of organized agonistic behaviour patterns. Individual animals vary in their sensitivity to agonistic cues and the responsiveness of an individual can alter on a number of time scales. In this chapter we look at studies whose aim is to explain these features of conflict behaviour in terms of neural mechanisms, seeking answers to the following questions.

1. How do the sensory pathways in the brain process the cues impinging on an animal during an agonistic encounter?
2. How are discrete, agonistic responses produced and integrated by the nervous system?
3. What neural mechanisms are responsible for generating differences in aggressive responsiveness?

The last of these questions is in some respects the most interesting (and the most difficult to answer) but a complete understanding of the role of the nervous system in the control of agonistic behaviour requires answers to all three.

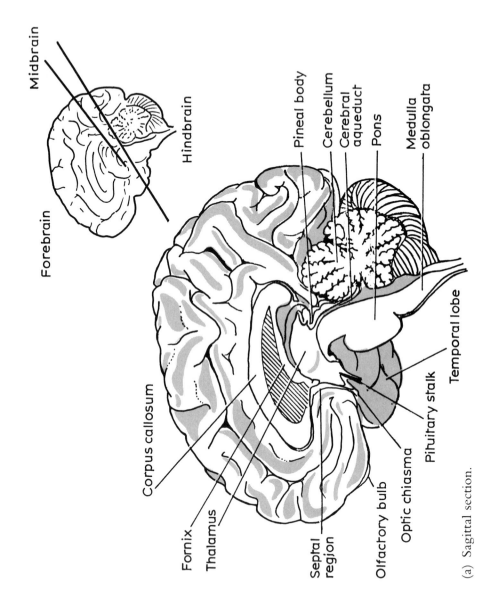

Forebrain

Midbrain

Hindbrain

Corpus callosum

Fornix

Thalamus

Septal region

Olfactory bulb

Optic chiasma

Pituitary stalk

Temporal lobe

Pineal body

Cerebellum

Cerebral aqueduct

Pons

Medulla oblongata

(a) Sagittal section.

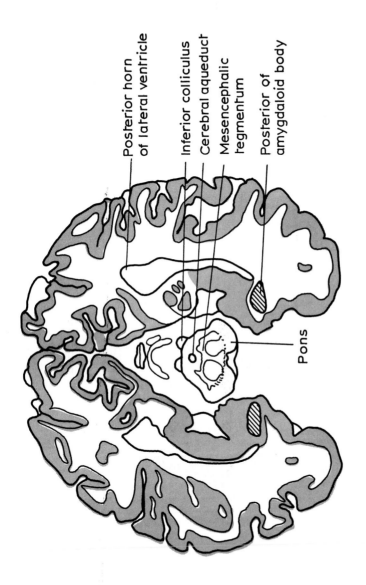

Posterior horn
of lateral ventricle

Inferior colliculus

Cerebral aqueduct

Mesencephalic
tegmentum

Posterior of
amygdaloid body

Pons

(b) Transverse section through the midbrain.

Figure 6.1 The brain of a mammal (*Homo sapiens*) (shading indicates grey matter). (Simplified from Niewenhuys, Voodg and Van Hurjzen, '1981.)

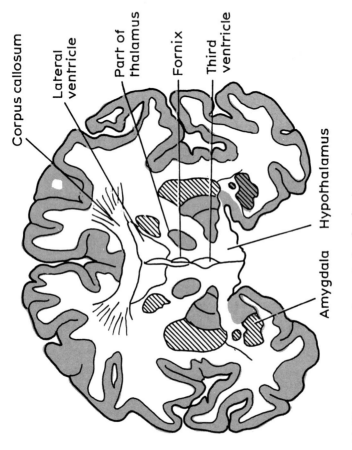

Corpus callosum

Lateral
ventricle

Part of
thalamus

Fornix

Third
ventricle

Hypothalamus

Amygdala

(c) Transverse section through the forebrain.

Figure 6.1 (*contd*)

In this chapter we assume a basic knowledge of the components of the nervous system and how these work, topics which are covered in any general textbook on biology. Figure 6.1 shows the gross anatomy of the mammalian brain, including the structures which will be discussed below. Table 6.1 lists some of the many chemical messengers (or neurotransmitters) of the mammalian brain and gives a brief description of their distribution. Most of the methods used to find out

Table 6.1 Some neurotransmitters of the mammalian central nervous system.

Chemical	Distribution	Physiological effects
Acetyl choline	Cholinergic nerves occur throughout the nervous system, especially in the autonomic ganglia.	Excitatory or inhibitory effects depending on nature of post-synaptic receptors.
Serotonin	Cell bodies in midbrain and lower brainstem project anteriorly to the hypothalamus, striatum and amygdala.	
Dopamine	Cell bodies lie in midbrain and project through brain stem, (especially in the medial forebrain bundle), including the hypothalamus, some fibres terminating in the ámygdala and septum but most in the striatum.	
Noradrenaline	Fibres arise in the pons and medulla and project via the brainstem to olfactory bulbs, hypothalamus, septum and amygdala and to the cerebral cortex.	
Glutamic acid	Abundant and widespread in the CNS.	Fast excitatory effect.
γ-Aminobutyric acid (GABA)	Abundant and widespread in the CNS.	Fast inhibitory effect.

how the nervous system performs the three tasks of processing agonistic information, producing agonistic action patterns and generating differences in agonistic responsiveness are variations of the following techniques.

- Relating activity in a particular neural structure to the perception of agonistic cues or to the performance of agonistic behaviour.
- Artificially activating a neural structure and observing the behavioural effects.
- Destroying or inactivating a neural structure and looking at the behavioural consequences.

For example, neurons in the lateral hypothalamus of cats (Fig. 6.1; see also page 145) are especially active during predatory attack. Stimulating this area with a small electric current triggers a natural-looking predatory attack in cats which do not normally kill rats. Destruction of the lateral hypothalamus eliminates predatory behaviour in rat killers. Each of these techniques has its limitations; for example, electrical stimulation induces an unnatural brain state and many other responses are absent following destruction of the lateral hypothalamus. However, in this case, all three point to the conclusion that the lateral hypothalamus is somehow involved in the control of predatory behaviour in cats.

Studies of brain–behaviour relationships have been greatly extended and strengthened by a number of recent neuroanatomical, neurophysiological and neuropharmacological developments. Special staining methods have identified a network of ascending and descending neurons which includes the lateral hypothalamic sites from which electrical stimulation elicits predatory attack in cats (Flynn et al., 1979). Biochemical stimulation has then been used to find out whether nerve cells lying in the lateral hypothalamus itself (as opposed to those in other parts of this network) are responsible for the attacks. Of many widely dispersed hypothalamic sites from which electrical stimulation evokes predatory attack in cats, just a few produce this effect when stimulated with the neurotransmitter glutamic acid. Unlike electrical stimulation, this chemical activates the cell bodies of neurons at the electrode tip but leaves nerve fibres unaffected. The cell bodies in the active sites, which are all clumped together, may be involved in the control of predatory attack (Bandler, 1982a).

6.2 NEURAL MECHANISMS AND THE PATTERNS OF ANIMAL CONFLICT

Of the various patterns of animal conflict outlined in Chapter 1, only a few have attracted the attention of neurobiologists. We know quite a lot about the neural mechanisms controlling production and recognition of mating calls in amphibia (Kelley, 1980; Walkowiak 1983) and of song in birds (Nottebohm, 1980). Both of these have a role to play in animal conflicts, but most research in this field concerns actual fighting. Many different patterns of fighting have been investigated; we have already given examples of the detailed work by Flynn and his collaborators on the neural control of predatory attack in cats. The so-called muricide response when rats attack and kill mice has also been intensively studied (Karli, 1981). These responses are sometimes referred to as aggression but the brain mechanisms that control predation and intraspecific fighting are at least partially distinct. Since intraspecific conflict is the main topic of this book, for the most part these cases are not discussed further here.

As we have stressed throughout this book, animals of the same species come into conflict with each other in many different contexts and over many different resources and outcomes. We should expect these different aspects of agonistic behaviour to have different neural bases, just as they depend on different hormones. Most of the research on neural mechanisms has concentrated on fighting between adult males but some other aspects of agonistic behaviour have been considered. The brain mechanisms underlying agonistic interactions between conspecifics have been studied in a wide range of species, from Spanish fighting bulls to ants and crayfish, taking in birds and fish on the way. The literature on this topic is enormous and our account is necessarily highly selective. We consider in turn each of the main functions that the brain must perform, namely analysing agonistic stimuli, initiating and integrating agonistic movements and generating shifts in agonistic motivation. The discussion concentrates on mammals, but work on other groups of animals is included wherever possible. Finally, we try to draw some general conclusions on the neural mechanisms which control agonistic behaviour.

6.3 ANALYSING AGONISTIC INFORMATION

As already stated, a complete understanding of the neural mechanisms that control agonistic behaviour requires information about how the sensory pathways analyse the information impinging on an animal during an agonistic encounter. Of all the aspects of brain function, sensory processing is perhaps the easiest to study; it is a relatively straightforward procedure to subject an animal to a specified stimulus and then to record the activity of sensory neurons. The pathways involved in the major senses are well defined (in mammals at least) and we know something about how they work. It is known, for example, that biologically relevant information is progressively extracted from the total complex of stimuli to produce a much edited version of the outside world. This selectivity can be explained in terms of synaptic connections at different levels of the sensory pathway (Camhi, 1984).

The sensory pathways must process the cues produced by a potential opponent, along with all the others an animal experiences. Some brain manipulations do seem to modify agonistic behaviour by altering neural activity in the sensory pathways. For example, Distel (1978) gives the following description of the behaviour of an iguana during electrical stimulation of the brain:

'After first displaying dewlap and body, the stimulated animal hissed, opened its mouth widely and engorged its tongue. While circling . . . it oriented body and head parallel to an imaginary aggressor, against whom it lashed its tail repeatedly,'

However, we know very little about exactly how the sensory pathways process agonistic cues. Two examples, involving hearing and vision, are briefly discussed in this section.

6.3.1 Auditory processing in birds

Song is important in triggering and directing agonistic responses in many birds (page 66). It is analysed by the auditory pathways (Fig. 6.2) for frequency, intensity and direction. In addition, successively finer discriminations are made at higher levels to provide information about socially relevant features such as the species, individual identity and motivational state (as reflected by call type) of the singer. Only at this point is the information passed to the nearby descending motor pathways to generate appropriate action.

There are two main pathways involved in auditory processing in song birds: one (via the magnocellular nucleus) computes the direction of the sound, by comparing input from the two ears; the other (via the angular nucleus) analyses the pattern of the sound. Different cells in this nucleus fire maximally in response to specific features of pure sound (pitch and intensity, for example). At the next stage, temporal patterns of intensity and frequency are analysed and units at the highest levels (in field L) fire only to specific categories of conspecific song (Leppelsack, 1983). For example, in the guineafowl, different field L units have been identified, each responding to just one type of call. Some units even differentiate between different individuals using the same call type (Schein et al., 1983).

6.3.2 Visual processing in crustacea

Sensory neurons in the crustacean optic nerve respond to specific patterns of visual stimulation and, here as well, by the time visual

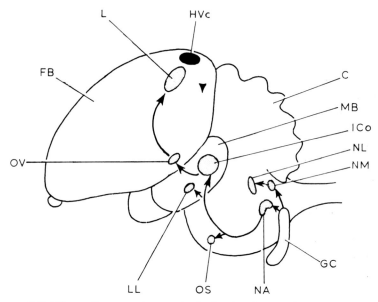

Figure 6.2 The auditory pathways in the brain of a songbird. C, cerebellum; GC, cochlear ganglion; FB, forebrain; ICo, intercollicular nucleus; L, field L; LL, lateral lemniscus; MB, midbrain; NA, angular nucleus; NL, laminar nucleus; NM, magnocellular nucleus; OS, superior olivary nucleus; OV, ovoid nucleus; HVc, hyperstriatum ventrale pars caudale, a major centre controlling song production (simplified from Leppelsack, 1983).

information reaches the brain a number of behaviourally important features have been extracted. For example, in crabs the so-called movement fibres habituate rapidly to stationary stimuli but fire vigorously in response to visual stimuli moving at a particular speed. The muscles that open the chelae (or claws) contract in response to visual stimuli (thus producing a component of the threat display) and this response habituates with a time course similar to that of the movement fibres. So the movement fibres may be involved in detecting potentially threatening stimuli (whether these come from a predator or an attacking conspecific) and in triggering an appropriate response (Glantz, 1974).

6.4 PRODUCING AGONISTIC MOVEMENTS

A considerable amount is known about general aspects of the control of movement in both vertebrates and invertebrates (Camhi, 1984), but these processes are only discussed here as they relate to the production of particular agonistic movement patterns.

6.4.1 Agonistic postures in crustacea

During agonistic encounters, advanced crustacea such as lobsters commonly assume a high body posture, with legs stretched, claws open and abdomen extended. The motor control of claw movements and abdomen extension has been studied in some detail. Male snapping shrimps make conspicuous snapping movements with their claws during a fight; the mechanism that produces this tiny but dramatic fragment of the agonistic repertoire is quite well understood. Whenever a shrimp is disturbed the neuron that activates the claw-opener muscles fires, causing the muscle to contract. When the claw is open, sensory hairs on the cuticle are stimulated. This sets up action potentials in a sensory neuron which, in turn, excite the motor nerves of the muscles closing the claw. When tension in the opener muscle (induced by contraction of the antagonistic claw-closer muscle) crosses a threshold level, the claw snaps shut, producing the characteristic sound and jet of water (Ritzmann, 1974; Malon and Stephens, 1979).

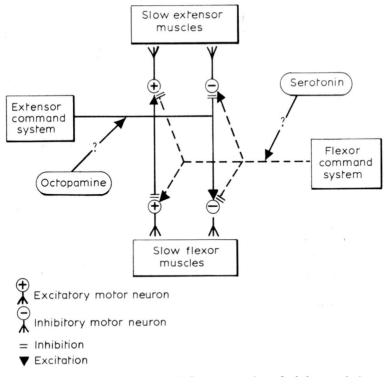

Figure 6.3 The slow extensor and flexor muscles of a lobster, their motor neurons and how these may be activated (electrically and biochemically) during agonistic displays (after Camhi, 1984).

Abdomen posture in crustacea depends on slow flexor and slow extensor muscles, each with a set of excitatory and inhibitory motor neurons (Fig. 6.3). The flexed abdomen typical of a victorious lobster can be produced by electrical stimulation of a system of neurons (called command neurons) in the central nervous system which activate the flexor muscles. Stimulation of the command neuron system for the extensor muscles produces the collapsed, extended posture typical of a defeated animal. Systemic injection of serotonin and octopamine (which are produced by the nervous system of lobsters) triggers the high and the collapsed body posture respectively (page 123), partly acting as neurotransmitters in the command neuron systems (Kravitz et al., 1983; Camhi, 1984).

6.4.2 Agonistic vocalizations in squirrel monkeys

The brains of many vertebrates, including fish (Demski and Knigge, 1971; Demski, 1973; Demski and Gerald, 1974), lizards (Sugerman and Demski, 1978; Distel, 1978, Tarr, 1982), birds (Akerman, 1966; Maley, 1969; Phillips and Youngren 1971; Vowles and Beasley, 1974) and mammals (Brown and Hunsperger, 1963; Roberts et al., 1967; Delgado, 1979) have been investigated by electrical stimulation. The subjects of these studies were relatively unconstrained animals in the presence of a range of external stimuli, including social companions.

The results indicate that the brain mechanisms that put together agonistic sequences in vertebrates are organized along hierarchical lines. Stimulation at some sites (predominantly in the hind brain) elicits behavioural fragments (in cats, erection of the fur, forward rotation of the whiskers, extension of the claws or hissing, Brown and Hunsperger, 1963). At others (in the hind- and mid-brain) the same fragments are combined into complete behaviour patterns, including threat postures. The elicited behaviour is very rigid in nature, is independent of any cues in the environment and starts and stops abruptly as the stimulating current is turned on and off. The agonistic behaviour induced by stimulation of a third category of sites (in the forebrain and particularly the hypothalamus) is more complex and variable, comprising flexible, motivated behavioural sequences. The subjects show restlessness in the absence of an appropriate environ-mental stimulus, and complete, well directed agonistic responses when companions are present.

We are concerned here with the first two levels of this hierarchy and give an example in which the muscles, nerves and motor control mechanisms generating agonistic behaviour patterns have been identi-fied. Squirrel monkeys produce a rich repertoire of vocalizations during agonistic encounters. The monkeys groan and shriek when

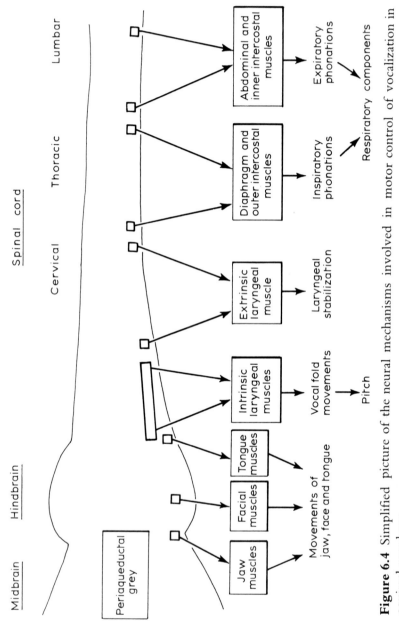

Figure 6.4 Simplified picture of the neural mechanisms involved in motor control of vocalization in squirrel monkeys.

defending themselves against an attack, they chatter during group mobbing of outsiders and dominant animals growl as a threat (Jurgens, 1983). The respiratory components of vocalization depend on the muscles which move the abdomen, ribs and diaphragm; the pitch and patterning of the sounds depend on the muscles which open and close the glottis and move the larynx; while articulation depends on the muscles of the face and jaw. The motor nerves which make these muscles contract are well known and are shown in Fig. 6.4.

Isolated fragments of unnatural sound, but never complete calls, can be elicited by stimulation in various points in the hindbrain and spinal cord, which are thought to generate the patterned motor output necessary for vocalization. Small lesions in the midbrain periaqueductal grey (see Fig. 6.1) eliminate particular vocalizations, and stimulation in this region elicits a variety of natural-sounding calls, depending on the precise location of the electrode. Via identified descending pathways to the patterning mechanisms of the hindbrain and spinal cord, the midbrain central grey is able to trigger production of particular vocalizations (Jurgens, 1983). Equivalent regions of the midbrain play a similar role in birds, reptiles and fish (Schein et al., 1983; Demski and Gerald, 1974).

6.5 PRODUCING CHANGES IN RESPONSIVENESS

6.5.1 Identifying changes in aggressive motivation

The neural mechanisms that generate long-term differences and short-term shifts in agonistic responsiveness are much more difficult to study than those involved in sensory processing or motor output. The following kinds of observation suggest that such mechanisms are being activated.

'The attacks consisted of rushing towards or chasing the non-stimulated bird and grabbing, pushing or tugging at the feathers of the neck, back, breast, tail and wings . . . If the non-stimulated bird fought back, the birds fought in the breast-to-breast posture until . . . one turned away.' (Maley, 1969, referring to attack produced by electrical stimulation in the forebrain of a mallard.)

'After the onset of stimulation, the animals became aroused . . . and kept an arched back position with moderate piloerection . . . (and) . . . intermittent bouts of vocal threat . . . (the partner animal) regularly became the target for short attacks . . . If (it) slashed back, fights inevitably occurred lasting until the stimulation had been turned off.' (Lipp and Hunsperger, 1978, referring to marmosets.)

Squirrel monkeys will learn a simple task if, as a result, they receive electrical stimulation in hypothalamic regions which elicit growling. 'In subjective terms this *means that hypothalamically-induced growling is accompanied by pleasurable . . . emotional states.' (Jurgens, 1983).*

This sort of information, together with precise behavioural measurement and analysis, is essential if we are to understand what is going on, and yet many studies of the effects of neurophysiological manipulation fail to provide the necessary detail. Often the experimental animals can show just one very limited aspect of agonistic behaviour. In one type of study, mice which have been paralysed by sectioning their spinal cord are given electric shocks, and the pressure exerted on a rubber hose placed between their teeth is uses as a measure of aggressiveness (Renfrew and Hutchinson, 1983). In other cases, rats and mice are isolated for a time before a test or given electric shocks during it, and in both cases fighting is commonly observed. However, shock-induced fighting, in rats at least, is predominantly defensive in nature and isolation induces a state of general hyper-responsiveness of which agonistic behaviour is just one part. In addition, unless the subjects are given the opportunity to do other things as well, there is a risk of mistaking non-specific effects of brain manipulations (for example on overall activity levels) for a specific influence on agonistic behaviour.

Even when the experimental animals can potentially show a range of responses, in many studies the effect of a physiological manipulation is assessed crudely, perhaps by counting the percentage of animals fighting before and after treatment or by using a single aggression rating. Simple measures of aggression may well be sufficient for studies primarily concerned with physiology, but with behaviour analysed at this level, many important features of brain function escape detection. The need for precise behavioural information is now recognized and, increasingly, neurophysiological studies are being carried out in conditions that approximate as nearly as possible to the subjects' natural social structure and that allow a full range of behaviour to be displayed and measured (Wiepkema, Koolhaus and Olivier-Aardema, 1980; Benton, 1981; Blanchard and Blanchard, 1981).

6.5.2 Different kinds of change

Such detailed behavioural analyses of the effects of brain manipulation make it clear that changes in aggressiveness can come about in many different ways (Table 6.2). Some are non-specific in nature, influencing an animal's readiness to show all kinds of different behaviour, with agonistic behaviour being just one reflection of this change. Brain stimulation and lesion can interfere with release from the pituitary

Table 6.2 Routes by which the nervous system can generate changes in agonistic responsiveness.

---

1. By altering production of the chemicals that cause release of pituitary hormones, especially gonadotropins.
2. By effecting a behavioural change of a general nature:
   (a) general activity,
   (b) general responsiveness to social and non-social cues,
   (c) learning processes.
3. By a specific effect on behaviour which indirectly influences agonistic behaviour.
4. By a specific and direct effect on agonistic motivation.

---

gland of the hormones that alter a whole range of reproductive behaviour patterns, including agonistic behaviour. Other brain mechanisms influence just one specific aspect of the way an animal behaves, but their influence on agonistic behaviour is indirect, for example via activation of competing responses. Yet other brain mechanisms exert direct and specific effects on agonistic behaviour. All these mechanisms, whether general or specific, potentially contribute to naturally occurring variation in agonistic responsiveness, so examples of each are discussed here. The limbic system has a special place in the study of the neural bases of agonistic behaviour and, until recently, most studies of drug-induced behavioural changes have been carried out with little reference to the brain mechanisms involved. So these are both discussed separately, even though their behavioural effects are often of a non-specific nature.

6.5.3 Non-specific effects

A number of brain manipulations alter agonistic responsiveness, together with many other aspects of behaviour, as part of a very general change in behavioural state. For example, the common taming effect of amygdalectomy in mammals (page 150) is accompanied by sluggishness, reduced responsiveness to social stimuli, hypersexuality and voracious eating (many of these effects may in turn depend on memory deficits, see below and Isaacson, 1982, for a review). Lesion of the olfactory bulbs greatly increases general reactivity in rats, which is manifested in increased levels of defensiveness among other behavioural changes (see Albert and Walsh, 1984, for a review). Severing the connections between the frontal lobes of the cortex and other brain regions (frontal lobotomy) reduces aggressiveness in various species of mammal (Dimond, 1980), but it also produces a general decrease in responsiveness, persistent behavioural stereotypies and many other behavioural abnormalities.

Changes in the production and metabolism of brain neurotransmitters can profoundly influence the way an animal behaves in an agonistic encounter, but this is often just one aspect of a non-specific behavioural effect. For example, drugs that block catecholamine synthesis in mice and rats impair many aspects of motor function, including aggressiveness. The increased shock-induced fighting seen in mice with depleted brain serotonin levels may be a secondary effect of a rise in general levels of arousal (see Eichelman, Elliot and Barchas, 1981, and Miczek, 1983, for reviews).

6.5.4 Changes in learning processes

Past experience of victory or defeat is an important determinant of what happens in a fight, so brain mechanisms that influence learning (particularly avoidance learning) could generate differences in agonistic responsiveness. The amygdala in mammals is implicated in various aspects of fear-motivated learning (Isaacson, 1982; Sarter and Markowitsch, 1985) and it has been suggested that learning and memory defects may be responsible for the reduced emotionality seen in rhesus monkeys following amygdalectomy (page 151; see also Aggleton and Passingham, 1981). The fact that subordinate vervet monkeys with bilateral amygdalectomy ignore dominants may reflect such an effect (Kling, 1972). Electrical stimulation in the forebrain (in a part called the hypostriatum) produces defensive attack in ring doves, possibly because this brain region is involved in the formation of conditioned emotional responses (Vowles and Beasley, 1974).

6.5.5 Specific effects

The previous sections have shown that many brain manipulations that modify aggressiveness, when looked at carefully, turn out to do so by means of behavioural effects that are not specific to the agonistic context; thus they may alter general arousal levels, overall social responsiveness or learning ability. Other effects of brain manipulation are more specific but influence agonistic behaviour only indirectly. For example, electrical stimulation of the preoptic region in sunfish suppresses aggression, but does so by activating an incompatible response, namely sexual behaviour (Demski and Knigge, 1971). Only where such non-specific or indirect effects have been ruled out can we ascribe to a neural system a specific agonistic function. For example, aggression induced in rats by electrical stimulation of the hypothalamus (page 148) is not accompanied by an increase in general activity. Far from being enhanced, both sexual behaviour and feeding are suppressed during stimulation-induced attack, just as they are by natural aggression. Like the natural behaviour, stimulation-induced attack is facilitated if the subject is in its own territory, but suppressed if the opponent is large and dominant (Koolhaas, 1978; Kruk *et al.*, 1983). All these observations suggest that stimulation has induced in the rats a specifically aggressive state.

6.5.6 The role of the hypothalamus

The hypothalamus, a collection of discrete neural structures (or nuclei) lying along the base of the third ventricle at the top of the brain stem, is one of a number of interconnected forebrain structures collectively referred to as the limbic system (Fig. 6.5). Receptors for

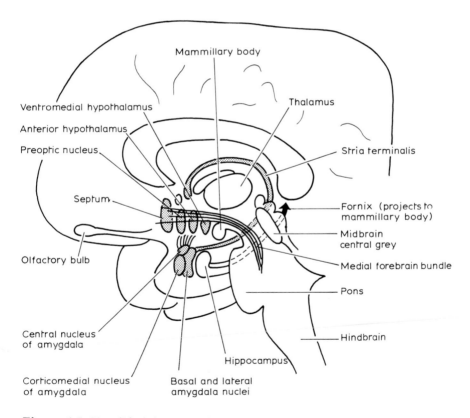

Figure 6.5 Simplified diagram of the components of the limbic system of a mammal (*Homo sapiens*; after Niewenhuys, Voodg and Van Huijzen, 1981).

steroid hormones, in particular for oestrogens and androgens, are found in many parts of the hypothalamus, but are most abundant in the ventromedial nucleus and in the preoptic area. The hypothalamus has direct neural connections with the pituitary gland and also secretes into local blood vessels the chemicals that promote release of pituitary hormones. Since hormones (and in particular those of the pituitary–gonadal axis) influence agonistic behaviour, this is one route by which the hypothalamus can potentially influence how animals fight. For example, lesion in the anterior basal hypothalamus of the male green anole lizard results in drastic testicular collapse and eliminates all aspects of reproductive behaviour, including sexual and agonistic responses. These behavioural deficits can be reversed by androgen injections, so the lesion must have destroyed those parts of the hypothalamus that, via production of pituitary releasing factors, normally trigger gonado-tropin production (Farragher and Crews, 1979).

Major descending projections run from the hypothalamus to the brain stem and spinal cord, especially to those parts that give rise to the autonomic nervous system, controlling among other things the physiological aspects of emotional arousal. Because of its connections with the pituitary, with the other limbic structures and with the autonomic nervous system, the hypothalamus is well placed to inte-grate the physiological, behavioural and endocrine components of emotional responses, including aggression. Traditionally this part of the brain has been ascribed an important role in the control of motivation and emotion. For example, Stellar (1954) suggested that each functional category of behaviour (such as feeding, drinking and fighting) is controlled by a pair of hypothalamic centres, one excita-tory and the other inhibitory. The relative activity of these two centres, which, according to this theory, depend on information about the internal state of the animal and on stimuli from the environ-ment, determine how likely it is to show the behaviour concerned. The idea of precisely localized centres with specific and simple effects on behaviour cannot be supported in the light of more recent evi-dence. However, as we shall see, the hypothalamus is clearly impli-cated in the control of agonistic behaviour and it does seem to contain systems with opposite facilitatory and inhibitory effects.

An intact hypothalamus is essential for the spontaneous expression of agonistic behaviour and, from fish to monkeys (Table 6.3), sti-mulation of the hypothalamus can produce (among other responses) offensive behaviour, defensive behaviour and flight, depending on the location of the electrode and the species concerned. The brain regions controlling attack, threat and flight are anatomically distinct in some species (for example, the marmoset) but overlap in others (for exam-ple, the cat, Lipp and Hunsperger, 1978).

Table 6.3 Examples of the effects of hypothalamic manipulations on agonistic behaviour.

Species	Observation	Reference
Sunfish, cichlids	Stimulation of periventricular region and inferior lobe elicits feeding and attack.	Demski and Knigge (1971); Demski (1973)
Green anoles	Lesion of the anterior hypothalamus and preoptic region reduces aggressive behaviour independent of testosterone levels. Testosterone implants restore aggression and then sexual behaviour.	Crews (1983)
Iguanas, fence lizards	Electrical stimulation in lateral periventricular regions and medial forebrain bundle elicits defensive displays.	Sugerman and Demski (1978); Distel (1978)
Pigeons	Electrical stimulation in the anterior hypothalamus and preoptic region elicits aggression.	Akerman (1966); Maley (1969)
Mallards	Posterior hypothalamic stimulation elicits agonistic behaviour.	Akerman (1966); Maley (1969)
Hamsters	Lesions in the anterior hypothalamus decrease low intensity aggression. In females, medial preoptic lesions decrease aggression.	See Siegel (1985) for review
Mice	Lesion of medial preoptic region abolishes intermale aggression.	Shivers and Edwards (1978)
Rhesus monkeys	Stimulation in the anterior hypothalamus and lateral hypothalamus elicits aggression in subordinate males and can reverse dominance.	Robinson *et al.* (1969)
Squirrel monkeys	Stimulation in the anterior hypothalamus and lateral hypothalamus elicits growling and shrieking.	Jurgens (1983)

The role of the hypothalamus in the control of agonistic behaviour is best understood in rats, where behavioural effects of both lesion and stimulation have been very carefully characterized (for example, Koolhaas, 1978; Wiepkema *et al.*, 1980; Kruk *et al.*, 1983). Stimulation of a continuous and fairly extensive region of the hypothalamus (including the anterior and lateral hypothalamus and parts of the ventromedial nucleus) in male rats produces co-ordinated attack sequences with the properties of natural territorial aggression; the stronger the stimulating current, the more intense the response. Feeding, drinking or sexual behaviour are suppressed and the stimulated animal attacks subordinate male intruders, but not larger or dominant males or oestrus females (Fig. 6.6(a)). After losing a stimulation-induced fight, rats will not attack their victorious opponent, either spontaneously or when they are stimulated in the identical spot. So experience during stimulation-induced fights has long-term effects similar to those observed as a consequence of spontaneous fights; in other words they apparently feel the same to the animal involved. Weak stimulation of the ventromedial hypothalamus terminates agonistic behaviour.

Lesions in the lateral hypothalamus abolish fighting (and many other aspects of behaviour; Adams, 1979) and rats with lesions in the anterior ventromedial hypothalamus are hard to handle and show greatly enhanced defensive fighting. In contrast, small lesions in the posterior ventromedial hypothalamus (and in the nearby premammillary nuclei, Van der Berg, ter Horst and Koolhaas, 1983) specifically increase offensive behaviour in territorial males, without rendering them hyperdefensive. Male but not female intruders are attacked, sexual behaviour is normal and the overeating associated with larger lesions of the ventromedial hypothalamus is absent (Olivier, Olivier-Aardema and Wiepkema, 1983).

Just like natural territorial behaviour, aggression elicited by stimulation of the lateral hypothalamus or by lesion of the posterior ventromedial area is absent in castrated male rats but is reinstated by systemic testosterone injections (Fig. 6.6(b)). So the brain manipulations depend on circulating testosterone and are not simply 'imitating' the effects of this hormone. Since the lateral hypothalamus at least does not contain many androgen receptors, the hormone must be detected elsewhere.

The medial anterior hypothalamus and the preoptic region, on the other hand, are rich in androgen receptors and the activity of the nerve cells in this region is altered by direct administration of testosterone. Destruction of the front end of the preoptic area rapidly and completely abolishes intermale aggression in rats; lesions further back eliminate sexual behaviour. Both effects persist when the subjects are given testosterone injections, so the behavioural effect is not the result of impaired release of luteinizing hormone from the hypothalamus.

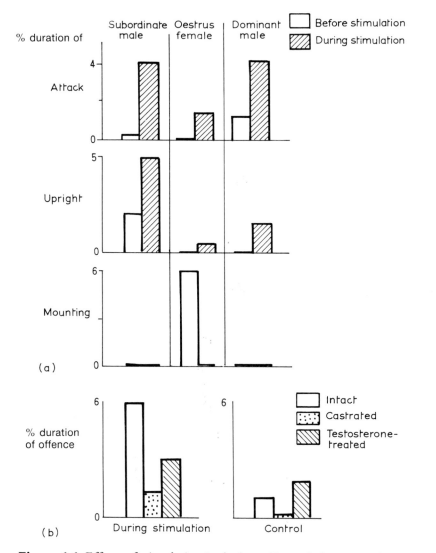

Figure 6.6 Effects of stimulation in the lateral hypothalamus on the percentage of time spent in various components of agonistic behaviour in male rats. (a) Response to different categories of opponent (after Koolhaas, 1978); (b) response after castration and testosterone replacement (after Wiepkema *et al.*, 1980).

Bilateral testosterone implants in the front of the preoptic region restore aggression but not sexual behaviour to castrated rats (and several other species). So testosterone-induced activity in the anterior preoptic region regulates aggressive behaviour in some way (see Bermond, 1978 and references therein).

Thus separate systems within the hypothalamus of the rat control defence and offence. Offensive behaviour depends on two agonistic systems; one of these (predominantly in the lateral hypothalamus) promotes the offensive behaviour typical of a territorial male and the other (situated more ventrally) has the converse effect. Neither system is effective in the absence of circulating androgen. This sounds remarkably like Stellar's antagonistic centres in the hypothalamus (page 146), without the precise anatomical localization. Even if this scheme approximates to reality, we still do not know how the hypothalamus might exert these effects. During predatory attack in cats, head turning, jaw opening and biting are elicited by tactile stimulation in a defined area around the mouth; striking movements with the paws are elicited from a sensory area on the forelegs. These two sensory fields are enlarged during stimulation of hypothalamic sites which evoke predatory attack (Bandler, 1982b). Whether similar mechanisms are involved during changes in agonistic motivation remains to be seen.

6.5.7 The role of other limbic structures

The limbic system (Fig. 6.5) is traditionally regarded as the part of the brain most closely involved in the generation of emotional states, with the hypothalamus acting, as it were, to implement its decisions. This view has received some experimental support, but it is clear that the role of the limbic system is complex and certainly not restricted to a specific effect on agonistic behaviour (see Isaacson, 1982, for a review). The two structures in the mammalian limbic system which have received most attention in this context are the amygdala and the septum.

(a) The amygdala

The amygdala consists of several groups of neurons (or nuclei) in the temporal lobe and is connected with the hypothalamus via the stria terminalis and the ventral amygdalofugal pathway (see Fig. 6.5). Manipulating the activity of the amygdala alters the threshold for stimulation-induced attack in the hypothalamus in various mammals, so these two regions are functionally as well as structurally interconnected. In an early experiment, bilateral destruction of the temporal lobe (including the amygdala) in monkeys produced a complex of behavioural changes known as the Kluver–Bucy syndrome, many aspects of which may stem from memory deficits. They included compulsive eating, hypersexuality, increased activity and loss of both fearful and aggressive reactions. Such a complex structure as the amygdala is likely to have many different behavioural roles, but more precise analyses have confirmed the suggestion that the amygdala is involved in the control of defensiveness in a number of species of mammal (see Ursin, 1981, and Albert and Walsh, 1984, for reviews). Stimulation (particularly in the basolateral region, which projects to the preoptic area and lateral hypothalamus) elicits defensive attack in cats and monkeys. Lesions of the amygdala make rats (Blanchard *et al.*, 1979), mice, cats (Ursin, 1981) and hamsters (Bunnell *et al.*, 1970, cited in Siegel, 1985) tame and easy to handle and impair shock-induced fighting.

The effects of manipulations of the amygdala on offensive behaviour depend on the species, past experience and social environment of the animals concerned. Following amygdalectomy, rhesus monkeys become more aggressive towards and/or less fearful of an experimenter when living alone; in a social context they often drop in the dominance hierarchy but this depends on the behaviour of other animals in the group (Rosvold, Mishkin and Pribram, 1954). Following destruction of the amygdala, guinea pigs are less likely to initiate attacks on companions and so lose status (Fig. 6.7(a)). This may be an indirect result of their impaired ability to defend themselves (Levinson, Reeves and Buchanan, 1980). In Java monkeys, stimulation of the amygdala increases aggression in dominant animals but increases defence and submission in subordinates (Van der Berken and Cools, 1980). Destruction of the amygdala reduces aggressiveness in previously dominant hamsters (without changing their status) but reduces submissive behaviour in subordinates (Bunnell *et al.*, 1970; see also Siegel, 1985).

Some of these effects of manipulations of the amygdala, both on offence and defence, could be a secondary consequence of the general behavioural deficits they produce. However, restricted lesions may have effects which are independent of some learning deficits at least. Territorial threatening and fighting in male rats (housed in a large cage with a female) is severely reduced (and exploration enhanced) following lesion of the corticomedial nucleus. This part of the amygdala receives input from the olfactory system and projects to the lateral and ventral hypothalamus and the premammillary nucleus. Basolateral lesions do not alter fighting ability but slightly impair the rats' ability to make use of previous fighting experience (Wiepkema *et al.*, 1980).

(b) The septum

Following septal lesions, many mammals show temporarily enhanced

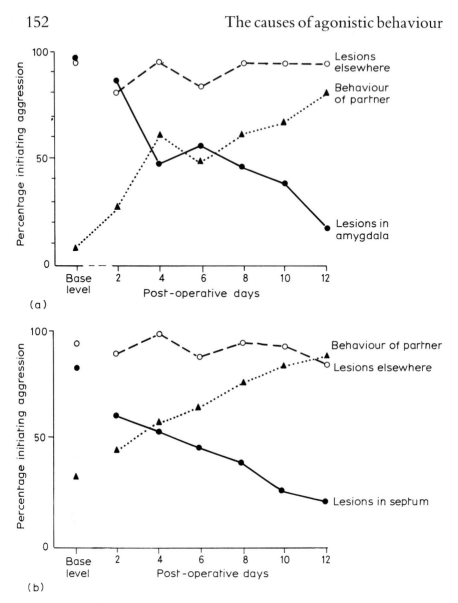

Figure 6.7 Effects of lesions in the (a) amygdala and (b) septum on fighting in dominant male guineapigs (after Levinson, Reeves and Buchanan, 1980).

irritability and exaggerated startle responses (the so-called septal syndrome). This is reflected in a greater tendency to fight when given electric shock. The full septal syndrome is not observed in hamsters, squirrel monkeys or rhesus monkeys (see Isaacson, 1982, and Albert and Chew, 1982, for reviews). Like the amygdala, the septum is a complex structure and complete lesions have a number of behavioural effects, some of a very general nature. The hyper-reactivity caused by septal lesion is often accompanied by increased general activity and gregariousness, by learning deficits and, depending on the species, by reduced aggression and impaired status. Dominant guineapigs and rats become less aggressive and fall in rank (Fig. 6.7(b); Lau and Miczek, 1977; Levinson, Reeves and Buchanan, 1980), perhaps as an indirect effect of their hyper-reactivity. On the other hand, in hamsters septal stimulation completely suppresses fighting, provided the amygdala is intact (Siegel 1985 and references therein).

6.5.8 The role of the forebrain in non-mammalian vertebrates

The available data suggest that the non-mammalian equivalents of the amygdala and septum may be implicated in the control of agonistic behaviour, but give little information about the precise behavioural nature of their effects. Lesions in the parts of the forebrain which are thought to be equivalent to the mammalian amygdala and nearby structures reduce fighting in caimans (Keating, Korman and Horel, 1970), lizards (Tarr, 1977) and mallards and pigeons (Phillips, 1964; Ramirez and Delius, 1979). They suppress courtship and challenge display in male green anole lizards; lesion in the equivalent of the septal region causes deficits in the performance of challenge display and aggression in males without altering testicular function (Crews, 1983 and references therein).

The forebrain of teleost fish develops in a very specialized way, and homologies with mammalian brain regions are hard to establish. However, in a number of species lesions in different parts of the forebrain have opposite, facilitating and inhibiting, effects on agonistic behaviour. For example, lesions placed near the back of the dorsal forebrain of male Siamese fighting fish eliminate biting and reduce frontal display while lesions further forward increase levels of both frontal and lateral display (see De Bruin, 1983, for data and a review).

6.5.9 Biochemical modulation of agonistic behaviour

Both naturally occurring and artificially induced changes in the production or metabolism of brain neurotransmitters (see Table 6.1) are known to influence agonistic behaviour. These changes are often part of a non-specific shift in behavioural state, perhaps in general activity or social responsiveness (page 143). Whatever their nature, they represent an important source of variation in agonistic responsiveness and so we need to know about them. In many studies, chemicals are administered systemically or assayed in the whole brain; they therefore provide little information about the site of action of the neurotransmitter concerned. This lack of specificity, together with the complexity of neurochemistry (for example, should we look at steady-state levels or at rates of metabolism?) makes the results hard to interpret. However, there are a few cases in which behavioural changes have been related to altered biochemistry in identified brain regions and this approach has great potential for the study of the neural mechanisms underlying agonistic behaviour.

(a) Whole brain biochemistry

Naturally occurring differences in behaviour can be related to differences in brain biochemistry. For example, the brains of mice from spontaneously aggressive strains contain lower levels of serotonin and higher catecholamine levels than do those of less aggressive animals (Serri and Ely, 1984). The activity of catecholamine-synthesizing enzymes is higher in strains of mice which readily become aggressive following isolation (Ciaranello, 1979). Aggressive, isolated mice have fewer serotonin binding sites in various parts of their brains than do non-aggressive, group-housed controls (Essman and Valzelli, 1984). These relationships suggest that differences in brain biochemistry may cause differences in agonistic behaviour. However, catecholamine levels and aggressiveness do not always change in parallel; although aggressiveness increases with isolation in some strains of mice, this is accompanied by a decrease in noradrenaline levels (Bernard et al., 1975). In addition, various aspects of fighting experience (including simply performing movements) can alter brain biochemistry (page 158), and cause and effect cannot be separated without experimental manipulation.

Systemic administration of neurotransmitters, or of substances which alter their activity, has been shown to alter a number of different aspects of intraspecific agonistic behaviour in mammals (Table 6.4; see also Eichelman et al., 1981, and Miczek, 1983, for reviews). Systemically administered acetylcholine has a rapid stimulating effect on activity in rats, cats and mice (and in Siamese fighting fish, Avis and Peeke, 1979) and this is reflected in increased levels of aggression, defence and flight, among many other changes. Anti-cholinergic drugs have the reverse effect. In mammals, experimentally reduced serotonin levels usually increase overall arousal and with it agonistic responsiveness (Eichelman et al., 1981). However, this effect depends on species and housing conditions. For example, rats placed on a tryptophan-deficient diet (which reduces brain serotonin levels) show enhanced shock-induced fighting; however, the same treatment decreases isolation-induced fighting in mice (Eichelman et al., 1981). Increased serotonin uptake reduces aggressiveness in isolated male

Table 6.4 A summary of the effects on agonistic behaviour of artificially increased levels of various neurotransmitters (from Miczek, 1983).

Behaviour	Acetylcholine	Serotonin	Noradrenaline	Dopamine
Shock-induced fighting in rats	↑	(↓)	↑	↑
Brain stimulation-induced defence in cats	↑	↓	↑↑	↑
Isolation-induced defence in mice		↕		
Isolation-induced offence in mice	↑	(↓)	(↑)	(↑)

↑ ↓ represents increases/decreases
( ) represent some conflicting results

mice, but has the opposite effect on group housed males (Essman and Valzelli, 1984). Serotonin also acts as a general inhibitory transmitter in invertebrates and artificially increased serotonin levels suppress agonistic responses (and general activity) in ants (Kostowski and Tarchalska, 1977).

The catecholamines noradrenaline and dopamine have a clear stimulating effect on defensive fighting in mice, rats and cats when administered systemically. This effect too is probably of a general nature (Schmidt, 1984), although catecholamine-induced aggression between wild-caught male mice may not be accompanied by an increase in general activity (Matte and Tornow, 1979). The effect of catecholamines on offensive behaviour depends on the past experience, present social conditions and motivational state of the subjects. In dominant rats, low doses of amphetamine (which, among other effects, promotes release of noradrenaline and dopamine) enhance aggression and high doses suppress it; subordinate animals respond to high doses with defensive behaviour (Miczek and Krisiak, 1981). Drugs which increase noradrenaline activity increase overall activity in Siamese fighting fish and (up to a point) increase attack rates. Dopamine agonists increase attack latency but also decrease flight (Ganzer, Schmidtmayer and Laudien, 1984).

(b) Biochemistry of specific brain regions

The best documented case of neurotransmitters influencing agonistic behaviour by known effects on specific neural structures come from the invertebrates. In crayfish, serotonin depresses electrical activity in the giant interneurons that mediate escape responses. Octopamine (which may be the arthropod equivalent of adrenaline) has the opposite effect. These two substances may be part of the mechanism by which crayfish respond differently to threatening stimuli. They are more responsive when moulting or injured (when perhaps they have lower serotonin and higher octopamine levels) but less responsive if hungry (when serotonin levels may be high and octopamine levels low, Glanzman and Krasne, 1983).

The behavioural effects of neurotransmitter activity in identified regions of the vertebrate brain are not so well understood. Serotonin, noradrenaline and dopamine are known to be concentrated in particular regions and pathways (Table 6.1), many of which have been implicated in the control of agonistic behaviour. Relating differences in agonistic responsiveness to altered activity of these neurotransmitter specific pathways would help us to understand how the brain controls this behaviour. To date there is rather little work of this kind. However, what there is suggests that shifts in readiness to fight are accompanied by changes in serotonin, noradrenaline and dopamine in the hypothalamus and limbic system. This is consistent with the picture of neural control of agonistic behaviour presented in earlier sections.

Neurons using serotonin run from the hind and midbrain to a number of limbic structures, including the hypothalamus, septum and the amygdala; those arising in a structure called the dorsal raphe project mainly to the amygdala. Destruction of the dorsal raphe induces defensive (shock-induced) fighting in rats, suggesting an inhibitory effect on the amygdala, which is implicated in the control of defensive behaviour (even if indirectly). Direct administration of drugs that enhance serotonin activity into the corticomedial amygdala (where lesion suppresses offensive behaviour, page 151) inhibits shock-induced fighting in rats with little effect on overall activity or pain sensitivity (Pucilowski, Plaznik and Kostowski, 1985, and references therein).

Several nuclei in the pons and medulla send noradrenergic projections via the median forebrain bundle to the hypothalamus, amygdala and septum. There is accumulating evidence of various kinds that these pathways are involved in the control of agonistic behaviour, though once again there is no evidence that their effects are specific to this aspect of behaviour. Attack elicited by hypothalamic stimulation in rats is facilitated by injections of catecholamines into this brain region (Torda, 1977, cited in Eichelman et al., 1981). Mice from aggressive strains have high noradrenaline levels in several brain regions, including the septum and preoptic region (Table 6.5(a)).

Table 6.5 Localized neurotransmitter levels and agonistic behaviour. NA, noradrenaline; DA dopamine; 5HT, serotonin.
(a) Concentrations (ng/mg protein) of catecholamines in specific brain regions of mice from an aggressive (NIH) and a non-aggressive (A/HeJ) strain (from Tizabi et al., 1979).

	Septum		Lateral pre-optic region		Hypothalamus	
	NA	DA	NA	DA	NA	DA
Aggressive strain	7.68	3.67	17.79	11.73	34.85	6.18
Non-aggressive strain	4.08 *	3.41	10.59 *	10.59	34.48	2.38 *

(b) Concentrations (ng/g wet weight) of monoamines in specific brain regions of rats following a bout of shock induced fighting, compared to shocked controls, (from Kantak et al., 1984).

	Brain stem			Amygdala			Hypothalamus			Caudate nucleus		
	NA	DA	5HT	NA	DA	5HT	NA	DA	5HT	NA	DA	5HT
Shocked controls	573	89	524	291	302	231	1764	919	812	426	14204	775
Fighting rats	491	87	415 *	363	304	209	1238 *	1182	632 *	668 *	8423 *	695

* denotes significant differences.

Noradrenaline levels are also higher in the septal region of mice that become aggressive as a result of isolation compared to those that isolation fails to render aggressive (Tizabi et al., 1979, 1980). Disturbance of noradrenaline levels in the amygdala induces deficits in fear-motivated learning, similar to those produced by surgical lesions of this area (page 151; Ellis and Kesner, 1983).

In mammals, the majority of dopaminergic neurons arise in the midbrain and project via the lateral hypothalamus to the striatum. So manipulations of the lateral hypothalamus which influence agonistic behaviour may do so, in part at least, by their effect on this pathway. Dopamine agonists administered systemically increase the readiness with which defensive threat can be elicited from the (ventromedial) hypothalamus in rats (Maeda et al., 1985). Dopamine uptake in the striatum is increased by a bout of shock-induced fighting (see below) and both steady-state levels and rates of dopamine turnover are higher in the striatum and hypothalamus of aggressive mice (see Table 6.5(a)), this may be a result of higher levels of motor activity in aggressive animals (Tizabi et al., 1979, 1980).

6.6 EFFECTS OF FIGHTING EXPERIENCE ON BRAIN BIOCHEMISTRY

As already mentioned, during an agonistic encounter, multiple and complex changes in neurotransmitter synthesis and metabolism occur in various parts of the brain. These are hard to interpret, because they could be specific effects of fighting or non-specific effects of active movement and because they may come about through alterations in production or utilization of the chemicals concerned, or both.

Brain stem levels of noradrenaline and dopamine initially decrease following a short bout of isolation-induced fighting in mice, but levels increase slightly in mice given daily fights (Welch and Welch, 1969, in Eichelman et al., 1981). After a brief shock-induced fight, rats and mice have reduced serotonin levels in their brain stem and hypothalamus, reduced noradrenaline levels in the hypothalamus (with increased levels in the caudate nucleus; Kantak et al., 1984; Table 6.5(b)). Shock-induced fighting increases the affinity of nerve cells in the striatum and nearby cortical regions for dopamine in mice (Hadfield, 1983).

The longer-term biochemical effects of engaging in a fight depend on its outcome. Dominant mice have lower brain stem noradrenaline turnover than do subordinates (Welch and Welch, 1969, cited in Eichelman et al., 1981). In rats and a number of other mammals, serotonin levels are higher following unsuccessful fights with a dominant animal. On the other hand, dominant male vervet monkeys have much higher (whole body) levels of serotonin than do subordinates and are much more sensitive to the effects of serotonin. Levels drop when the dominant ceases to be the recipient of submissive behaviour (McGuire et al., 1984). Subordinate pheasants (Phasianus colchicus) have lower dopamine levels in the neostriatum but do not differ from dominants in their noradrenaline levels (McIntyre and Chew, 1983). In trout as well, dopamine levels are lower in subordinates and are negatively related to the number of times they are attacked (McIntyre et al., 1979). Ants which have been involved in a fight have elevated serotonin levels but reduced levels of noradrenaline (Kostowski, Tarchalska and Markowska, 1975), so effects of fighting on brain biochemistry are not confined to vertebrates.

Whatever their origin, because brain biochemistry influences behaviour, these experience-induced changes in neurotransmitter activity are likely to alter subsequent agonistic behaviour. The short-term biochemical changes could make animals more active and aggressive during the early stages of a fight, when this behaviour is most adaptive. Where subordinates have high levels of serotonin and low levels of catecholamines, this is likely to suppress their subsequent aggressiveness, along with other behavioural changes, and so may prevent them from getting involved in repeated, hopeless fights. To speculate even further, the enhanced serotonin levels induced in a dominant male vervet by the behaviour of his subordinates may reduce his aggressiveness and keep him sweet.

As discussed on page 12, chronic exposure to unavoidable attack activates the endogenous opiate system, raising pain thresholds and suppressing pain-induced defensive fighting. Following prolonged agonistic encounters in male Mongolian gerbils, losers have increased levels of two endogenous opiates (β-endorphins and Met-encephalin) in their amygdalae, compared to victorious animals. The amygdala is implicated (directly or indirectly) in the control of defensive behaviour in rodents and implants of morphine in this region increase pain thresholds. So, opiate-induced suppression of amygdala activity may represent one mechanism by which hopeless counterattack is suppressed in chronically defeated animals given no opportunity to escape (Raab, Seizinger and Herz, 1985).

6.7 OVERVIEW

6.7.1 Have our questions been answered?

(a) Processing agonistic information

When an animal detects agonistic cues, like any other aspects of sensory input these are progressively analysed so that behaviourally relevant information is highlighted, until electrical activity in specific neurons signifies a very precise event in the environment. We can see this in guineafowl when area L cells fire in response to specific calls given by particular individuals and in crabs when the movement fibres monitor threatening visual events. These mechanisms, though by no means specific to agonistic behaviour, mean that certain kinds of signal will be particularly effective as agonistic cues.

(b) Production and integration of agonistic behaviour patterns

For decapod crustacea, performance of some components of agonistic postures can be explained in terms of precisely identified muscles, nerves and neurochemicals. This degree of detail is not available for vertebrates, although the motor systems responsible for production of agonistic vocalizations in squirrel monkeys have been

identified. The result of brain stimulation suggests that a hierarchical system controls the production of agonistic responses in
vertebrates. Lower level units (predominantly in the hindbrain) produce small-scale movements, while at higher levels (especially the
midbrain) increasingly complex movement patterns are put together.
The systems for offensive and defensive fighting are distinct although
they may be closely associated anatomically. Performance of particular agonistic movements must involve unique patterns of muscle
contraction, but higher level components of this system are involved
in other types of behaviour.

(c) Generating differences in agonistic responsiveness?

A combination of good physiological techniques and careful behavioural observation has identified a whole range of neural mechanisms
that alter the probability that an animal will fight (page 141). Some of
these are specific to agonistic behaviour while others are more general
in nature, but all could potentially contribute to naturally occurring
permanent differences and short-term fluctuations in aggressiveness.
   The components of the limbic system are involved, although we do
not really know how these work. Manipulation of the hypothalamus
produces relatively specific effects (both facilitatory and inhibitory) on
agonistic behaviour (page 145). Like real aggression, these depend on
social circumstances and on circulating androgen levels, which may be
detected by the preoptic region (page 148). Other structures in the
limbic system such as the septum and amygdala (page 150) have
antagonistic stimulatory and inhibitory effects, but these are of a much
more general nature (page 143). Changes in production and utilization
of neurotransmitters have profound, if indirect, effects on aggressive
responsiveness. Serotonin has a predominantly inhibitory effect and
acetylcholine and the catecholamines are generally facilitatory; the
effects of the catecholamines depend on dose levels and on the context
in which these are administered (page 155). Whatever their precise
behavioural effects, these various neural and neurochemical mechanisms will predispose animals to a certain level of responsiveness in an
agonistic encounter.

6.7.2 General principles

Beyond recognizing that where agonistic behaviour is concerned,
brain–behaviour relationships are particularly complex, are there any
valid generalizations to be made about the neural control of this aspect
of behaviour? Constructive use of the comparative method is made
difficult by the fact that scientists working on different animals tend to
ask different questions and to use different techniques to answer them.
However, there is a degree of uniformity in the answers that emerge.

1. Such information as we have shows that, as expected, the neural mechanisms responsible for the different patterns of agonistic behaviour among conspecifics are at least partially distinct. This is seen particularly clearly for defensive and offensive fighting.

2. Certain structures and systems are consistently implicated in the control of agonistic behaviour in vertebrates. These include midbrain regions involved in sound production (page 141), a special role for the hypothalamus (page 147) and antagonistic effects of the forebrain structures such as the amygdala and septum (page 153). Stimulatory effects of acetylcholine and inhibitory effects of serotonin are observed in various vertibrates and in invertibrates as well (page 155).

Although certain neural entities, in particular those in the hypothalamus (page 148), do seem to act specifically on agonistic responsiveness, in many cases, effects on this behaviour are just one part of a much more general change. Even the brain systems with specific effects are anatomically diffuse and form part of a complex, interacting network of brain structures which act as a whole to determine when and how animals fight. The concept of a single discrete neural centre controlling aggressive behaviour (a key concept in early studies) is thus an abstraction.

3. Fighting has both short-term and long-term effects on brain biochemistry in many kinds of animal, from ants to monkeys. Although the exact causes and consequences of these changes are far from clear, they potentially provide a mechanism for adapting agonistic behaviour to present needs (short-term changes in neurotransmitter activity may mediate the warm-up effect at the start of a fight) and past experience (modified brain biochemistry could render previous losers less likely to fight).

4. In many species the effects of brain manipulations depend on the experience and status of the subject and on the context in which the encounter takes place. These effects are particularly important in mammals such as the primates which have complex social systems.

5. Differences between species in the way the brain controls aggression can be interpreted in functional terms. The adaptive significance of the precise tuning curves of neurons in the auditory pathways of birds (page 136) is obvious. The slight degree of overlap between the hypothalamic regions controlling defence and flight in the marmoset may reflect the ineffectiveness of these small animals in defensive attack against larger animals (page 146). The rather different behavioural effects of manipulations of the septum in hamsters compared to other rodents (page 153) may be related to their intolerant, solitary way of life; high levels of aggressiveness have to be suppressed periodically to allow access to potential mates.

CONCLUDING COMMENTS ON THE CAUSES OF AGONISTIC BEHAVIOUR

The three levels of analysis discussed in this section (behavioural, endocrine and neural) interact with each other in complex ways. This means that studies which focus on just one can provide only a restricted picture of the control of agonistic behaviour and that, ideally, explanations should span all three levels. However, in our present incomplete state of knowledge, different questions about agonistic behaviour are most satisfactorily explained at different levels. For example, behavioural studies have gone a long way towards precise characterization of the way external stimuli influence agonistic behaviour. One of the great success stories of neurophysiology is the understanding of how the wiring of the sensory pathways generates such selective responsiveness. Behavioural analysis is also a powerful tool for identifying causally related actions, but to understand how these are generated and integrated, we need to look at the neural organization of motor systems. At present, rapid fluctuations in aggressiveness are most readily explained at the behavioural level, by means of motivational models. Although candidate hormonal and neural mechanisms have been identified, as yet we know little about how they work. In contrast, when it comes to permanent differences and longer-term changes in aggressiveness, behavioural studies can only provide trivial explanations; in this case, the most satisfactory explanations are in terms of endocrine physiology.

What does emerge clearly from all these studies of the proximate causes of agonistic behaviour is that the identified mechanisms reflect the need for animals continually to adapt their behaviour to present conditions and past experience. This can be seen in the way behavioural consequences alter the state of the control mechanisms, whether these are hypothetical motivational states, hormone levels or activity in particular brain regions. The selection pressures responsible for this state of affairs are considered in Chapter 11.

PART THREE
Genetic and environmental influences

Observations of animals in their natural habitats indicate that, if and when they do fight, all members of a given species and sex use the same basic repertoire of agonistic behaviour patterns. It is also clear that in some cases, animals are able to fight normally even though they can have had no opportunity to learn how to do so. This sort of observation has led to the suggestion that aggression is a genetically determined, instinctive behaviour pattern which is resistant to environmental modification.

However, development is the result of a continuous interaction between genetically coded biochemical events and the environment (inside and outside the animal) in which these events occur. Genes cannot make adult animals in a vacuum and environmental influences can only act on the material provided by the genome. Thus, every characteristic of an animal is genetically and environmentally determined at one and the same time and agonistic behaviour is no exception. It is important (for theoretical and practical reasons) to try to discover the relative strength of these two influences, and the way they interact to determine how animals fight. This is the subject matter of the next two chapters.

7
Genetics

7.1 INTRODUCTION

Agonistic behaviour, like any other trait, develops through a continuous interaction between genetic and environmental factors. So if
we wish to understand how diversity in agonistic behaviour arises we
need to investigate its genetic bases and how these interact with the
developmental environment. In addition, because inherited variation is the raw material on which natural selection acts, a study
of the genetics of agonistic behaviour provides a background to the
functional issues raised in Chapter 11. In this chapter, in which we
assume that the reader has an understanding of the principles of
classical genetics, as presented in any standard biology textbook, we
are trying to answer the following questions.

1. Which aspects of agonistic behaviour (its form, its frequency or its
   higher-level manifestations such as territoriality and status) are
   inherited?
2. What is the pattern of inheritance: for example, how many genetic
   loci are involved and on which chromosomes are these situated?
3. How do the genes exert their effects on agonistic behaviour?

It is worth pointing out at the start that although inherited differences in agonistic behaviour have been identified in many species, this does not mean that environmental factors have no role to play in the development of this trait.

Where differences in agonistic behaviour are so marked that the animals in question can be classified unambiguously as aggressive or peaceful, their inheritance can be studied by a simple crossing experiment. Young crickets of the species *Gryllus campestris* fight fiercely but young of the closely related *G. bimaculatus* do not. Fertile hybrids can be formed by crossing these two species and the distribution of aggressive and non-aggressive animals in the progeny (Fig. 7.1) strongly suggests that differences in aggressiveness are inherited. A pair of alleles segregating at a single locus, with the aggressive phenotype dominant over the non-aggressive form, is probably responsible (Hörman-Heck, 1957).

More often, however, animals show a whole spectrum of agonistic behaviour, some being peaceful, others very aggressive, but the majority lying somewhere in between. Such a continuous pattern of variation is typical of a trait that is controlled by alleles segregating at several different loci and/or that is susceptible to the environmental influences. To study the mechanism of inheritance of such continuous traits the techniques of quantitative genetics have to be used (Falconer, 1981; Hay, 1985). A very simple account of some concepts in quantitative genetics is given below.

The overall variability of a trait in a population (its total phenotypic variance or V_p) is the combined result of genetic differences between its individual members (the genetic variance or V_g) and of differences in the environment in which they develop and live (the environmental variance or V_e). Provided there are no gene–environment interactions (an unlikely provisio, see Falconer, 1981).

$$V_p = V_g + V_e$$

Some genetic variance (the non-additive genetic variance or V_{na}) is the result of interactions between alleles – dominance effects, for example. The gene associations responsible for such non-additive variance are broken up by chromosome segregation at gamete formation. The variance resulting from those effects of alleles which are independent of their interactions with other alleles is called additive genetic variance or V_a. So:

$$V_p = V_{na} + V_a + V_e$$

The proportion of the total phenotypic variance of a continuous trait which is the result of additive genetic effects is called the heritability (h) of the trait and indicates the extent to which its state can be predicted across generations. That is,

$$h = V_a/V_p$$

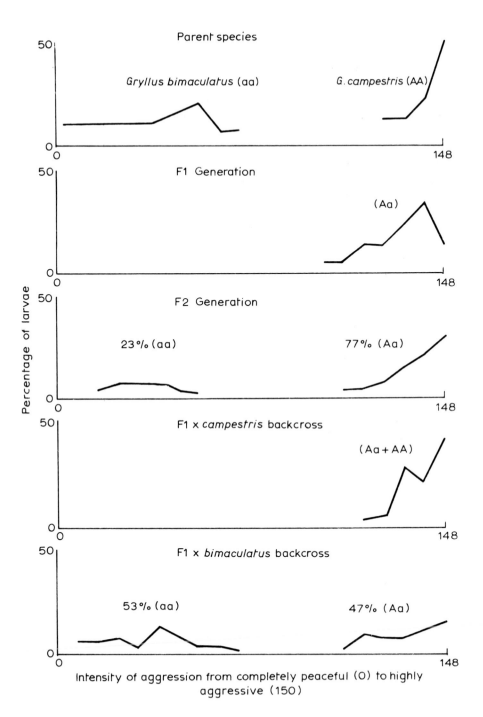

Figure 7.1 The percentage of larvae fighting in two cricket species and in the progeny of various experimental crosses (after Hörmann–Heck, 1957, from Huntingford, 1984).

One important statistic that quantitative genetic analyses attempt to measure is the proportion of the total variance in a given trait that depends on additive genetic effects. This summarizes the extent to which the trait can be predicted across generations and is called its heritability. Measures of heritability can suggest how selection may have operated in the past; strong directional selection in favour of extreme forms promotes genetic dominance (with heterozytes showing the favoured form); this increases the non–additive component of variance and lowers heritability values.

There are a number of ways in which the heritability of a continuously variable trait can be estimated, one being the use of artificial selection. If a significant proportion of the continuous variation in agonistic behaviour within a population is the result of genetic differences, then animals chosen for their fierceness will be well endowed with 'aggressive' genes. These will be passed on to offspring and, in a programme of selective breeding, an aggressive line will arise. The rate of change in aggression for a given strength of selection depends on the heritability of this behaviour, and so can be used to estimate it. Fig. 7.2 shows the results of three generations of selection for high and low levels of aggression in breeding male three-spined sticklebacks. The line selected for low levels of aggression diverges from the control line, suggesting that some at least of the variability in the original population depended on genetic differences between the fish. The heritability estimates based on the rate of this response is 50% (Bakker, 1986). The rapid response to selection for low levels of aggression suggests that natural selection has favoured aggressive males in the past.

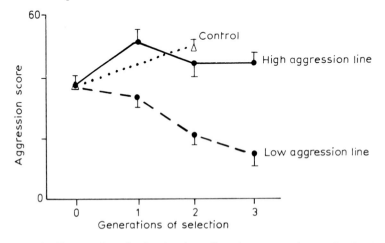

Figure 7.2 The results of selective breeding (means and s.e.) for high and low levels of aggression in male sticklebacks (after Bakker, 1986).

A number of other characteristics change in the selected lines along with the alteration in agonistic behaviour. By studying these we can get an idea of how the genes that influence this behaviour may be acting, although the characters might be associated by chance. For example, male sticklebacks in the line selected for high levels of aggression are more brightly coloured and have heavier kidneys than those in the low aggression line. These are both androgen–dependent characters, so selection may have acted to enhance androgen secretion, either directly or via increased gonadotrophic hormone production (Bakker, 1986). If this association between aggression and hormone-dependent characteristics is reliably found in every line selected for high male aggression and if it persists in crosses between high and low lines, then the conclusion that the behavioural differences between the selected lines is the result of inherited differences in endocrine activity would be supported.

7.2 GENETICS AND THE PATTERNS OF ANIMAL CONFLICT

Some studies have investigated non-violent responses to conflict; for example, the genetic mechanisms responsible for species-specific patterns of calling in crickets (which play a role both in attracting mates and in deterring rivals) have been analysed in some detail (Bentley and Hoy, 1972). Differences within a single species in the frequency of calling (leading to alternative strategies for attracting mates, page 53) are known to be inherited (Cade, 1981). In fruit flies, mutations have been identified which make males more effective at preventing females remating, one aspect of conflict between the sexes (Pyle and Gromko, 1981). However, most studies involve direct fighting, usually between adult males of the same species; furthermore, the great majority have been carried out on mice. While it is valuable to study one system intensively, the pattern of inheritance of agonistic behaviour may be quite different for other species, for females and even for males fighting in different contexts. Therefore in this chapter we also look at the genetics of agonistic behaviour in other species, in females as well as males and in males fighting in different contexts. We then try to identify general principles about the way agonistic behaviour is inherited.

7.3 IDENTIFYING GENETIC INFLUENCES

Crosses between behavioural variants and selective breeding pro-grammes are just two ways of studying the genetics of behaviour; Table 7.1 summarizes other sources of evidence. This battery of

Table 7.1 Evidence used to identify inherited difference in agonistic behaviour.

---

1. Crosses between behavioural variants.
2. Behavioural differences between chromosome variants.
3. Behavioural differences between mutants and normal animals.
4. Behavioural differences between inherited morphological variants.
5. Behavioural resemblances between relatives.
6. Selective breeding from extremes of the behavioural continuum.
7. Behavioural differences between geographic or inbred strains.

---

techniques has been used to identify inherited differences in agonistic behaviour in all kinds of animals. In this section we discuss some examples from the huge literature on rodents and then give a few examples from other groups of animals. Details of the pattern of inheritance are considered separately in a subsequent section, but in cases where only limited information is available, this is included here for the sake of simplicity.

7.3.1 Rodents

(a) Chromosome variants

Different races of wild mice may have different chromosome comple-ments; for example, in northern Italy, two forms are found – one with twenty-four and the other with twenty-six chromosomes. Male mice belonging to the race with twenty-six chromosomes are consistently more aggressive than those of the twenty-four chromosome race in laboratory encounters and appear to compete more successfully in the wild. So something about the possession of the extra chromosomes makes male mice fierce (Capanna *et al.*, 1984).

(b) Identified mutants

A number of mutant alleles, identified by their effects on morphology in mice, also influence agonistic behaviour, indicating that in some situations detectable differences in this trait can be caused by allele substitutions at a single locus.

- A mutant allele at a locus on the X chromosome produces a syndrome called testicular feminization in mice (and other mam-mals), which renders most male body tissues insensitive to androgens. Males with this trait neither attack other mice nor elicit attack from them, even when treated with androgen (Ohno, Geller and Lai, 1974).

- Mice bearing a mutant which makes the coat yellowish (at an autosomal locus) produce fewer ultrasonic vocalizations (given by the losers of fights) than do normal mice (D'Udine, Robinson and Oliverio, 1982).
- Swiss albino male mice (homozygous for an autosomal recessive mutant) are highly aggressive, perhaps because they are more sensitive to light than are normal mice (see Simon, 1979).
- Male rats rendered sterile by a single allele substitution are equally aggressive as wild rats but win more fights, perhaps because the condition also makes them larger (Christensen, 1977).

(c) Resemblance between relatives

Particular patterns of agonistic behaviour in male mice run in families, suggesting that these may have a genetic component.

- There is a marked similarity in aggressiveness towards other males among sibling male mice and between father and sons, even when the mice are all fostered by unrelated females (Cairns, MacCombie and Hood, 1983).
- In a study in progress, the sons of dominant male deermice tend to outcompete the sons of subordinate males even when they are the same size and were reared without contact with their fathers (D.A. Dewsbury, personal communication).
- Among the offspring of feral house mice, some males consistently fight for territories; others coexist peacefully, showing no evidence of territoriality and offspring behave like parents in this respect (Bisazza, 1982).

(d) Selection experiments

Artificial selection has produced changes in various aspects of agonistic behaviour in mice (Fig. 7.3), so at least some of the variation in agonistic behaviour in the original population must have been inherited. Selection for rapid attack by male mice on other males produces a marked effect, but selection for slow attack rates is less successful; heritability estimates are of the order of 30–40%. An increase in speed of attack on other males occurs in all young male mice as they approach maturity, but this takes place at a younger age in the males of the fast-attack line. Males in the fast-attack lines are also more aggressive than normal to their mates and offspring, so some of the same inherited factors must influence these different aspects of male agonistic behaviour (Lagerspetz, 1979; Cairns et al., 1983; Van Oortmerssen and Bakker, 1981). Selection for both fast and slow attack by female mice on another female is effective and heritability is

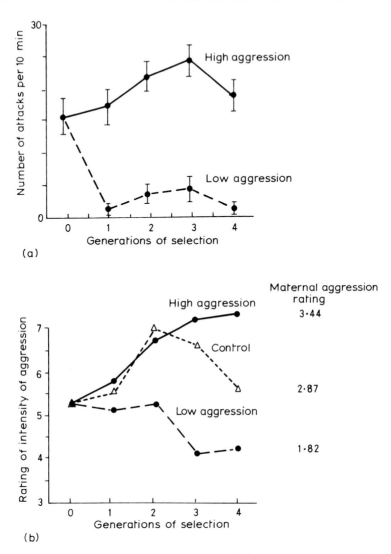

Figure 7.3 Selective breeding for agonistic behaviour in mice, and some correlated effects. (a) Attack (means and s.e.) by males on another male in a neutral arena (after Cairns *et al.*, 1983); (b) attack by isolated females on another female (after Ebert and Hyde, 1976).

estimated at about 40–50% (Hyde and Sawyer, 1979, 1980). Aggression in females is not enhanced in lines selected for male aggression, nor is the converse found. So fighting is inherited by separate genetic mechanisms in the two sexes.

(e) Strain differences

Behavioural differences are often found between geographically isolated strains of the same species (whose gene pools have almost
certainly diverged). Where these differences persist in animals reared
under identical conditions, genetic factors are implicated, although
maternal effects (acting both before and after fertilization) could
complicate the issue (Eleftheriou, Bailey and Denenberg, 1974; see
also Michard and Carlier, 1985). For example, two subspecies of
deermice (*Peromyscus maniculatus bairdii* and *gracilis*) differ in the
aggressiveness with which mothers protect their young, *P.m.gracilis*
being much fiercer. This difference persists in animals reared in a
uniform laboratory environment, suggesting that it may be inherited
(King, 1958).

Mice from different inbred strains (produced for medical purposes)
have different genotypes and also vary in many aspects of their
agonistic behaviour; the differences often persist when maternal
effects (postnatal at least) have been ruled out (Fig. 7.4 and Table 7.2).
So there must have been a great deal of genetically determined
variation in agonistic behaviour in the wild mice from which the
strains were derived. Which strains behave most aggressively depends
critically on the context in which fighting is studied and on how the
behaviour is measured. However, there are some robust effects; for
example males of the strains DBA/1 or /2 are usually more aggressive
to other males than are those of the strains C57BL/6 or /10. In these

Table 7.2   Some differences in agonistic behaviour between inbred strains
of mice, (from Michard and Carier 1985; Ogawa and Makino, 1984).

Behaviour	Ranking of strains
Agonistic behaviour of adult males	
Percentage of victories in cross/ strain pairings	C57 BL/J > BALB/cJ
Latency to first attack	BALB/C > C57BL/10
Number of attacks	DBA/2J = Swiss > C57BL/6 = C57BL/10
Agonistic behaviour of adult females	
Pregnant and lactating against adult males	AKR/J > DBA/2J > C57BL/6 > BALB/C
Group housed against lactating female	Swiss > C57BL/6 > C3H

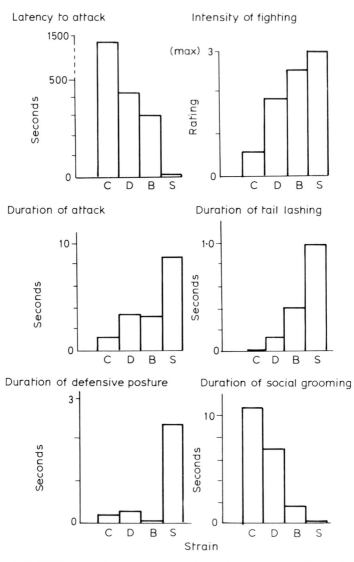

Figure 7.4 Differences between strains in various aspects of agonistic behaviour. C, C57Bl/10J; D, DBA/2J; B, BALB/cJ; S, SJL/J (after Hahn and
Haber, 1982).

strains at least, aggression in lactating females (but not attack by
females towards lactating intruders) covaries with that of the males.
Strains also differ in responsiveness to early environmental influences.
For example, early handling makes males of the non-aggressive
C57BL/10 strain more aggressive but has no effect on the behaviour of

aggressive (DBA/1) mice (See Hay, 1985, and Michard and Carlier, 1985, for reviews).

7.3.2 Other animal groups

(a) Crosses between behavioural variants

Hybrids between beagles and coyotes fight using recognizable behaviour patterns inherited from their parents; no intermediate forms are observed so individual items of the agonistic repertoire seem to be inherited intact. However, individual hybrid animals may use a combination of complete acts, some typical of their beagle parent and others typical of the coyote one. It looks as though the various complete items of the agonistic repertoire are inherited independently. All the hybrids, regardless of the actual behaviour patterns used during agonistic encounters, are both timid and aggressive, like their coyote parent. The overall readiness of animals to engage in a fight seems to be inherited independently of the actual behaviour patterns used (Fox, 1978).

Cave-dwelling fish (*Poecilia sphenops*) show a dramatic reduction in the frequency with which they perform a variety of agonistic behaviour patterns, compared to closely related river-dwelling forms. F1 hybrids produced by crossing these two forms show very little agonistic behaviour, suggesting that the original behavioural difference depended on genetic factors and that the low aggressive condition is dominant. Further crosses indicate that the more of the cave-dwelling genome a fish has, the less aggressive it is, suggesting that a number of loci are involved. Levels of the different elements of the agonistic repertoire remain correlated in all crosses, so these must all be inherited by the same genetic factors (or by a set of closely linked factors; Parzefal, 1979). Lake char are much less aggressive than brook char and hybrid crosses show that this behavioural difference is inherited. In this case, though, F1 hybrids resemble brook char, suggesting that the aggressive condition is genetically dominant (Fig. 7.5). Although, on the whole, the different components of the agonistic repertoire covary, certain differences in their pattern of inheritance suggest that they depend on separate, though probably linked, genetic mechanisms (Ferguson and Noakes, 1982, 1983).

(b) Chromosomal variants

In a number of viviparous fish, all-female species exist in which genetically identical eggs are produced without a reduction division. In some cases, these females mate with males of closely related species

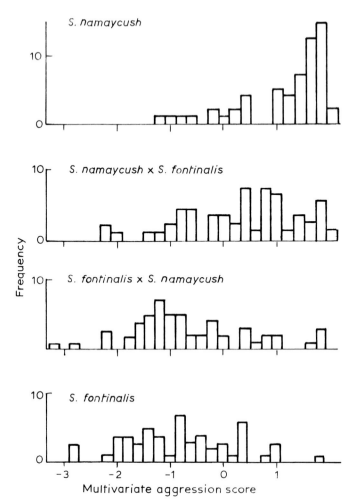

Figure 7.5 Compound aggression scores in brook char (*Salvelinus fontinalis*) and lake char (*S. namaycush*) and in their reciprocal hybrids (after Ferguson and Noakes, 1982, from Huntingford, 1984).

producing what are called hemiclones. The all-female species *Poeciliopsis monacha-lucida*, getting its maternal genes from *P. monacha* and its paternal genes from *P. lucida*, consists of at least eighteen different, naturally occurring hemiclones. Females from these different hemiclones were mated in the laboratory to males from an inbred line of the species *P. lucida* to produce forms which differed only in the genes inherited from their mother. Females in one of the experimental lines were markedly more aggressive than the others, often killing male fish and chasing and ramming both males and females. So differences in the maternal genome of these fish may give rise to marked differences in agonistic behaviour, although prenatal maternal effects are possible. Geographical distribution of the hemiclones suggests that the most aggressive form (together with the next most aggressive one) is able to prevent the others from colonizing certain favoured areas (Keegan–Rogers and Schultz, 1984).

(c) Identified mutants with behavioural effects

Chickens bearing the so-called dwarf mutant (which decreases body size) engage in more frequent and sometimes more intense agonistic encounters than do normal birds (Marsteller, Siegel and Gross, 1980). Male fruit flies (Drosophila melanogaster), homozygous for the vestigial mutant, lack the wings used in the normal courtship of this species but make up for this deficiency by energetically chasing females about and interfering with the courtship of normal males (Rendell, 1951).

(d) Inherited polymorphisms with behavioural effects

Animals of various species show inherited variation in their morphology; for example, some male ruffs are light coloured and others are dark. In some cases, these morphological differences are accompanied by differences in agonistic behaviour (light coloured ruffs do not defend territories, van Rhijn, 1973), suggesting that the genetic factors that cause the morphological difference also influence behaviour. Of four different morphs of the cichlid fish Pseudotropheus zebra (identified on the basis of their body dimensions) one is more aggressive and defends larger territories than the others (Schröder, 1980). White-throated sparrows come in two colour forms (which also have different chromosome types); one form, with a white head stripe, initiates more attacks in the spring and the other, with a tan head stripe, initiates more attacks in the autumn (Ficken, Ficken and Hailman, 1978; Watt, 1984).

(e) Resemblances between relatives

Such resemblances suggest that different patterns of agonistic behaviour may have a heritable component in a variety of species.

- Aggressiveness and dominance run in families in red grouse reared in standard conditions in the laboratory (Moss et al., 1982).
- Variability in the frequency with which sticklebacks attack each other is greater between broods than within broods (Goyens and Sevenster, 1976).

- Worker honey bees establish dominance hierarchies and are able to produce parthenogenetic eggs. Variability in social dominance and in the amounts of queen substance (which suppresses reproductive activity in other workers) is much greater between the broods of different workers than within broods; heritability values are about 40% and 89% respectively (Moritz and Hillesteim, 1985).
- Sibling lobsters are more similar than are unrelated individuals in latency, duration and frequency of fighting. Heritability values are significant but small (Finley and Haley, 1983).
- The marine snail *Nassarius pauperatus* comes in two forms; twisters (but not non-twisters) swing the shell from side to side when they come in contact with other snails and gain a competitive advantage (page 46). Non-twister snails produce young which behave in the same way; twisters produce both forms (McKillup, 1983).

(f) Selective breeding

Various breeds of domestic animals have been selectively bred for behavioural traits and sometimes for different levels of aggressiveness. That such selection has been possible shows that some of the behavioural variation in the original stock from which these different breeds were produced must have had a genetic basis. Among dog breeds, terriers are bred for fierceness and beagles for their peaceable nature. (Fox, 1978). Selection for sporting purposes has produced fighter crickets, fish, cocks, cows and bulls.

Other, scientific, programmes of breeding for differences in agonistic behaviour provide additional information. Silver foxes can be selectively bred for docility and this is associated with reduced production of adrenal hormones in response to stress (Belyaev and Trut, 1975). Systematic selection for aggression and dominance produces a positive response in both directions in chickens and in grouse (Cook and Siegel, 1974; Moss *et al.*, 1982). Selective breeding for aggressiveness in young sticklebacks of both sexes (average heritability = 45%) and in adult females (heritability = 48%) produces a response in both directions (Fig. 7.6). Females selected for adult aggression show equivalent changes in juvenile aggression, so some of the genes affecting these two aspects of agonistic behaviour must be the same. In adult males, selection for territorial aggression and, in separate lines, for the ability to dominate another fish in a pairwise encounter is effective in the downwards direction only (Fig. 7.2). Selection for dominance does not bring with it a change in territorial aggressiveness, so these must be at least partly distinct in their genetic bases (see also Francis, 1984), but males selected for low levels of aggression in a territorial context are less aggressive to females during courtship (Bakker, 1986).

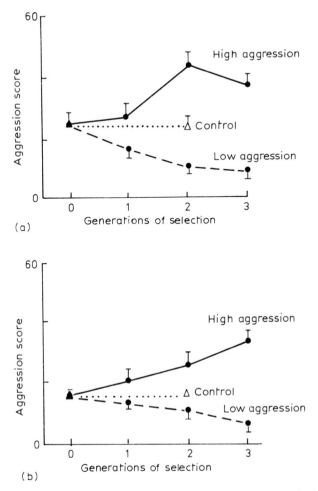

Figure 7.6 The effects (mean and s.e.) of three generations of selection for high and low levels of aggression in sticklebacks. (a) Juvenile males; (b) adult females (after Bakker, 1986).

(g) Differences between strains

Specially bred strains of bees are available for apicultural purposes, and these differ in the readiness with which they will attack a variety of opponents. Crosses give a heritability value of about 60% and suggest that the aggressive form is dominant. Spiders (*Agelenopsis aperta*) collected from desert sites are much more aggressive towards conspecifics than are those from more productive habitats. These differences persist in animals reared under identical conditions in the laboratory, suggesting that they may be inherited (Maynard Smith and Riechart 1984).

7.4 PATTERNS OF INHERITANCE

The previous sections have demonstrated genetically determined variability in many aspects of agonistic behaviour in species ranging from snails to dogs. In this section we look in more detail at the way in which these differences are inherited. A full understanding of the genetic architecture of agonistic behaviour (or any other trait) requires knowledge of how many loci are involved, where these are located and the pattern of genetic dominance (which trait is expressed in the heterozygote). Complete information is not available for any species, but a certain amount is known about how differences in agonistic behaviour are inherited in mice.

7.4.1 Rodents

Carefully designed crosses between inbred strains have gone some way towards elucidating the genetic architecture of agonistic behaviour in mice (Table 7.3). Certain aspects of agonistic behaviour (maternal aggression in females and, in males, short accumulated attack time and offensiveness to a non-aggressive opponent) show low heritability and strong directional dominance for the aggressive condition. In contrast, overtly offensive behaviour by males against fierce opponents has a high heritability and low directional dominance. Interpreting these figures in terms of the probable effects of selection in the past, extreme levels of maternal aggression and offensiveness to a non-aggressive opponent have been favoured by selection but offensiveness to an aggressive opponent has not, presumably because this will result in escalated fights (Hahn and

Table 7.3 Heritability estimates and directional dominance for various aspects of agonistic behaviour in male mice (from Singleton and Hay, 1982).

Behaviour	Opponent	Heritability (%)	Directional dominance
Number of jump attacks	Aggressive	91	None
	Non-aggressive	85	None
Accumulated attack times	Aggressive	58	For low expression
	Non-aggressive	78	For low expression
Number of bites	Aggressive	56	For low expression
	Non-aggressive	75	For high expression

Table 7.4 Numbers of aggressive mice in the progeny of crosses between an aggressive (Balb/cJ) and a non-aggressive strain (Balb/cN) (from Ciaranello, Lipsky and Axelrod, 1974).

	Fighters	Non-fighters
Balb/cJ	6	0
F1	2	4
Balb/cN	0	6

Haber, 1982; Singleton and Hay, 1982; see also Michard and Carlier, 1985).

Several studies (often with small sample sizes) have looked at the behaviour of the progeny of crosses between inbred strains of mice which have markedly different levels of inter-male aggressiveness. In some cases, the results suggest that the original behavioural difference depended on one autosomal locus, with the non-aggressive phenotype dominant (Table 7.4; Ciaranello, 1979; Orenberg et al., 1975; Kessler et al., 1977). A more complex study looked at the behaviour of a series of recombinant inbred lines set up from the progeny of crosses between the two parental strains, (see Hay, 1985). The results suggest that the difference in aggressiveness between the original strains depends on alleles segregating at two loci, with the aggressive form dominant (Eleftheriou et al., 1974). Whether alleles at one or a few loci have major effects in the natural genome of wild mice remains to be seen.

Identification of the chromosomal location of genetic factors influencing aggression in male mice has been most successful for the sex chromosomes (Maxson, 1981; Selmanoff and Ginsburg, 1981). When mice of two strains are crossed the male progeny of reciprocal crosses are genetically identical (having half their autosomes from each parental strain) except for the source of their X and their Y chromosome (which they get from their mother and father respectively). The male progeny of reciprocal crosses between an aggressive strain (such as DBA/1) and a non-aggressive strain (such as C57BL/10) resemble their father in terms of their aggressive behaviour, whatever the strain of their mother (Table 7.5). This suggests that at least some of the original difference in aggressiveness between the strains depends on genetic factors on the Y chromosome. The genome of the mother makes some contribution, since males with a C57 father are more aggressive if their mother was DBA (see Table 7.5; Selmanoff et al., 1975). The aggression-enhancing effect of the Y chromosome of aggressive (DBA) strains depends on the autosomal chromosomes

Table 7.5 Aggressive behaviour in male offspring of reciprocal crosses between aggressive (DBA) and peaceful (C57) strains of mouse (from Selmanoff *et al.*, 1975).

| | | Father's strain | |
		DBA	C57BL
Mother's strain	DBA	35.4*	10.7
	C57BL	19.8	7.3

* Compound aggression score, max. = 180

with which it is associated. There is some evidence that differences in overtly offensive actions are inherited by Y chromosome factors while autosomal factors control defensive behaviour (Maxson *et al.*, 1982).

7.4.2 Other animal groups

There are few precise data available about the mechanisms by which differences in overtly agonistic behaviour are inherited in other groups of animals. Male killifish with the chromosome complement YY are more aggressive than are XY males, suggesting that, in fish as in mice, the Y chromosome may make a special contribution in aggressiveness. This is certainly suggested by the results of crosses between two inbred strains of guppy differing in the frequency of gonopodial thrusting (a way of fertilizing females without their co-operation) and in the level of aggression. The state of both these characteristics in male progeny resembles that of the strain from which they inherit their Y chromosome (Farr, 1983).

The funnel web spider *Agelenopsis aperta* shows geographic variation in aggressiveness which can be related to local food levels (page 79). The F1 progeny of crosses between the two forms fight extremely fiercely. A motivational model devised to explain this observations was discussed on page 84; the genetic model to which it gives rise is presented in Table 7.6. According to the model, the levels of aggression and fear are controlled by alleles segregating at two independent loci, with the two parental strains of spider being homozygous and with dominance in favour of high levels of aggression and low levels of fear. F1 hybrids are heterozygous at both loci, which makes them just as aggressive but not as fearful as their aggressive parent, so they fight even more fiercely (Maynard Smith and Riechert, 1984).

The ability to release poisons, which characterizes the killer trait in ciliate protozoa such as *Paramecium* (page 14), is inherited. A single

Table 7.6 Proposed mechanism for the inheritance of agonistic behaviour in spiders (*Agelenopsis aperta*) (simplified from Maynard Smith and Riechert, 1984).

Parents	Geographic strain			
	Desert grassland		*Desert riparian*	
Genotype	AA	bb	aa	BB
Phenotype	High aggression	High fear	Low aggression	Low fear
	F1 hybrids			
Genotype	Aa		Bb	
Phenotype	High aggression		Low fear	

allele is responsible for this trait, with the killer phenotype dominant over the non-killer, but its effects depend on particles (possibly symbiotic bacteria) in the cytoplasm. These produce the killer substance and can only be maintained if the host carries the killer allele (Dini and Luponni, 1982).

7.5 THE ROUTE FROM GENES TO BEHAVIOUR

There are a great many ways in which a genetic difference can produce a difference in agonistic behaviour. In the first place, the precise behavioural nature of a gene-induced change in aggression can be very different. Some genetic differences result in a change in the form of the movements used during fights (for example, in coyotes, page 177); others alter an animal's ability to produce agonistic cues (for example, scents in mice, page 172) or its sensitivity to relevant environmental stimuli (for example, albino mice, page 173). In addition to such effects on motor patterns and on production and reception of agonistic cues, genes can change the readiness with which fighting occurs. This may be the result of changes which are not specific to the agonistic context; for example genetic factors modify the readiness with which mice learn to avoid unpleasant stimuli (see Hay, 1985) and so influence their ability to modify their agonistic behaviour in the light of experience. However, in other cases genes probably produce specific effects of agonistic responsiveness (for example in mice and spiders).

There are a number of ways in which a genetic difference might generate these various kinds of behavioural change. Whatever the route from genes to behaviour, in each case an inherited difference produces adult animals which respond differently in an agonistic context. So the mechanisms by which genes influence agonistic behaviour are considered briefly here. The categories used are not mutually exclusive; for example a gene could alter central nervous system structures which control the agonistic behaviour of an adult animal by a change in hormonal environment during development.

7.5.1 Patterns of development

Many of the inherited changes that accompany domestication (reduced or absent weapons, lower levels of aggression) can be interpreted as the result of a reduction in the rate of development in domesticated forms; traits which are typical of juvenile animals in the wild form persist into adulthood in the domesticated form. Some of the differences in aggressiveness between different strains of mice may arise in an equivalent way. For example, in male mice selectively bred for slow attack rates as adults, the sharp, transitory increase in attack frequency which normally occurs as they reach sexual maturity is smaller and occurs later (Cairns et al., 1983). Similar results are obtained for sticklebacks (Bakker, 1986). Mice from strains that are more mature at birth are less susceptible to early environmental effects, including the aggression-enhancing consequences of handling (Oliverio, 1983).

7.5.2 Body size and appearance in adults

Inherited differences in body size, described for a whole range of animals (Falconer, 1981), are likely to influence fighting ability, fighting behaviour and the cues an animal sends out during an agonistic encounter (page 70). Great tits which have inherited a large size from their parents are more successful in fights than the offspring of smaller birds (Garnett, 1981) and large, genetically sterile male rats win more fights than normal animals (page 173). Several examples of genetically controlled differences in agonistic cues have been described in the previous sections and these too will alter the course of a fight. Some examples are colour in sticklebacks (page 171) and sounds (page 173) and smells (page 172) in mice.

7.5.3 The endocrine system

Some inherited differences in agonistic behaviour can be related to differences in the hormones of the adrenal gland, which are known to influence agonistic behaviour (page 118). For example, selection for

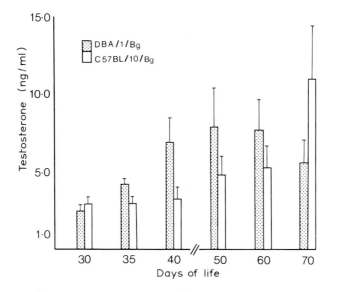

Figure 7.7 Changes in testosterone levels (mean and s.e.) with age in aggressive (DBA/1/Bg) and non-aggressive (C57BC/10/Bg) strains of mice (after Selmanoff *et al.*, 1977b).

docility in domestic animals is often accompanied by reduced pituitary–adrenal activity (page 180). On the other hand, aggressive male mice tend to have enhanced adrenal activity (Lagerspetz, Tirri and Lagerspetz, 1968; Ciaranello, 1979).

The reproductive hormones have been implicated in the control of agonistic behaviour in a number of species and variation in production or metabolism of these hormones may provide the link between genetic and behavioural variation. Production of gonadotropic hormones may be enhanced in sticklebacks bred for juvenile aggression and of gonadal hormones in those bred for male territorial aggression (Bakker, 1986; see page 171). Aggressive chickens show enhanced sensitivity to androgens (Cook and Siegel 1974), while the testicular feminization mutant (page 172) reduces aggression in mice by making the body tissues insensitive to androgens (Olsen, 1979). Male mice from a strain selectively bred for fast attack towards territorial intruders have higher blood androgen levels than those from slow attack lines. They are also more sensitive to the aggression-promoting effects of testosterone administered after castration; this effect may be the result of different hormonal environments during development (Van Oortmerssen, in press).

Males of non-aggressive strains of mice (such as C57BL/10/Bg) show a weaker pubertal surge in production of luteinizing hormone than do those of aggressive strains (such as DBA/1/Bg) and so have lower circulating testosterone levels around days 35–40 (Fig. 7.7). In reciprocal hybrids, patterns of hormone secretion as well as levels of aggression in males reflect the strain from which they inherit their Y chromosome (Selmanoff, Goldman and Ginsburg, 1977; see also Maxson, Shrenker and Vigue, 1983 for a review).

7.5.4 The peripheral nervous system and sensory pathways

There are many examples of genes that act on muscles and sense organs and some inherited differences in aggressiveness probably depend on such effects. Selection for low rates of impulse conduction in peripheral nerves in mice slows the leg withdrawal reflex in response to pain (Hegner, 1979). This is likely to influence the way they respond in an agonistic encounter. The aggressiveness of Swiss albino mice may be the result of differences induced in the visual system which render the animals more sensitive to bright light (Simon, 1979).

7.5.5 The central nervous system

Differences in agonistic behaviour in adult animals can be ascribed to genetic factors that alter the structure or function of the central nervous system. For example, some inherited differences in aggressiveness in mice are associated with reduced brain size (Simon, 1979). The effects of the testicular feminization mutant on aggression in mice (page 172) are mediated via a small group of cells in the hypothalamus (page 145). Mutant-induced androgen insensitivity in these cells is apparently responsible for the absence of aggression in affected mice (Ohno et al., 1974).

Many differences in brain biochemistry have been identified among inbred strains of mice (Ingram and Corfman, 1980), and some of these are reliably correlated with variation in agonistic behaviour. For example, aggressive strains often have higher brain noradrenaline levels and lower serotonin levels than do non-aggressive strains, as well as higher levels of the enzymes that synthesize noradrenaline (Lagerspetz et al., 1968). When two inbred strains of mice and their progeny are compared (although using small samples, and making the comparison at the level of groups and not individuals, see Michard and Carlier, 1985), high levels of activity of the enzymes that synthesize catecholamines are associated with well developed shock-induced fighting (Ciaranello, 1979). These findings make sense in the light of the facilitatory and inhibitory effects of catecholamines and serotonin on aggressive behaviour in mice (page 153).

7.6 OVERVIEW

7.6.1 Have our questions been answered?

(a) Inherited variation in agonistic behaviour

The preceding examples show that in every group for which data are available, inherited differences in agonistic behaviour exist. Many features of agonistic behaviour (from the form of the movements and the readiness with which these are used to patterns of territoriality and dominance) are influenced by genetic factors. In crickets (page 168), cave fish and char, and in canids (page 177), variation in genetic material has been shown to be responsible for differences between closely related species. In a whole range of animals from protists (page 184), snails and lobsters (page 180) to red grouse (page 179), mice (page 172) and dogs (page 180), naturally occurring within-species differences in agonistic behaviour have been shown to have some genetic basis. However, environmental effects are always superimposed upon these genetic influences; this is reflected in the fact that even in laboratory tests, though heritabilities may be high, they never even approach 100%

(b) Genetic architecture

In some cases, naturally occurring behavioural differences are known to be the result of alleles segregating at one or a few loci. The killer trait in protists (page 184) and fighting in juvenile crickets (page 168) provide examples. On the restricted background of the inbred strain, single allele substitutions can produce marked differences in aggressiveness in male mice (page 183). However, most naturally occurring differences in agonistic behaviour depend on a number of loci distributed throughout the genome. Some of these are found on the autosomal chromosomes (page 173), but in mice and guppies (page 183) the Y chromosome appears to play a special role. In crickets (page 168), bees (page 181), char (page 177), lactating mice and male mice responding to a non-aggressive opponent (page 182), the aggressive phenotype is dominant over the non-aggressive form. In cave fish (page 177) and in male mice responding to an aggressive opponent (page 182), the non-aggressive condition is dominant.

(c) Mechanisms of gene action

The key mechanisms by which genetic variation generates differences in agonistic behaviour may be a change in external appearance (page

186), in the peripheral nervous system or sensory pathways (page 186) or in the structure, neurophysiology or biochemistry of the central nervous system. Perhaps the best documented route from a genetic to a behavioural change is some sort of modification to the endocrine system, whether this involves a difference in production, subsequent metabolism or reception of a particular hormone. These various kinds of genetic effect are not mutually exclusive; thus an inherited difference in androgen production may alter both body size and brain activity, while a genetically determined change in brain biochemistry may alter production of the endocrine system.

7.6.2 General principles

1. The different aspects of agonistic behaviour (by different sexes and ages, in different contexts and to different categories of opponent) are often inherited independently. Thus selection for aggression between male mice does not alter that shown by females in the selected lines (page 174). In some cases, the different patterns of agonistic behaviour covary, as when both adult males and lactating female DBA/2 mice are very aggressive (page 176) and when male sticklebacks selected for low aggression to males are also less aggressive to females (page 180).
2. Considering fighting within one particular context, in lobsters (page 180), char, cave fish (page 177), and mice (page 176), but not in coyotes (page 177), the various components of the aggressive response tend to be inherited together, although they can sometimes be dissociated. They may therefore depend on the same genetic factors, or perhaps on closely linked ones. This system of linked inheritance for the different components of agonistic behaviour means that, in general, an animal will inherit the ability to perform the full adaptive complex of responses. The fact that the linkage is not complete means that there is some limited potential for producing new combinations of action patterns on which natural selection may act.
3. Inherited differences in the overall tendency to perform agonistic behaviour have been identified in many animals which nevertheless use the same motor patterns when they do fight. How an animal fights and how much it does so are inherited by independent mechanisms; potentially, therefore, these two aspects of agonistic behaviour could be independently modified by selection to suit local circumstances.
4. The mechanisms by which differences in agonistic behaviour are inherited, and particularly the patterns of genetic dominance, suggest how selection may act on animals in an agonistic context. Genetic dominance for aggressiveness (seen in juvenile crickets and char, in lactating mice, and in male mice matched with a non-aggressive

opponent) suggests that these traits have been subjected to strong directional selection. This is what we might expect because competition for food, protection of young and the ability to press home an advantage are critical determinants of fitness. On the other hand, the non-aggressive form is sometimes dominant, suggesting that extreme aggressiveness can be disadvantageous. Animals make complex assessments of probable costs and benefits of fighting and of their chances of winning (page 279) and this is reflected in the way fighting behaviour is inherited.

8

Development

8.1 INTRODUCTION

Somewhere between conception and maturity the capacity for agonistic behaviour appears. The course of its development is dictated by two major factors; one is the need for animals to compete successfully in the conflicts they experience when young. Another is the need for them to develop the skills they require to fight effectively when they become adults.

Because childhood is a period of both vulnerability and rapid growth, the need for resources such as shelter and food is particularly strong. Young animals often come into conflict both with adults and with their contemporaries over these resources. For example, survival in young anole lizards is critically dependent on the ability to find shelter from predators; shelters are in short supply and are fiercely disputed (Stamps, 1983a and b). Conflict among young animals is particularly intense because a small advantage at an early age can result in improved competitive ability for the rest of the animal's life. In groups of young rats, animals which initially have a small size advantage over their companions become dominant; this gives them access to more food, so that the initially small size difference becomes accentuated. Even after several months of separation from their dominant companions, the rats that were subordinate when young show less aggression to other males (Lore and Stipo-Flaherty, 1984). The status attained by a red deer stag (and thus his reproductive success) depends on his rate of growth in the first year of life; this is at least partially a consequence of his success in early disputes over food (Clutton-Brock *et al.*, 1982).

The outcome of early conflicts can have marked effects on morphological as well as behavioural development. In beetles, for example, males that initially grow fast become large, develop horns and fight for burrows. Those that grow slowly when young follow a different developmental route, never becoming large, failing to develop weapons and apparently wandering rather than defending burrows (Eberhard, 1979). Development of male morphology and behaviour is suppressed in subordinate paradise fish, who therefore become females (Francis, 1984).

With a few exceptions, agonistic behaviour is a complex business. An animal needs not only a complete repertoire of fighting techniques, but also the ability to deploy these effectively. During a fight, agonistic actions must be directed towards appropriate animals, and continually adapted to what the opponent does, to the probability of victory and to the costs and benefits of fighting. In the longer term, agonistic behaviour must fit into the social circumstances (territoriality, dominance hierarchy or whatever) in which the animal lives.

To see how these two requirements (of competing for resources when young and of assembling effective adult behaviour) are met and to understand how variability in adult agonistic behaviour arises, we address the following questions.

1. When and how do co-ordinated agonistic behaviour patterns arise during ontogeny?
2. How do these change in form and frequency with age? How do they come to be elicited by the right stimuli and in the right context?
3. What factors are responsible for the differences in form and intensity of adult agonistic behaviour that are observed between species, sexes and individuals?

8.2 THE AGONISTIC BEHAVIOUR OF VERY YOUNG ANIMALS

Very young animals often have the capacity to display at least part of the agonistic repertoire typical of their species on the very first occasion that they fight. Table 8.1 summarizes the results from a number of species; here we consider just a few of the better documented examples.

- Immediately after hatching, young spiders (*Trite planiceps*) fight with competitors. They show a full repertoire of movements and postures similar to that of adults and, as in adults, the larger of two opponents wins. Although there are differences (for example, young spiders make more use of stationary displays than do adults), this may just reflect the costs involved. Small moving objects elicit predatory attacks, so moving displays would expose young spiders to the risk of cannibalism (Forster, 1982).
- Young lizards (*Anolis aeneus*) live in clearings, where they are reasonably safe from predators. Suitable living space is in short supply and they compete fiercely for what is available, using the same set of displays as do fighting adults. Because the advantages of a good living site are so great, juveniles are extremely aggressive and, unlike adults, direct their attack towards all opponents, regardless of size (Stamps, 1983a and b).
- Newly fledged great tits fight with their peers over food, pecking each other violently, sometimes with lethal consequences. Larger

Table 8.1   Examples of effective agonistic behaviour in very young animals.

Species	Observation	Reference
Crabs	Young fight with species-typical actions straight after metamorphosis.	Jacoby (1983)
Cockroaches	Newly hatched young fight over refuges with species-typical actions.	Deleporte (1982)
Fish (mollies)	Young show the full species-typical repertoire in adult-like sequences from a few days after hatching.	Parzefal (1969)
Chameleons	Hatchlings fight over resting position with butting, chasing, bobbing and wrestling.	Burghardt (1978)
Golden eagles	Bite and stab at siblings.	Ellis (1979)
Pikas	Scent marking and species-typical agonistic behaviour shown by very young animals.	Whitworth (1984)
Peccaries	Agonistic behaviour appears soon after birth in disputes over suckling access.	Byers (1984)

birds dominate smaller ones and success at this stage improves their chance of survival (Garnett, 1981).

8.3 THE DEVELOPMENTAL ORIGIN OF AGONISTIC MOVEMENTS

Even when animals perform complete agonistic actions when they first fight, these movements have developmental precursors. From similarities in form and from the time course of development, it seems that in some species co-ordinated agonistic behaviour patterns originate in the weak, undifferentiated movements of very young animals. Sporadic, unco-ordinated movements of the eyes, fins and operculae occur in young blue gouramis by the second day of hatching. Co-ordinated mouth and opercular movements are observed by day three, when they begin to be used during active feeding, which is also accompanied by arching of the body. The same movements are also directed towards the dark eyes of companions as well as at food, with the result that the fish engage in head-to-head encounters. These usually end in flight by both participants. By three weeks, a dark tail spot develops and head-to-head encounters give way to bites (as if feeding) at the tail of other fish. Arching, previously shown while observing prey, develops into part of the agonistic response to other fish (Tooker and Miller, 1980).

In other cases, agonistic behaviour patterns develop from actions used by the young animals in other contexts, for example from interactions with parents. In young orange chromides (a cichlid fish), agonistic postures develop from movements used when feeding on the mucus secreted by their parents (see Fig. 8.3). Forward, head-on movements such as frontal display develop from the micronip (a forward feeding movement) and sideways-on postures such as lateral displays develop from glancing (brief sideways contact with their parent's side, apparently to stimulate mucus production; Wyman and Ward 1973). The active submission posture of adult canids (dogs and their relatives) arises from the sideways rooting movements of suckling pups and from the licking used by older pups to solicit food. Young pups lie on their backs in a supine position while their mother licks them to stimulate urination; this probably represents the developmental origin of the passive submission posture of adult canids (Fox, 1971).

Adult male jungle fowl fight by leaping at each other with neck feathers raised, while striking with the legs and pecking (Fig. 8.1(a)). Unoriented hopping in young chicks (Fig. 8.1(b)) appears at a very early age, initially with no aggressive function. Hopping is contagious, and simultaneously hopping chicks often bump into each

(a)

(b)

Figure 8.1 Adult fighting and its precursors in chicks. (a) Fighting in adult cockerels. (b) Hopping in young jungle fowl (reproduced with permission from Kruijt, 1964).

other. By about seven days, hopping is directed at other chicks, though it still occurs during general activity. By eight to nine days, chicks briefly stretch out their necks at the end of a hop, thus performing a juvenile version of the frontal threat. This is accompanied by raising of the neck feathers, as soon as the chicks have any to raise. Later, hopping and threatening are followed by directed leaping towards another chick, although the legs are not thrown forward so the chicks simply bump chests. By twenty-one days, pecking is observed, frequently at the partner's head. Thus 'genuine little fights' develop from an assortment of movements used by the chicks in other contexts (Kruijt, 1964).

8.4 PLAYFUL FIGHTING

In most mammals (and perhaps some birds, Ficken, 1977) the agonistic movements and postures used by adults are first observed during playful fights between young animals (see Fagen, 1981, and Smith, 1984, for reviews). The play movements themselves arise from other actions; for example, play soliciting in puppies is derived from the successive paw raising and upward head movements seen during suckling (Fox, 1971).

Play is a very conspicuous activity, but is not easy to define with precision (Chalmers, 1984). During playful fights, animals make harmless use of actions which will eventually be used in real fights. These are put together into unpredictable sequences of exaggerated actions, with the participants often rapidly exchanging roles. During playful fights, which are often initiated by special postures and facial expressions, the participants appear to be trying to gain an advantageous position without themselves being put at a disadvantage. Table 8.2 gives some examples of playful fighting in mammals.

The form of playful fighting in a given species reflects its adult agonistic repertoire; for example, young horned ungulates head butt and wrestle when playing but members of the pig family use playful bites and slashes. Differences between adults of two species of hyrax (*Procavia johnstoni* and *Heterohyrax brucei*) in the relative frequency of several components of agonistic behaviour correspond exactly to differences in the same actions during playful fights (Caro and Alawi, 1985). In species such as the bighorn sheep, where adult males are more aggressive than adult females, young males indulge in more playful fighting (Berger, 1980). Adults of monogamous carnivore species such as the bush dog are equally aggressive and in these cases young males and females play with equal enthusiasm (Biben, 1983).

These observations, together with the obvious physical similarity between playful and real fights, suggest that playing may allow young animals to practise sophisticated fighting skills in a context where mistakes will not have serious consequences (Symons, 1978). Non-contact play usually appears later than contact play; and effective attack sequences may develop as young animals gradually learn that less and less intense actions can be effective in bringing about a desired endpoint (Fagen, 1981). However, playful fighting cannot be essential for the development of efficient adult agonistic behaviour in mammals, since species such as hamsters and pikas (a relative of the rabbit) which do not play extensively fight perfectly well (Hole and Einon,

Table 8.2 Examples of playful fights in mammals.

Species	Observation	Reference
Yellow-bellied marmot	40% of social interactions between juveniles playful; play fighting involves wrestling. Males play fight more than females.	Nowicki and Armitage (1979)
Rat	Play fights appear at about seventeen days, initiated by high pouncing and involve wrestling, biting and chasing.	Poole and Fish (1975); Meaney and Stewart (1981)
Polecat	Open-mouthed face used during play fights. Biting is inhibited; intimidating and immobilizing action absent. Roles often reversed.	Poole (1966)
Bear	Open-mouthed, puckered lip face used during play fights, involving head butting, wrestling, biting and chasing.	Henry and Herrero (1974)
Rhinoceros	Subadults frequently show playful butting, biting and horn wrestling.	Owen-Smith (1975)
Deer	Fawns play by butting, kicking and chasing. Males play more than females. Decreased during food shortage.	Müller-Schwartz et al. (1982)
Ibex	Butting, chasing, clashing play fights. Males prefer to fight with other males, especially with relatives of a similar age.	Byers (1980)
Peccaries	Biting, slashing, kicking and chasing play fights. Seen in adults as co-ordinated herd activity associated with scent marking.	Byers (1983)

1984). Nor can the development of fighting skills be the only function of play, because adults (peccaries, for example) quite often play.

Playful fighting is a particularly conspicuous feature of development in many primates, although vervets and squirrel monkeys never play much nor do langurs growing up in unfavourable environments (Poirier and Smith 1974; Baldwin and Baldwin, 1977, Lee, 1983). A variety of signals (special gaits, postures and, in particular, the open-mouthed play face) are used to indicate playful intentions. Playful fights in young rhesus monkeys involve chases and wrestling match-es; vocalization, piloerection and aggressive threat are absent. Each participant behaves as if trying to achieve a position from which he or she can bite without being bitten, although bites are inhibited and rarely hurt. Young males play fight more frequently and more energetically than do young females, but playful fights between females are more likely to turn into serious fights. As the monkeys get older, playful fights become less mutual and more one-sided, as one animal consistently attacks and the other defends (Symons, 1978).

8.5 CHANGES WITH AGE

Once agonistic behaviour has appeared in the repertoire, it changes in a variety of ways as the animals concerned get older. These changes are the result of a number of rather different processes (Table 8.3), all of which are illustrated in the following sections. Some of these apparent changes with age are in fact the result of differential mortality in young animals. For example, although the head bobbing displays of young lizards (*Anolis aeneus*) are very similar to those of adults, between-individual variation in one (the jerkbob) is markedly reduced as the animals get older. This could be because animals that do not perform this display in the standard way fail to survive (Stamps, 1978).

However, in most cases, differences in agonistic behaviour in animals of different ages are the result of developmental events in the animal concerned. Some of these simply reflect the process of matura-

Table 8.3 Some factors that cause differences in agonistic behaviour in animals of different ages.

---

1. Selective mortality of behavioural variants

2. Changes due to maturational processes
   Development of muscles
   Development of nerves
   Growth of body and/or weapons
   Development of agonistic cues
   Endocrine changes

3. Changes due to experience
   Practice of performing movements
   Experience of the consequences of fights (see Table 8.5)

---

tion. Improved overall muscle co-ordination brings with it the ability to perform controlled movements, including agonistic actions. Aggressiveness may increase with body size and weapon development as the animals become less vulnerable and more effective fighters (in cockroaches and lobsters). Animals of many species become more aggressive with the approach of physical maturity, when they start to compete for a very valuable resource, namely mates. In some cases, this change depends on enhanced production of gonadal hormones (page 113), but in hamsters, for example, aggression increases in castrated males as well as intact ones (Whitsett, 1975).

Finally, changes in agonistic behaviour during development may be the result of specific experience at previous stages. For example, the large major claw which lobsters use as a weapon fails to develop normally if young lobsters are prevented from moving their claws normally (Govind, 1985) and young animals of many species learn to use and respond to threats as a result of experience in early agonistic encounters.

8.5.1 Changes in form and frequency

Agonistic behaviour patterns often change in form with age, perhaps because of a general improvement in co-ordination. Young fiddler crabs begin to perform the waving and rapping displays (which in adult males attract females and elicit attack from rival males) some six months after hatching. As the crabs get older, the frequency of incomplete waves drops while the duration of the wave and the height to which the claw is raised increase. Larger crabs are able to produce longer, louder, more rapid rapping sounds which are more effective as signals (Hyatt, 1977).

Some changes in the form of agonistic behaviour patterns can be related to changes in the development of fighting equipment. Agonistic interactions over food and space in larval rock lobsters include an exchange of threats. These initially involve sweeping the short antennae forward and back over the carapace, a movement which produces a rasping sound. As the larvae get older, their antennae lengthen and are used to lash opponents during fights. Encounters are won by the animal with the longest antennae (Berril, 1976).

In young lobsters, escape movements are the predominant response to a threatening stimulus; as they get older, lobsters switch to defensive threat, with extended claws. Various morphological changes accompany these behavioural shifts; the abdomen becomes relatively less well developed and the giant nerve fibres which control escape become relatively smaller, so that the escape response is slower and less effective. In contrast, the claws increase in size, which makes

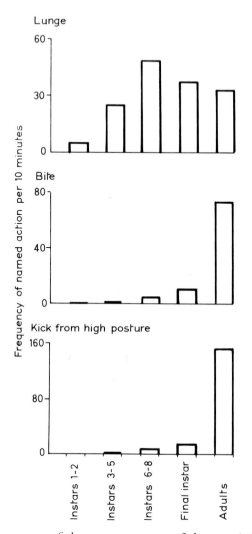

Figure 8.2 Frequency of three components of the agonistic repertoire in cockroaches of different ages (after Gorton and Gerhardt, 1978).

the extended claws an effective threat (Lang *et al.*, 1977).

The frequency and intensity of fights also change with age, reflecting changes in the requirements of the young animals and in their vulnerability to injury during fights. Young mollies (*Molliensia velifera*, Parzefal 1969) and lizards (*Anolis aeneus*) become less aggressive as they get older; in the case of the lizards, their need for shelter becomes less pressing (Stamps, 1983b). The cave-living cockroach *Eublaberus posticus* is a social species, feeding on the abundant food generated by bat droppings. In this species, the intensity and frequency of fighting decreases with age (Gorton and Gerhardt 1978). In general, though, animals fight more often and more intensely as they get older; this is the case, for example, in another cockroach species (*Shawella couloniana*; Fig. 8.2; Gorton and Gerhardt, 1978).

Overall, fights become more frequent and intense with age in Dungeness crabs, but young crabs fight particularly aggressively in instars 3–6, an age at which in the wild they would be migrating into open water and assuming a solitary way of life (Jacoby, 1983). Young pikas live at high latitudes where the summer is very short. Levels of aggression rise very quickly between the age of four and six weeks, as the young pikas compete for the feeding territories they need if they are to survive their first winter (Whitworth, 1984).

8.5.2 Expanding the repertoire

Although pikas fight more often as they get older, they make greater use of threat displays rather than overt attack. A shift from overt attack to the use of threats is seen in many animals (blue gourami, Tooker and Miller, 1980; mollies, Parzefal, 1969), so that the repertoire of agonistic behaviour patterns expands with age. In young rainbow trout, chasing and nipping are first observed at the time the young emerge from the gravel in which the eggs hatch. Threats (involving head down posture and dorsal fin erection) appear after a few days and by the time the fish are a year old, many encounters are settled by an exchange of threat display. The young fish probably learn to associate the sight of a displaying rival with the pain of being bitten (Chizar, Drake and Windell, 1975; Dill, 1977; Cole and Noakes, 1980).

Early agonistic movements may multiply into new forms by changing the emphasis on existing features or by adding new components. In orange chromides, a repertoire of new agonistic behaviour patterns develops by accentuating different features of the two parent-oriented, feeding movements (Fig. 8.3; Wyman and Ward, 1973). As jungle fowl chicks experience the pain of being attacked, they begin to show fear responses when confronting a rival. The agonistic repertoire is then expanded by addition of a number of displays which are thought to depend on simultaneous attack and flight motivation (Kruijt, 1964).

8.5.3 Changes in motivational bases

The agonistic behaviour patterns used by very young animals are often under the control of causal systems other than aggression or fear.

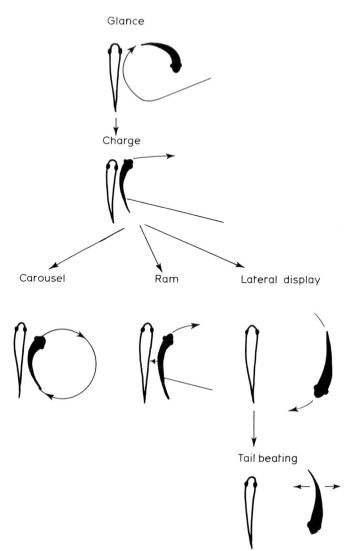

Figure 8.3 Development of five elements of agonistic behaviour from the glancing movements of young orange chromides (after Wyman and Ward, 1973; from Huntingford, 1984).

The separation of these actions from their early causal bases and the appearance of the systems of aggression and fear found in adult animals (page 60) is an important aspect of the development of fighting. Attack and feeding are motivationally linked in a number of young fish. In some species, this relationship persists into adulthood, but in orange chromides the feeding component of the early agonistic actions starts to disappear when the young become independent of their parents. However, until quite a late stage quiver, charge and mouth fight occur more frequently after a bout of feeding (Wyman and Ward, 1973).

In jungle fowl chicks, hopping attacks and pecking are linked to general activity and to feeding, but gradually become emancipated from these systems. Overt signs of aggression disappear within a few seconds of a fight in young chicks, but older birds remain aggressive for some time, suggesting that a sustained motivational state is aroused. Fear responses appear as the chicks learn the consequences of being attacked and the mutually inhibitory interactions between aggressive and sexual behaviour develop in older birds as a result of social experience with birds of both sexes (Kruijt, 1964).

A major behavioural reorganization occurs at weaning in kittens as the aggressive and predatory behaviour of adult cats develops from the actions used during play. Some play items drop out of the repertoire, but others are retained in adults. Two play items, face off and stand off, occur at the start of agonistic sequences in adult cats as they do in playful fights. The play component of arching develops into the arched back posture with raised fur of adult fights, and in adults as in young this is often followed by chasing. Wrestling (Fig. 8.4(a)) remains in the repertoire after the kittens cease to play, being used (with an added kicking component) both during fights with other cats (Fig. 8.4(b)) and when bringing down prey. Approach and bite gradually switch from siblings to prey, forming part of the predatory system of adult cats (Barrett and Bateson, 1978; Caro, 1981).

8.5.4 The development of territoriality

As they get older, animals increasingly base decisions about whether and how to fight on the context in which an encounter occurs. One consequence of this change is often the appearance of territoriality or dominance relationships. As we have seen, some species (such as anole lizards, page 195) show territoriality from a very young age; in others it appears gradually. Sometimes young animals establish territories by gradually spacing themselves out from an area where they live communally. Juvenile Richardson's ground squirrels, and especially young females, tend to remain near to their mother and siblings; at about nine weeks they gradually space out and establish their own core areas, from which they repel unrelated animals (Michener, 1981).

In other cases, young animals gradually concentrate their aggression in one limited area of their range. In young blue gourami fights

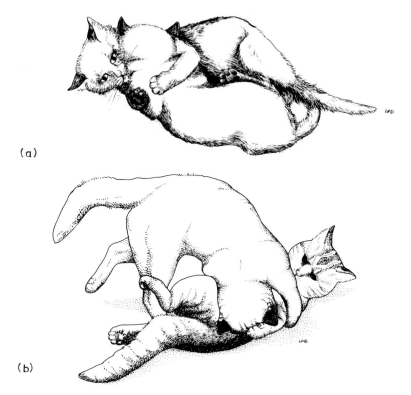

(a)

(b)

Figure 8.4 Wrestling during play in kittens (a) and fighting in adults cats (b).

occur in the absence of food, but are more intense when food is about. Larger fish gradually concentrate their attacks in specific areas where food is abundant. At first they move off when the food is finished, but by seven to eight weeks they persist in the defence of good feeding sites even when food is temporarily absent. Neighbouring fish exchange agonistic displays and attacks and a permanent system of territories develops (Tooker and Miller, 1980). Aggressive interactions between young cichlids (*Haplochromis burtoni*) intensify in fish of between five and six weeks, with the most active males gradually establishing exclusive rights to specific areas. Successful defence of a territory coincides with the appearance of particular colour patterns, including the black eye bar and orange spots that play a role in adult agonistic interactions (page 71; Fernald and Hirata, 1979).

Yet other animals establish territories by gradually encroaching on existing territories. In a mouse colony, females and juveniles move freely between the territories of adult males. As young males reach maturity, they form a base near some shelter within an existing territory and gradually push back the boundaries until they have a small territory of their own (Mackintosh, 1970). Young male red-winged blackbirds establish territories in a similar way, and are more successful if they attempt to do so in the area in which they spent the previous year. In this case, they receive fewer attacks from existing territory owners, who may have learned to ignore the presence of the young bird (Yasukawa, 1979).

8.5.5 The development of dominance relationships

Clearly defined (if unstable) dominance–subordinance relationships are observed during agonistic interactions between cockroaches almost as soon as they hatch (Deleporte, 1982). More commonly, such relationships only appear in older animals. Outside the breeding season, in red deer females and males of less than five years old (when they reach their maximum size, Fig. 8.5), the older of two animals is almost always dominant; above this age, rank in males depends on size. For the most part, sexually mature males who have not yet reached their full size may obtain surreptitious matings but stags only begin to hold harems successfully when they reach maximum size. Success in maintaining a harem peaks at an age of about eight years, after which both body condition and weight decline (Fig. 8.5; Clutton-Brock *et al.*, 1982).

Status differences often arise because less able fighters learn that if they fight a particular individual they will be defeated and so refrain from doing so; successful fighters learn the opposite. By the time they

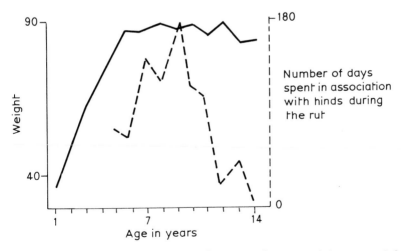

Figure 8.5 Reproductive success as a function of age in red deer stags (after Clutton-Brock *et al.*, 1982).

are thirty-five days old, jungle fowl show avoidance responses when attacked. Not until some time later do they habitually avoid specific individuals and thus show the beginning of dominance relationships. Rank orders start to stabilize at about 50–80 days, and intense fights are confined to encounters between closely ranked birds (Kruijt, 1964).

Small early behavioural differences can become entrenched as permanent 'personality traits' as a result of social encounters with siblings. In quail and chickens, rank within the brood is a good predictor of the outcome of later encounters with non-brood mates (Boag and Alway, 1980; Rajecki *et al.*, 1981). In rats, adult dominance relationships are gradually established during playful fights as these become less mutual and more one-sided (Meaney and Stewart, 1981). A similar behavioural change is seen in young rhesus monkeys, but in this case development of status is complicated by a system of aid giving among relatives (page 215).

8.6 ENVIRONMENTAL INFLUENCES: IDENTIFYING THE RELEVANT FACTORS

The movements and postures used during fights are fairly resistant to environmental influences during development. However, the efficiency and frequency with which these are used and the social structure that results are more labile. Differences between adult animals in these aspects of agonistic behaviour are likely to depend, in part at least, on the environment in which development occurs.

The usual procedure for finding out whether and how a particular factor influences behavioural development is to manipulate that factor and see how the experimental animal behaves when it grows up. If its actions are normal then the manipulated factor is not essential for the development of the behaviour in question. If adult behaviour deviates from normal following the treatment, it may play a role in the natural developmental process. For example rhesus monkeys reared in social isolation get involved in many fights, because they persist in eye contact after being threatened by a rival, rather than looking away which is the normal submissive response. Some aspect of early social interactions is necessary for development of the ability to respond appropriately to agonistic signals (Symons, 1974).

The results of this sort of treatment (called a deprivation experiment) are often difficult to interpret (Bateson, 1981). For example, the manipulation may be so unnatural as to produce highly disturbed animals whose behaviour is irrelevant to what normal animals do. In addition, development is often buffered against environmental changes in the sense that if one relevant factor is absent, another can act in its place. This state of affairs, the advantages of which are discussed below (page 219), means that even if behaviour is more or less normal in animals reared without a particular experience, we cannot conclude that this has no role in normal development.

8.7 ENVIRONMENTAL INFLUENCES: KINDS OF IDENTIFIED EFFECT

A variety of environmental factors influence the way agonistic behaviour develops (Table 8.4); these are all exemplified in the following sections. Aspects of the non-social environment (such as background nutrition level) are important, as is the social environment. Learning ('. . . that process which manifests itself by adaptive changes in individual behaviour as the result of experience'; Thorpe, 1963) represents just one of many different kinds of environmental influences acting on the developing animal; however, it is clear that it has a very important role to play. We make no attempt to summarize the huge literature on this subject, but simply draw attention to the various patterns of learning as they relate to the development of agonistic behaviour (Table 8.5).

For example, imprinting determines the stimuli to which aggressive responses are directed in ants, fish and mice; habituation makes it easier for birds to establish territories in familiar areas and classical or Pavlovian conditioning promotes the development of avoidance response to threat postures. Many young animals learn by trial and error how to use agonistic responses effectively, as these bring positive or negative consequences (retreat of rivals or pain). Young chaffinches learn what to sing by imitation and young vervet monkeys learn by observation how and when to fight.

Animals are not infinitely plastic in what they learn; some strains of

Table 8.4 Sources of environmental influence on the development of agonistic behaviour.

---

1. From the non-social environment
      Food availability
      General stimulation

2. From the social environment
      Interactions with parents
      Interactions with unrelated adults
      Interactions with siblings
      Interactions with unrelated age peers

Table 8.5 A highly simplified classification of the types of learning process which influence the development of agonistic behaviour.

Practice:	feedback from performing movements alters later performance.
Habituation:	responses wane after repeated exposure to the same stimulus without adverse or favourable consequences.
Imprinting:	social responses which can initially be elicited by a wide variety of objects come to be elicited by a specific stimulus through exposure to that stimulus during development.
Classical conditioning:	a stimulus which does not initially elicit a response comes to do so through being repeatedly paired with one that does do so.
Trial and error learning:	animals learn to repeat actions which in the past have led to pleasant experiences (or positive reinforcement) and to avoid those with unpleasant consequences (negative reinforcement).
Observational learning:	animals learn by observing the consequences for other animals of behaving in a particular way.

mice are particularly resistant to early environmental influences (see Hay, 1985). In addition, animals may be predisposed to learn certain things. They are particularly good at associating certain cues given out by an opponent with the pain of being attacked, but some cues may be more effective than others. For example, male three spined stickle-backs which have never seen the red and blue breeding colours of their species learn more readily to avoid a model fish which attacks them if the model is red underneath like a real stickleback than if it is red on top (Chauvin-Muckensturm, 1979).

8.8 ENVIRONMENTAL INFLUENCES: THE NON-SOCIAL ENVIRONMENT

Mice that experience a short period of undernourishment shortly after birth are more likely to initiate an attack on another mouse, but being smaller are less likely to win fights (in some strains at least, Tonkiss and Smart, 1982). The development of the mutual intolerance which disperses young spiders (*Coelotes terrestris*) from their egg cocoon is delayed when an ample food supply is available (Horel, Roland and Leborgne, 1979). Rats that have an opportunity to dig burrows when young are more aggressive as adults than those denied this opportunity (Nikoletseas and Lore, 1981). Gentle handling when young makes mice and rats larger, less emotional and less reactive to pain as adults, all of which are likely to influence responses to conflict (Levine and Broadhurst, 1963).

8.9 ENVIRONMENTAL INFLUENCES: THE SOCIAL ENVIRONMENT

Social experience is often not essential for the development of the actions used in fights, but it does influence the way these are used; that is how fiercely and efficiently animals fight and how well they are able to deliver and respond to subtle cues such as threats. Early relations with parents, with siblings and with unrelated animals are all important in this context.

8.9.1 The effects of social isolation

Many invertebrates and vertebrates are able to show efficient agonistic behaviour (often the complete adult repertoire) even when they have been raised in social isolation (Table 8.6). This makes sense in animals with little or no parental care and in which juveniles compete for resources from an early age. Although isolates (for example, spiders, cockroaches, hermit crabs, cichlid fish) tend to be particularly aggressive, this abnormality can disappear quickly with social experience (Hazlett and Provenzano, 1965; Wyman and Ward, 1973; Apsey, 1975; Deleporte, 1978).

However, even though animals reared in isolation are able to perform species-typical agonistic actions, their behaviour is often abnormal in complex and subtle ways. The following examples illustrate the nature of such deficiencies.

- Gouramis reared in isolation display all the agonistic responses typical of normal fish of the same age, but the co-ordination of the movements and the way these are linked into sequences are impaired, especially during the initial phase of a fight when visual signalling normally occurs. Dominant isolates show abnormally high levels of overtly offensive actions and few threat displays, subordinate isolates show fewer appeasement gestures and, in both cases, the agonistic repertoire is impoverished (Table 8.7; Tooker and Miller, 1980).
- Male and female rats reared in isolation fight with normal behaviour patterns but are markedly more aggressive towards other rats, both as juveniles and as adults. They also elicit more attacks and fewer

Table 8.6   Examples of the effects of social isolation on the development of agonistic behaviour.

Species	Observation	Reference
Hermit crabs	Isolates fight with normal actions but are unusually aggressive.	Hazlett and Provenzano (1965)
Cockroaches	Isolates fight with normal repertoire but are more aggressive.	Deleporte (1978)
Wolf spiders	Isolates use the same leg waving displays and dominants and subordinates behave differently as in adults.	Apsey (1975)
Mollies	Fish reared in isolation show normal repertoire of agonistic behaviour patterns but are abnormally aggressive.	Parzefal (1969); Franck, *et al.* (1978)
Cichlids (*H. burtoni*)	Black eyebar elicits attack in males reared in isolation as it does in normal males. Development of territorial behaviour is accelerated.	Fernald (1980)
Jungle fowl	Isolates show full agonistic repertoire but are intensely and indiscriminantly aggressive.	Kruijt (1964)
Mice	Isolates initiate more attacks and attack non-aggressive companions.	Cairns *et al.*, 1985
Gerbils	Isolated gerbils are fearful and show reduced territory marking and aggression.	Berg *et al.* (1975)
Cattle	Calves reared without social contact lose agonistic encounters, being overinclined to retreat.	Broom and Leaver (1978)

affiliative responses from their opponents. Lack of the opportunity to play with peers may well make an important contribution to these effects of isolation; rats reared in isolaion but given a brief daily encounter with another pup behave normally as adults, provided that their companion is able to play. Rapid alternation of roles during play may provide practice in inhibiting inappropriate acts (Day *et al.*, 1982; Hole and Einon, 1984, and references therein).

Table 8.7   Agonistic behaviour in group-reared and isolation-reared gouramis (from Tooker and Miller, 1980).

Rearing conditions	Approach	Chase	Lateral display	Tail beat	Carousel	Behavioural diversity
			Frequency of			
Isolated	92.1	42.4	7.9	2.2	0.2	1.95
In groups	59.6	21.7	27.3	9.3	3.7	2.21
	*	*	*	*		*

* Represents significant difference at $p < 0.05$ or less.

- Isolated rhesus monkeys are capable of producing most of the gestures, facial expressions and vocalizations of the normal repertoire; for example, they show open-mouthed threat when frightened. However, they are extremely aggressive, do not use agonistic actions in the appropriate context and fail to react appropriately to the gestures and expressions of their companions (Symons, 1974).

8.9.2 Experience with parents

In species with parental care, interactions between an animal and its father or mother have an important influence on the development of agonistic behaviour. For example, grasshopper mice are more aggressive if they are reared by both parents rather than by their mother alone (McCarthy and Southwick, 1979). Similar results are found for mice, even when physical contact between father and young is prevented. Young mice become imprinted on the smell of an adult male and this becomes a more effective stimulus for eliciting attack (see page 76; Wuensch and Cooper, 1981). Cichlids (*Haplochromis burtoni*) which have been reared with an adult male and female of another species (*Cichlasoma nigrofasciatum*, which looks very different) direct their agonistic responses almost exclusively to males of the foster species (Fig. 8.6). This preference is reduced but never completely abolished by subsequent experience with their own species (Crapon de Caprona, 1982). The presence of a father can influence aggression in other ways; adult male rats join in playful fights with their developing young and generally win. This delays the onset of typical adult fighting in the young males (Drews *et al.*, 1982).

The importance of mother–offspring interactions in the development of agonistic behaviour is demonstrated by the fact that chicks

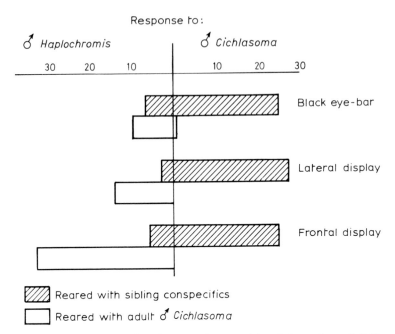

Figure 8.6 The effects of early experience with adult males of different species on the stimuli to which agonistic behaviour patterns are directed in cichlids (*Haplochromis burtoni*, after Crapon de Caprona, 1982).

that have been brooded are less aggressive than those that have not (Fält, 1978) and that mice reared by rat mothers are less aggressive as adults towards both mice and rats than are mice reared by their own species (Lagerspetz and Heino, 1970). The relationship between a young primate and its mother is the earliest and most intense that it experiences; even short-term separation from the mother has profound consequences, making the infant both fearful and aggressive (Sackett, 1967). A correlation exists between the level of aggression shown towards a young rhesus monkey by its mother and its own subsequent aggressiveness, so conflict between a young primate and its mother enhances aggressiveness (Chamove, 1980). Young squirrel monkeys are able to learn the appropriate responses to agonistic signals by observing their mothers fighting (see Baldwin and Baldwin, 1977 and references therein).

8.9.3 Aid giving and alliances in primates

Adult female primates frequently harrass and injure young females,

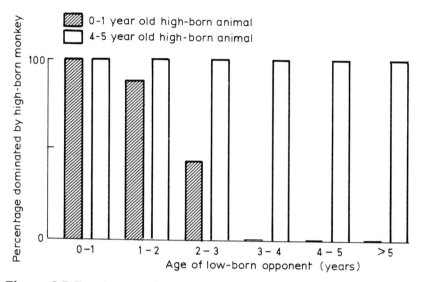

Figure 8.7 Dominance of animals of different ages born to low-status female rhesus monkeys by those born to high-status females (from figures in Datta, 1983).

especially those unrelated to themselves. Mothers (and other adult female relatives) give support to their daughters when they are attacked in this way, and when their offspring compete with peers. In both rhesus macaques and vervets, high ranking females are more successful in this respect (Berman, 1980; Horrocks and Hunte, 1984; Fairbanks and McGuire, 1985) and a young animal's short-term success and ultimate status both depend on the rank of its female relatives (Fig. 8.7).

In species in which young males leave the natal troop, their eventual rank cannot be directly influenced by that of their mother, although there may be an indirect effect, since the offspring of high-ranking females are likely to get better food and grow larger. For both sexes, but particularly for females, the end result of the complex system of aid giving is that inherent differences between individuals are of less importance in determining their ultimate position in society than is the status of their close relatives.

8.9.4 Experience with other conspecifics

Adults other than parents (both relatives and unrelated individuals) can influence the development of agonistic behaviour. For example, the first few weeks of life of a young worker ant are an important time for learning (by a process similar to imprinting) how to recognize friend and enemy by smell (Jaisson, 1985). Similarly, sweat bees accept into their colony other bees with whom they have been raised but aggressively reject those that smell differently (Buckle and Greenberg, 1981).

The way aggression develops can be influenced by early interactions with siblings. If a brood of red grouse contains more than one member of the same sex, one dominates the other(s) and this early status is reflected in their later interactions with non-siblings (Boag and Alway, 1980). Male mice reared in litters containing a number of brothers show less intense aggression as adults than those reared with sisters (Namikas and Wehmer, 1978), presumably as a result of similar kinds of early experience.

8.9.5 Experience during adult life

The environment does not cease to act on agonistic behaviour when an animal reaches sexual maturity, but continues to produce adaptive changes. Thus winners of fights behave more aggressively than do losers in future fights in slugs (Zack, 1975), crickets (Alexander, 1961), fish (Frey and Miller, 1972) and mice (Lagerspetz, 1979). In some cases, these changes are the result of altered hormonal activity (page 116) but learning processes are also involved. Male rats (but not females) show suppressed aggression after a defeat. This change depends on the presence of testosterone, because it is absent in castrates. However, it is not the result of altered testosterone production, because castrates given a fixed dose of testosterone also modify their subsequent behaviour following a defeat (Van de Poll *et al.*, 1981).

8.10 OVERVIEW

8.10.1 Have our questions been answered?

The examples discussed in the preceding sections go some way towards answering the questions posed in the introduction to this chapter, although the answers are partial and information is available only for a small fraction of animal species.

(a) Developmental origins

For the most part, co-ordinated agonistic behaviour patterns arise from the unco-ordinated movements of very young animals (gouramis) or from behaviour patterns used in other contexts (orange chromides, junglefowl and canids, page 196).

(b) Changes with age

Once they have appeared, these behaviour patterns may change in form or frequency; this sometimes reflects altered chances of winning and/or risks of injury, but the changes also depend on the value of the resource in question. Animals usually fight more often and more intensely as they get older, and especially as they approach sexual maturity, but the fights between young lizards over shelters are as fierce as any seen in adults (page 195). The agonistic repertoire often expands with age, by accentuating different aspects of earlier movements or by adding new components to existing patterns (jungle fowl, page 203). In particular, threat displays appear in the repertoire and so overt aggression ceases to be the predominant agonistic response (page 203). Territoriality develops as animals gradually exclude conspecifics from a given area (ground squirrels, page 205) or restrict fights over resources to predictably rich patches (gouramis, page 206). Changes in status with age may reflect an increase in size (red deer, page 207) but may also be the result of learning (junglefowl, page 208).

Many of these developmental changes reflect morphological and physiological maturation in the animal concerned; co-ordination improves, body size increases, weapons develop or gonads begin to produce sex hormones. Others are the result of experience, ranging from the effects of practice (lobsters, page 201) to an influence of non-specific environmental factors such as prenatal stress (page 211) and specific learning experiences during encounters with conspecifics (pages 203 and 209).

(c) Origins of differences between adult animals

The overall form of the behaviour patterns used during fights is quite resistant to environmental influences. On the other hand, many kinds of environmental influence modify the development of higher order aspects of agonistic behaviour, such as the stimuli which control it and how ready animals are to resort to aggression. These sometimes generate differences between species (some differences between mice and rats disappear when they are fostered to females of the other species) between the sexes (young male and female primates experience very different social environments and this influences their patterns of aggression as adults) and between individuals of the same species and gender (male cichlids direct attacks to opponents that look like fish they encountered when young (page 213), and monkeys whose mother attacks them become very aggressive themselves, page 214).

Some of these environmental effects are simply alterations in the rate of processes which will take place anyway; enhanced food levels and social experience delay but do not prevent the appearance of aggression in spiders (page 210) and cockroaches (page 212). On the other hand, they may have a drastic effect on how aggression develops (rats, page 193), even to the extent of directing development along quite different routes (paradise fish, page 194).

8.10.2 General principles

In general, it is clear that in many animals the actual motor patterns of fighting are relatively resistant to environmental modification, but that the way they are used is susceptible to all kinds of influences from the social and non-social environment. To understand why this should be so, we need to consider how easy it is to preprogramme (using an analogy from computing) different aspects of behaviour, whether or not opportunities for environmental modification present themselves and whether such modification is adaptive.

(a) Is preprogramming possible?

Producing a recognizable movement requires that a specific collection of muscles contract with particular time relationships. This is not a very complicated physiological feat and it is not too difficult to see how an embryo could be preprogrammed to produce particular patterns of agonistic behaviour. In contrast, the way these actions are put together into effective sequences depends on the ability to respond to present and expected future behaviour of an opponent and to several features of the context in which the encounter occurs (page 79). It is hard to see how these complex abilities of assessment, prediction and rapid response could be programmed to develop in the absence of any experience.

(b) Is environmental modification possible?

Some life styles and patterns of social organization (those of many insects, for example) provide little opportunity for agonistic be-haviour to be modified by early social interactions. In this case, efficient fighting has to develop without social experience. On the other hand, young animals of many species (particularly vertebrates, and particularly mammals within the vertebrates) develop alongside conspecifics of various ages. The details of the social structure will dictate the extent to which agonistic behaviour is open to modification by interactions with particular categories of social companion (mother, father, related or unrelated contemporaries, and so on). In some cases, evolution has taken the initiative, as it were; the evolution of play was probably partly the result of the need to create an environment in which animals could learn how to fight when direct practice would be unsafe.

(c) Is environmental modification adaptive?

Each species of animal has a certain body form, sometimes a given battery of weapons and specific regions which are more or less vulnerable to attack. These place strong constraints on the kind of movements which are effective but safe for use in fights and limit the value of any developmental adjustments of the broad mode of fighting. On the other hand, the ability to adapt higher level aspects of agonistic behaviour (overall aggressiveness, which animals are attacked and whether territoriality or dominance exist) to local environmental conditions is an important determinant of fitness (see Chapter 11). The fact that these aspects of the agonistic repertoire are accessible to environmental modification (i.e. they are developmentally open) allows them to be modified in accordance with local, unpredictable conditions of the social and non-social environment.

8.10.3 Self-regulatory processes in development and the costs of openness

This sort of developmental openness has its costs. There is the risk that an essential environmental factor may be missing so that effective agonistic behaviour fails to develop. The chances of this happening are reduced by the fact that animals that avoid the attention of experimental scientists mostly grow up in an environment containing all the necessary features (mother, father, siblings, and so on). In addition, development is often well buffered in the sense that if one environmental factor is absent, another can take its place. For example, orange chromides reared in isolation direct at inanimate objects the movements they normally show to other fish. Playful fights are important for the complete development of effective agonistic behaviour in squirrel monkeys. However, when this behaviour is not possible, for example, when the young animal has no companions of the same age, some agonistic skills can be learned in their first real fights.

Developmental flexibility opens up the possibility that animals might manipulate the behavioural development of potential rivals. Rank-based differences in maturation rate and the ruthless persistence with which some young mammals force playful fights on to their

companions may represent examples of such a process. So the buffered nature of development may also be the means by which animals
protect themselves against abuse of their developmental system by
rivals.

GENES, ENVIRONMENTS AND AGONISTIC BEHAVIOUR

The examples discussed in Chapters 7 and 8 have confirmed that both an animal's genes and its developmental environment have multiple and complex effects on the way agonistic behaviour develops and thus on how adult animals fight. They have gone further than this, though, in characterizing some of these influences. For example, we know that genetic variation can produce differences in agonistic behaviour by many different routes. Inheritance is usually polygenic, but single allele substitutions can have major effects and the way in which agonistic behaviour is inherited reflects the complex costs and benefits of fighting. We also know that the motor patterns used during fights are resistant to environmental modification, but that the details of how, how often and to whom these are deployed can be modified by various aspects of the physical and social environment. A number of gene–environment interactions have been identified. For example, animals with different genotypes differ in how susceptible they are to environmental influences and animals may be predisposed to learn certain things. The end result is an adaptive combination of rigidity in some aspects and flexibility in others.

PART FOUR
Consequences, fitness and evolutionary change

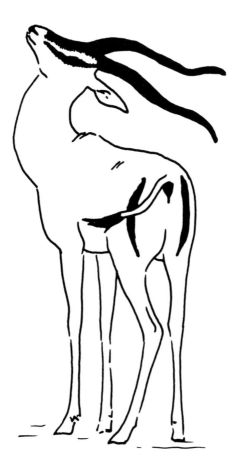

So far we have considered animal conflicts and the events leading up to them. But agonistic behaviour has consequences as well for the individual animal and for the population and ecological community to which it belongs. Such consequences (which are discussed in Chapter 9) acting over many generations determine the way in which agonistic behaviour and the structures used in fights change during the evolutionary history of the species concerned; these issues are considered in Chapter 10. The end result of these changes is that today's animals use agonistic behaviour that is finely adapted to their needs, and this is the subject of Chapter 11.

9

The consequences of animal conflict

9.1 INTRODUCTION

We saw in Chapters 5 and 6 how fighting and other conflict situations induce neuroendocrinological changes in both the winner and the loser. But rapid responses to fights can have longer-lasting effects. Conflicts do not occur in a biological vacuum. Each combatant is a member of a population of its own species and of an ecological community made up of many species. Whatever happens to losers and winners of disputes may therefore have ramifying consequences for populations and species both within and between generations. These potential consequences of animal conflict are the topic of this chapter. Three broad questions will be considered.

1. What happens to the winners and losers of disputes after their encounters?
2. The repercussions of agonistic behaviour can be viewed on more than one timescale: do they affect populations and ecological communities as well as the individuals concerned in the agonistic encounter?
3. Does agonistic behaviour influence the evolution of species?

9.2 CONSEQUENCES FOR INDIVIDUALS

The physiological changes that result from conflict may drastically affect an individual throughout his or her life, especially within the context of a stable group structure, by influencing maturation, reproduction, mortality and emigration.

9.2.1 Maturation and reproduction

Subordinate animals are often less successful at breeding than dominants (see page 287). This is sometimes because subordinates are poor competitors, but there are many instances of dominant individuals in a group directly interfering with a subordinate's breeding attempts. Individuals can also be prevented from breeding by excluding them from suitable areas. An excluded individual may remain passively in or near the territories as part of a 'floater' population, or may stay on a territory and help the territory owners in their breeding attempt (page 314). A dominant animal may still interfere with the helper, however, to prevent him or her attempting to breed. In some cases a subordinate's reproduction may be reduced rather than totally suppressed. Individuals are sometimes forced into less suitable habitats where they breed less well. Dominant cotton rats, for example, use the most preferred habitat and are more often in reproductive condition than are the subordinates (Spencer and Cameron, 1983). Other examples follow.

- Among rodents and carnivores, subordinates living in a group led by a dominant animal or pair fail to breed even though they are sexually mature (Ayer and Whitsett, 1980; Hrudecky, 1985).
- Adult male prairie deer mice inhibit the maturation of juvenile males, causing them to have smaller body weights, testes and seminal vesicles than juveniles reared away from adult males (Bediz and Whitsett, 1979).
- The production of reproductive hormones is suppressed in subordinate male mice after fighting (Bronson, 1973).
- A pheromone produced by adult male mice causes an increased frequency of chromosomal abnormalities in the sperm of younger mice (Daev, 1981).
- Among pied kingfishers, primary helpers at the nest (male offspring of the breeding pair of the previous year) have lower testosterone levels than the breeding male or unrelated helpers probably because of their subordinate position (Reyer et al., 1986).
- Among fish, adult platyfish and reef fish Amphiprion inhibit maturation of juveniles (Borowsky, 1978; Sale, 1980).

- In fish in which males arise by sex reversal of females (such as *Labroides dimidiatus*) the presence of dominant males inhibits this process (Sale, 1980).

These effects of agonistic encounters on maturation and reproduction can come about in several ways (see review in Christian, 1980). The increased secretion of adrenal steroids, oestrogen and progesterone inhibits the release of gonadotrophins and thus suppresses reproduction. Immatures are especially sensitive to this inhibition. Social subordination in mice decreases the activity of a pheromone that accelerates the maturation of females. Adrenocorticotropic hormone directly inhibits reproduction in rodents, possibly via an effect of associated endorphins (page 121) on decreasing luteinizing hormone and increasing growth hormone and prolactin secretion; in tree shrews the weight of the epididymus and prostate decline in subordinates due to an increase in prolactin output (Heinzeller and Raab, 1982).

9.2.2 Mortality and emigration

Agonistic behaviour could increase mortality or the emigration of groups of individuals. In field studies it is often difficult to distinguish between these two possibilities and they are often considered together.

The physiological consequences of fighting reduce an animal's resistance to disease and so increase its likelihood of dying. Glucocorticoids (produced during stressful situations) adversely affect the immunological response and the ability of wounds to heal (Christian, 1980). Defeated voles have reduced, involuted thymuses (which produce lymphocytes which help the body fight an infection; Clarke, 1953) and fighting in bandicoot rats decreases blood counts of lymphocytes, with the changes being greater in the losers than the winning animals (Ghosh, Sahn and Maiti, 1983). Male mice defeated in fights show a reduced immune response to injected sheep red blood cells (Beden and Brain, 1982). Stressed animals are thus at higher risk from pathogens, disease and parasites.

Heavily attacked mammals suffer a variety of physiological disturbances that can lead to death (Barnett *et al.*, 1975; Rasa and Vandenhoovel, 1984). Stress has also been implicated as the cause of death in cockroaches (Ewing, 1967). In a study of voles, subordinate males and small females were most frequently attacked and were most likely to die. The catecholamines released during stressful encounters increased glycogenesis in the liver and this led to the females suffering from low blood sugar. In the males, the increased output of adrenal hormones constricted the capillaries in the kidneys and caused them to fail (Rasa and Vandenhoovel, 1984).

Consequences, fitness and evolutionary change

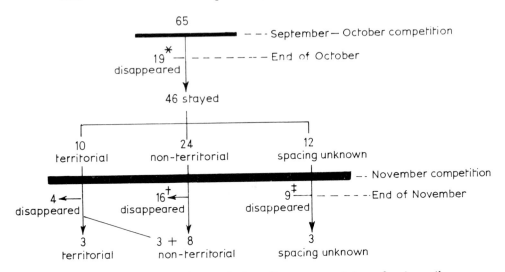

Figure 9.1 The fates during the breeding season of sixty-five juvenile grey plovers colour marked in September–October at Teesmouth, north-east England: *, one to northern France; †, one to western France; + three returned to Teesmouth in February; ‡ one returned to Teesmouth in February (after Townshend, 1985).

The effects of stress incurred during agonistic behaviour depend to a large extent on the prevailing social and environmental circumstances, so laboratory studies may not be relevant to the situation in the wild. However, physiological changes are certainly seen in wild populations. Adrenal weights and renal pathology of mammals increase at high population densities when social stress is likely to be high. In the dasyurid marsupial *Antechinus stuartii* males die soon after the start of the breeding season. This may be linked with their elevated levels of plasma glucocorticoids (Bradley, McDonald and Lee, 1976).

As well as these fatal physiological effects, other consequences of agonistic behaviour can result in death. Fighting animals can die from fatal injuries (page 43) or because of increased conspicuousness to predators. There are also serious energy costs of agonistic behaviour which can weaken an animal (page 277).

Losers of fights may also die because they are excluded from a vital resource or from a territory which serves as a long term supply of a resource (page 295). In red grouse, the territory provides protection from predation and an exclusive supply of food; most territorial birds thus survive the winter but non-territorial ones rarely do (Watson, 1977). After fledging, great tits are aggressive towards each other and it is the larger, dominant individuals that survive the best (Garnett, 1981). However, although large animals usually win fights and gain access to resources, they also require more food to survive so may be at a disadvantage when food is scarce. In species where males are larger than females, for example, they often suffer higher mortality (Clutton-Brock and Albon, 1985).

Some, but by no means all, studies suggest that agonistic behaviour, especially that involving females, is a cause of emigration. Emigration in female ground squirrels, for example, is related to high levels of aggression among siblings (Pfeifer, 1982). Among birds such as great tits and partridges, too, emigration is associated with high levels of aggression in the population (Green, 1983; De Laet, 1985).

Individuals differ in how much they are affected by aggressive encounters; poor competitors and low ranking animals, especially juveniles, are most likely to be forced out. Among small mammals, for example, the individuals that emigrate are often subordinates (Gaines and McClenaghan, 1980). Where juveniles are concerned, most usually move away from their natal site to a different wintering and/or breeding area, but the dominant individuals may still be able to get possession of the nearest suitable sites and resources, forcing others to take less favoured sites. Late-fledging great tits, for example, move further away from their parents' nests than do the older and hence larger and dominant ones (Dhondt and Hubler, 1968). Similarly, young grey plovers that fail to obtain a territory on a particular estuary over the winter continue their migration from the breeding grounds (Fig. 9.1; Townshend, 1985). A long migration may itself contribute to an increased risk of dying for subordinates (Pienkowski and Evans, 1985).

9.3 CONSEQUENCES FOR POPULATIONS

9.3.1 How is population size regulated?

The number of individuals of a species living in a particular area (the population size) is not static. The population size can vary from year to year, sometimes fluctuating considerably, sometimes varying only slightly, and sometimes regularly peaking then declining over a period of years (a phenomenon known as cycling). Despite these variations, however, populations in the wild neither continue to increase indefinitely nor do they habitually crash and become extinct. An upper limit to numbers is set by the resources that are available but other factors can act on populations before the supply of resources becomes critical. An important ecological distinction can be made between those factors that are independent of the density (numbers per unit area) of the population and those that act more strongly as the density of the population increases. Severe weather is sometimes a

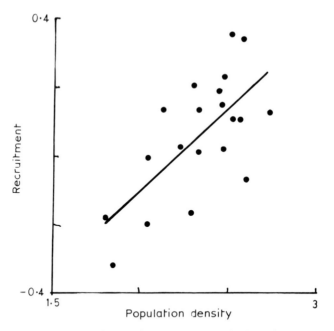

Figure 9.2 Recruitment of juvenile great tits into the breeding population in relation to population density. Recruitment is assessed from field data on population dynamics. Population density is assessed by $\log_{10}$ established juveniles in midwinter (after McCleery and Perrins, 1985).

cause of deaths that acts in a density-independent way. Many of the consequences of agonistic behaviour, such as mortality, emigration or exclusion from breeding, can be density-dependent.

The relative importance of density-independent and density-dependent factors in controlling the size of populations is a matter of controversy, but probably varies widely between taxa. Bird populations tend to be sensitive to the effects of density-dependent factors, whereas populations of small mammals, fish and insects in variable environments are often affected by density-independent factors. For example, recruitment of great tits into the breeding population is density-dependent (because of territorial behaviour) and so this can regulate the population size; density-independent mortality is still an important factor, however, in determining population fluctuations (Fig. 9.2; McCleery and Perrins, 1985). In damselfish, on the other hand, territorial individuals do not determine the number of young fish that obtain territories; the effects of weather and predators on recruitment seem to play a greater role (Doherty, 1983). In those situations where density-dependent factors are crucial in determining the size of the population, however, agonistic behaviour, in particular territorial exclusion, is likely to exert considerable influence. Even when mortality is due to a density-independent factor such as severe winter weather, it can be mediated by density-dependent behaviour such as interference with feeding or territorial exclusion. Even a weak density-dependent winter mortality can have a considerable impact on the size of a population (Goss-Custard, 1980).

The way a population grows, whether or not it reaches an equilibrium size, and how much the population size fluctuates depend in part on the population's per capita rate of increase at a given density. This in turn depends in part on the reproductive, mortality and emigration rates of the different classes of individuals making up the population. Agonistic behaviour plays a role, of varying importance, in these population processes.

9.3.2 Density-dependent effects of agonistic behaviour

(a) Reproduction

The number of individuals that obtain breeding territories can depend on population density and on territory size, the spatial variation of aggression and the settlement pattern (Patterson, 1980, 1985). In blue grouse, for example, some individuals are excluded from breeding territories in a density-dependent manner; productivity varies between years but mortality rates are constant and the population is stable (Zwickel, 1980). Repression of sexual activity in young subordinate animals may also increase at high densities (for example, house mice, Van Zegeren, 1980).

(b) Mortality

During the breeding season, mortality of dependent young may increase at high densities because of increased disturbance of the mother from intruders (for example, house mice, Van Zegeren, 1980). Resources are often shared unequally between individuals in a population, dominants or territory owners getting the most and thus being likely to survive a shortage of the resource. Individuals excluded from winter territories or good feeding areas may be more likely to die as we have seen. This mortality can be density-dependent, for example, in immature willow tits over the winter (Ekman, 1984). Consequently, the size of the population is likely to remain stable. If each member of a population gained only a part of the resource, more animals would die and the size of the population would oscillate. Thus territorial or despotic behaviour stabilizes populations (Sibly and Smith, 1985; Sutherland and Parker, 1985).

(c) Emigration

Among small mammals, numbers moving out of a population are generally positively correlated with population density and the rate of increase in density (Gaines and McClenaghan, 1980), suggesting that they have left because of an increased frequency of encounters with more dominant individuals. In voles, juvenile survival and recruitment are negatively correlated with the density of females because the latter are aggressive to juveniles (Boonstra, 1978, 1984; Redfield, Taitt and Krebs, 1978).

9.3.3 Cannibalism and mortality

Cannibalism is a major cause of mortality in several species. Entire age classes can be eliminated, resulting in large fluctuations in recruitment and skewed age/size distributions. Cannibalism is often a function of population density, either because alternative sources of food are scarce at high densities or because crowding increases encounter rates and territorial disputes. It reduces density not only by removing individuals from the population, but also by interfering with the feeding success and growth rate of other individuals. In the waterbug *Notonecta*, juveniles avoid the cannibalistic adults and are less successful at feeding as a result (Murdoch and Sih, 1978). So cannibalism can regulate the size of a population especially as it is both a means of gaining a food supply and a means of eliminating competitors (Polis, 1981).

A long-term study of pike in Lake Windermere, England, suggested that cannibalism is an important factor in determining population size and growth (Kipling, 1983). One-year-old pike eat young of the year and cannibalism can be intense when other prey of a suitable size, mainly young perch, are scarce. Between 1969 and 1978, when there were few small perch, pike year classes measured at two years were weak while adult pike were larger than usual, suggesting that cannibalism on young-of-the-year was particularly high.

9.3.4 Agonistic behaviour and cycling populations

Agonistic behaviour may be involved in the regulation of populations around an average density, but can it cause a population decline and population cycling? There are many theories to account for annual fluctuations and cycles, in mammals and birds, based on everything from food to predation. Some of these theories involve the effects of agonistic behaviour. Thus a decline in numbers may be due to

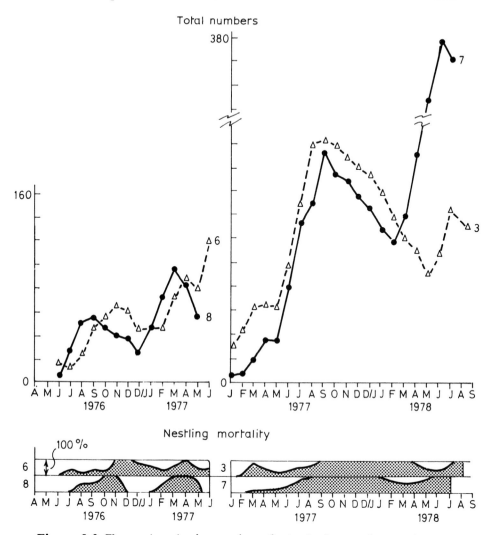

Figure 9.3 Fluctuations in the number of mice in four study populations, censused at four-weekly intervals over a period of more than one year, together with nesting mortality for the same populations. Thirteen populations were studied and all gave comparable results (after Van Zegeren, 1980).

subordinates being forced out by aggressive individuals at a time when (for genetic or environmental reasons) aggressive behaviour is increasing in frequency (Christian, 1971; Krebs 1978).

The role of aggressive genotypes in causing population cycles has been a matter of considerable dispute. Residents and emigrants are

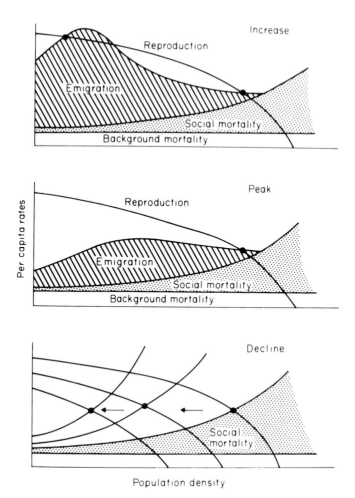

Figure 9.4 A schematic model to explain population cycles in voles. According to this model, voles come in two broadly different types: docile individuals can sustain high rates of reproduction and emigrate readily while aggressive voles fight rather than emigrate and thus cause high social mortality and low reproductive rates. In the low and increasing phase of the cycle (top graph) docile voles predominate and losses are primarily the result of emigration. Reproductive rates are high and the population increases but docile voles are gradually replaced by aggressive ones, which compete more successfully for breeding space as this becomes in short supply. The population in the peak phase (middle graph) thus consists predominantly of aggressive voles, so that emigration is low. Socially induced mortality is high and reproductive output low, so that the population goes into a decline (bottom graph). Competitive ability ceases to be at a premium, docile voles increase in frequency again and the cycle starts again. Hatched and stippled areas indicate the curves relating emigration and social mortality (respectively) to population density. (Reproduced with permission from Krebs, 1985).

sometimes of different genotypes (Gaines and McClenaghan, 1980) but aggressive individuals are not always most common during a population increase as theory predicts (for example, red grouse, Moss and Watson, 1985). The preponderance of aggressive genotypes may thus be a result, rather than a cause of, a decline. It may not be necessary anyway to invoke different genotypes to explain a decline since at high densities a general increase in sexual repression, disturbance of litters and aggression could have the same effect (Van Zegeren, 1980; Fig. 9.3).

Several recent studies point to a combination of factors inducing population declines and cycles, and it is likely that different mechanisms operate in different species. Aggression seems to be important in inhibiting reproduction and contributing to mortality and dispersal in voles (Krebs, 1985; Fig. 9.4). On the other hand, although abundant food cannot prevent a decline it does allow numbers to increase and reach high densities. Similarly, improved cover in the habitat can improve survival but cannot prevent a decline (Taitt and Krebs, 1983; Taitt, 1985). So a combination of food and spacing behaviour seems to be the main force behind rodent cycles rather than socially induced aggression alone. The spacing behaviour of females, rather than that of males, may have most influence, since females defend areas for breeding, adjusting the size according to the food supply and available cover, whereas males compete for females not for 'space'. In red grouse cycles, spacing behaviour may also play a role, although changes in aggressive behaviour accompany the cyclic changes in population size rather than causing them (Moss and Watson, 1985).

Although most studies of population cycles have considered small mammals and gamebirds, regular fluctuations in numbers do occur in other groups. The social wasps, *Vespula vulgaris* and *V. germanica* in Britain show a two-year and possibly a seven-year population cycle (Archer, 1985). Agonistic behaviour may drive the two-year cycle since, following a year of increasing numbers, queens are numerous in the spring and there is a great deal of fighting over the ownership of nests resulting in a considerable mortality of queens. High levels of competition for resources during a year of increase may also result in queens being of a poor quality and producing unsuccessful nests. Over a period of seven years the population fluctuations dampen to equilibrium, but the effect of weather on the success of colonies then becomes more important.

9.4 CONSEQUENCES FOR ECOLOGICAL COMMUNITIES

An animal's lifestyle within a community is called its niche. There are many dimensions to a niche, such as what an animal eats and in which habitat it feeds. Competition within and between species can change the size of this niche.

9.4.1 Competition between species

When two species with similar ecological requirements coexist both may contract their niche thus reducing niche overlap. However, more commonly one species behaves as dominant over the other, and experiences less niche contraction relative to the subordinate. The behaviourally dominant species is usually the larger of the two, but several individuals of a small species can gang up to defeat the larger one or may affect the dominant by depleting resources, especially when the subordinate species is common relative to the dominant. As a result, the dominant species may contract its niche and the normally subordinate species may expand its own. The vole *Microtus montanus* is dominant to *M. pennsylvanicus* in the laboratory; in the field the former is found on dry sites and the latter on wetter areas when they are equally abundant. However, when *M. pennsylvanicus* is numerically the more abundant species it is able to expand its range into the habitat of *M. montanus* (Stoeker, 1972).

Which species is dominant may also depend on the reproductive condition or age of the individuals concerned, since females are generally more aggressive when breeding and older individuals may be larger and more able to be dominant than younger ones.

An example of territorial exclusion is provided by great tits and chaffinches, which defend exclusive interspecific territories on Scottish islands but not on the mainland. Chaffinches are the larger birds and occupy most woodland types on the islands whereas the great tits are restricted to hazel scrub, which the chaffinches do not enter. On the mainland, great tits can be found in all types of woodland and their territories overlap those of chaffinches. Reed (1982; Fig. 9.5) removed chaffinches from an area of oak scrub and found that the great tits expanded their range into this habitat. Island woodland is structurally simpler and smaller than on the mainland so niche divergence may not be possible for these species here. In addition, the great tit is a recent arrival, breeding on Eigg since 1957 (with breeding attempts in 1939 and 1952), whereas the chaffinch has been on the island since at least 1880, leaving little time for the species to diverge. So competition and aggressive encounters may be intensified in such circumstances, in this case reducing the habitat types used by the great tit.

Conflict is likely to be most intense or most frequent between closely related species using similar limiting resources, because their

Removal zone before removals   Removal zone after 7 days

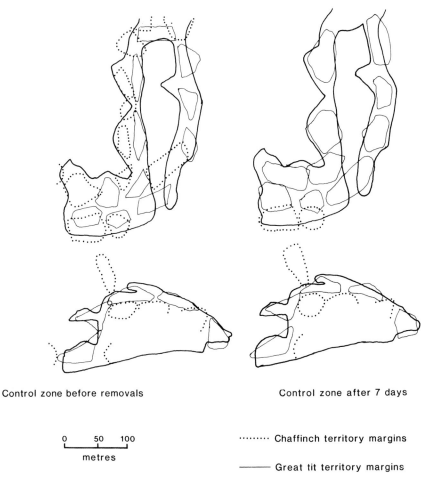

Control zone before removals   Control zone after 7 days

0   50   100   ········ Chaffinch territory margins
└─┴─┘
metres   ──────── Great tit territory margins

Figure 9.5 The distribution of great tit and chaffinch territories on control and removal zones in the isle of Eigg before the start of a removal experiment and after seven days (after Reed, 1982).

recent histories are similar. But even widely separated taxa can come into conflict. Sunbathing areas are disputed by vultures and howler monkeys; stingless *Trigona* bees can prevent hermit hummingbirds from feeding at passionflowers (Gill, Mack and Ray, 1982). If the bees are experimentally excluded the hummingbirds make more use of the flowers.

Figure 9.6 Herring gulls fighting over food on a rubbish tip. Gulls fight over individual food items and dominant birds exclude subordinates from favoured sites. (Courtesy of Neil Metcalfe.)

9.4.2 Competition within a species

Whereas interspecific conflict usually results in niche reduction, intraspecific conflict expands a species' niche since individuals diverge to avoid competition. For example, subordinates within a species sometimes use different feeding areas to dominants, because they either avoid, or are driven away by, them. For example, young oystercatchers leave mussel beds because of interference from older, dominant birds (page 291).

In some species, males tend to be dominant over females and thereby maintain permanent differences in foraging behaviour. In the three-toed woodpecker, males and females share territories but the smaller females use a wider range of foraging sites and techniques than the males. In the genus *Dendrocopus*, however, males and females are similar in size with more overlap in their foraging niche than in the three-toed woodpecker; males and females have exclusive territories with males dominating females (Hogstad, 1978).

Intraspecific conflict can alter the distribution of a population. Individuals may aggregate if agonistic encounters are low, but if such encounters are frequent some sections of the population, such as subordinates, often juveniles, are forced out to different areas (Sutherland and Parker, 1985). Thus, in both oystercatchers feeding on mussel beds and herring gulls on rubbish tips (Fig. 9.6), dominants feed in different areas to young subordinate birds (Monaghan, 1980; Goss-Custard *et al.*, 1984), extending the area used by the population as a whole. Dominant–subordinate relationships and territorial exclusion may also be involved in determining the distribution of coastal birds in winter, the most dominant migrating the shortest distance from the breeding grounds (Pienkowski and Evans, 1985, and inclusive references). Similarly, in the dark-eyed junco winter habitats are different for the two sexes, the smaller, lower ranking, females migrating further south than the males, probably as a result of male dominance (Roberts and Weigl, 1984).

9.4.3 Agonistic behaviour and predator–prey interactions

The combatants involved in agonistic behaviour are not the only individuals affected. Aggression within and between species has repercussions in the whole community.

Agonistic behaviour between predators usually favours the prey population because it reduces the time available for searching for prey and thus reduces the searching efficiency. It may also result in the dispersal of the predators, which in turn reduces the pressure on the prey species. However, the prey sometimes suffers. For example, pollination of flowers may be decreased by the aggression of *Trigona* bees against their pollinators, the hermit hummingbirds (Gill *et al.*, 1982; page 239).

Although the relevance of this interference between predators has been little studied in the wild, it has been demonstrated in laboratory experiments with insect predators and parasitoids. Searching efficiency of these insects decreases with increasing predator densities because they react to the presence of others by moving away from the food, sometimes because of direct aggression (as in the ichneumonid wasp *Rhyssa persuasoria*). As a result, the population sizes of the parasitoid and the prey are stabilized (Hassell, 1978; Strong *et al.*, 1983).

Social behaviour in the prey species could also affect the predator population. Predators tend to forage in the most profitable patches where prey density is high, and the time spent feeding in a patch depends on the abundance of prey within it, relative to other patches. When prey are widely dispersed, perhaps because of the presence of territories, the searching time of the predator may be increased and its efficiency at finding the prey decreased. But, if there is a floater population of non-territorial individuals, this would be in the predator's favour.

9.4.4 Agonistic behaviour and community structure

As we have seen (page 238) one effect of competition between individuals can be a change in niche breadth both within and between species. So predation pressure on another species may be increased as these are introduced into the diet, or reduced if dropped from the diet. Small humbug damselfish, for example, eat smaller prey in the presence than in the absence of large individuals (Coates, 1980).

Territorial behaviour may also reduce the number of predators feeding on a single species. The diversity of species and the community structure within the territory may even be changed. Herbivorous fish on coral reefs can, by excluding feeding competitors, increase the abundance of algae at the expense of corals (Sammarco and Williams, 1982). The damselfish *Hemiglyphidodon plagiometopon* maintains a high and characteristic diversity of algae in its territories (Sammarco, 1983). In experimental plots, diversity was lower in ungrazed and freely grazed sites than on territories. In freely grazed areas, blue-green algae predominate, in ungrazed areas red algae are most common, but in the damselfish territories there is a mixture of both types. This change in algal type also affects the local chemistry since blue-green algae fix nitrogen.

Thus, because, it changes niche size and predator–prey relations, competition moulds the structure of animal and plant communities, determining the relative abundance and distribution of the species within them. Despite many studies of competition, however, its prevalence and ecological importance (relative to other factors such as predation) in nature is still disputed. A detailed discussion of this is not possible in the space available here but see, for example, Schoener, 1983a; Strong *et al.*, 1984).

9.5 EVOLUTIONARY CONSEQUENCES

9.5.1 Agonistic behaviour and evolutionary mechanisms

Because an individual's reproductive success depends on what other animals do, behaviour is likely to be an important selective force in the evolution of populations and species. In addition, the agonistic interactions of animals help determine the structure and growth of the population and hence its genetic composition and the direction and speed of evolutionary change. Agonistic behaviour in particular influences gene flow between populations and gene frequency within populations via the population size and the extent of in-breeding and random mating.

(a) Gene flow

Movements between populations help to mix different genotypes, but agonistic behaviour may prevent such mixing occurring. Pikas, for example, are very aggressive to strange individuals but allow related animals to live close by (Smith and Ivins, 1983). Consequently there is little genetic variability within a population of pikas (Glover *et al.*, 1977). In contrast, genetic variability between isolated groups of animals may be high; this is seen in the pocket gophers who aggressively defend a group territory from strangers (Patton and Feder, 1981).

However, gene flow can occur even between apparently isolated groups, such as one finds in house mice, for example. Mice live in fiercely defended territories, containing a male with his females and their offspring, but females at least are able to move between groups (Baker, 1981). In many species gene flow is also promoted by the movements of young animals of one or other sex or both.

(b) Breeding population size

We have seen that the actual size of populations and density of individuals are influenced to a large extent by agonistic behaviour such as territoriality, and that the size of the breeding part of the population is also influenced in this way. Often there are individuals which do not breed in any one year because they are excluded by others as a consequence either of territoriality or of dominance hierarchies. In extreme cases only the most dominant male or pair in a group may reproduce and this could greatly restrict the genetic variability of the next generation (page 290). However, shared paternity in wild populations may be more frequent than has been supposed from studies conducted on enclosed groups. In one study of rabbits in the wild, for example, dominant males sired 37–74% of kittens (Daly, 1981) so subordinates do obtain some matings even though behavioural observations had suggested the dominant male did almost all the copulating.

(c) Random mating

Animals often do not mate randomly, but select a partner according to a variety of characteristics, including size or the effectiveness of a display. Agonistic behaviour may influence the development of these characteristics; for instance, large animals are often the winners in fights. The mating system used also affects the randomness of gene combinations, some animals mating with many individuals, others with just one.

(d) Speciation

Non-random mating serves as a mechanism that isolates populations of the same or similar species and thus promotes speciation. Individuals of closely related species may recognize each other by song, scent or other cues and avoid interbreeding. Aggression can also be important in this discrimination. Blind mole rats, for example, include four chromosomal species, morphologically identical, but females can choose mates of their own species on the basis of vocalizations and olfaction. Aggressive behaviour in this mole rat is greater between than within species, and is also greater between species whose distributions are contiguous than between non-contiguous ones (Nevo, Naftall and Guttman, 1975). So in this case aggressive behaviour prevents interbreeding between the species and thus contributes to the speciation process.

9.5.2 Competition for mates

Competition for mates is a strong selective force on the sex that competes (a process which is referred to as sexual selection). It can influence the evolution of characteristics such as large size, sometimes leading to striking differences in appearance between males and females (Fig. 9.7). Competition for mates is obviously important in the breeding season and selection pressure at this time may be strong but selection operates throughout the year and may not always be in the same direction. Thus a large male may obtain females during the breeding season but could lose out if food was scarce during the winter. The costs and benefits of such selection pressures are discussed in Chapter 11.

Conflict that may lead to larger males has been recorded in a wide range of animals from crustaceans and insects to birds and mammals. Some of the evidence is circumstantial. Within a group of related species, males are often the larger sex in species where a male mates with several females (where inter-male conflict is likely to be high, as in lizards, Dugan and Wiewandt, 1982) or in species where inter-male combat is known to be frequent (for example, turtles, Berry and Shine, 1980). Some experimental evidence is also available, however. In natural and staged fights between males it is the larger male that usually wins, suggesting that larger male size would be selected for. In the milkweed beetle large males win fights and pair with the females. In laboratory experiments the benefits of being large were much greater when males were allowed to fight over females than when a male could have a female to himself (McCauley, 1982). So male–male conflict can have a potentially important effect on male characteristics.

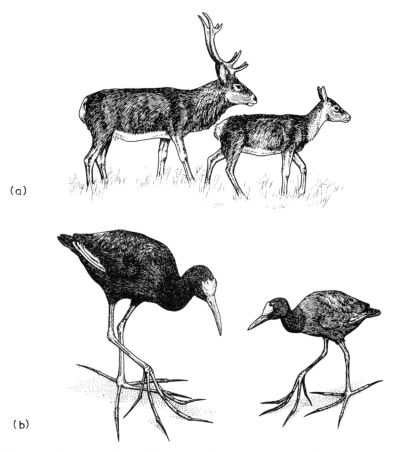

(a)

(b)

Figure 9.7 Sexual dimorphism. In red deer (a) males are larger than females and during the breeding season sport weapons (antlers) that are absent in females. Male and female jacanas (b) are similar in form, but the females can be up to 75% larger than the males.

Large body size is not the only result of conflict over mates. Weapons such as the larger, robust horns of antelope (Packer, 1983), and colouration such as the blackened chelae of snapping shrimps (Knowlton, 1980) have also probably evolved in response to male combat and the need to signal during agonistic encounters.

Males are not always larger than females. In some species sexual selection results in small males or large females. In invertebrates with discrete generations the winner of competition for a resource may be the one that develops fastest in his generation. So small male adult insects that emerge first may have the first choice of females and could avoid direct combat (Singer, 1982). Sexual selection also acts on females, since they may fight over nesting sites and large females are most likely to win (Johnson, 1982). In some species such as phalaropes (wading birds) females are brightly coloured and compete for access to the drab males.

Males that obtain many matings tend to be favoured but for females the ability to rear good quality offspring is often more important and this rather than or in addition to competition between males for mates, may cause differences between the sexes. Thus the horns of male antelope are large and robust, designed for clashing fights whereas those of females are thinner and more pointed, designed for stabbing predators (Packer, 1983). Also, females may be large because this enables them to produce more offspring (e.g. amphibia) or else larger offspring with a better chance of survival (e.g. bats, Ralls, 1977). Alternatively, small females may be favoured. For example, female weasels are smaller than males, probably because they are more efficient at using energy and thus in rearing their litters (Moors, 1980).

9.5.3 Competition for other resources

Mates are just one resource for which there is competition between individuals within a species. Conflict over any resource could result in differences between individuals since competition favours the evolution of structures that increase an animal's fitness. For example, over-winter mortality increases differences in size between the sexes in the house sparrow (Johnston and Fleischer 1981), males being larger than females. In severe weather, large females and small males are at a disadvantage in the feeding flocks. The largest individuals displace small birds from food so waste little time fighting; the smallest individuals avoid encounters so also spend little time fighting and they also need little food anyway. It is the intermediate sized birds that lose time and energy in fighting and die as a result.

Differences in morphology, colouration and behaviour, as well as in size, can sometimes be seen in populations. Theoretically, these differences could be maintained by one or more of several factors, such as frequency-dependent predation, but competition between individuals can play a role. The importance of competitive ability can depend on the environmental circumstances. In the marine snail *Nassarius pauperatus* there are two behavioural morphs: twisters that swing their shell from side to side on contact with another snail and non-twisters (Fig. 9.7; McKillup, 1983). Twisters are apparently at an advantage when food is both in short supply and found in discrete patches because they can push their way between other snails to reach the food.

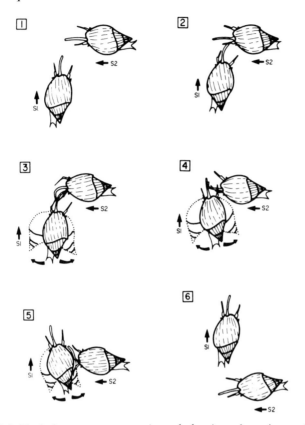

Figure 9.8 Twisting movements in a behavioural variant of the snail *Nassarius pauperatis* (reproduced with permission from McKillup, 1983). 1, Snail 1 and 2 approaching; 2, snail 1 contacts snail 2; 3, snail 1 commences twisting, snail 2 contacts snail 1; 4, both snails remain in contact, snail 1 continues twisting, snail 2 does not twist; 5, snail 1 ceases contact with snail 2 but continues twisting; 6, snail 1 ceases twisting.

9.5.4 Competition between species

As we have seen, competing species frequently have slightly different niches and over evolutionary time they may diverge until competition is much reduced or eliminated. Such divergence of characters is difficult to demonstrate, however. The marine snail *Hydrobia* provides one possible example (Fenchel, 1975). One species, *H. ulvae*, is larger than another, *H. ventrosa*, where they coexist but not where they are allopatric, suggesting that competition has produced a divergence in size.

Where interspecific aggression is eliminated, as occurs when one only of two or more competing mainland species reaches an island, selection pressure for diverging characters is relaxed. Island birds often have broader niches and more variable bill sizes than their relatives on the mainland (van Valen, 1965).

Mimicry is a possible case of character convergence due to conflict between species. On the islands of the Australasian region orioles, friarbirds and the honeyeater (*Pycnopygius stictocephala*) form a guild of birds with similar diets and similar feeding sites. There is a high frequency of fighting among the members of the guild with the large friarbirds being dominant to the smaller orioles; the honeyeater is the smallest and most subordinate species on New Guinea. Orioles are usually brightly coloured, but when they are sympatric with a friarbird they are brown and similar in plumage to the friarbird. The friarbirds tolerate both conspecifics and the oriole mimics more than they do other species in the guild. The mimics may also benefit by dominating other species. On New Guinea, the honeyeater mimics the oriole (Diamond, 1982).

Apart from morphological convergence or divergence of competing species there is also experimental evidence for a species' competitive ability changing over a number of generations (Arthur, 1982). In the real world, the relative competitive abilities of sympatric species are theoretically of great importance in the continuation, extinction or replacement of species over evolutionary time. There is some evidence of species replacement due to interference competition. In modern times, barnacles of the Balanoidea group outcompete *Chthamalus* species. *Chthamalus* is evolutionarily the older taxon; it radiated early then declined, whereas the Balanoidea are more recent and today are more successful because they have porous shells which allow them to grow faster than *Chthamalus* which has a solid shell wall (Stanley and Newman, 1980). Because the former are faster growing they can undercut, smother and crush *Chthamalus* which now only exists in dry refugia since it is more resistant to desiccation and extreme temperatures.

9.6 OVERVIEW

9.6.1 Have our questions been answered?

(a) What happens to the winners and losers?

The physiology of an individual can be affected markedly by the presence of a more dominant animal, a territory owner or as a result of a fight. Some aspects, such as tissue growth and the production of reproductive hormones, are often suppressed; others, such as glucose metabolism, increase sometimes to dangerous levels. The effects continue until the individual concerned either escapes from the situation, perhaps emigrating from the population, or dies. The act of engaging in agonistic behaviour is itself likely to lead to a higher risk of mortality.

(b) Consequences for populations and species

All the important parameters that influence the dynamics and distribution of a population (mortality, emigration and reproduction) are under the influence of agonistic behaviour.

Not just one individual but whole subsets of a population are affected. Populations are made up of many kinds of individuals of different age and sex classes and also of different behavioural classes. Often juveniles are forced to leave an area occupied by more dominant adults. Even within an age class or sex class there is a great deal of individual variation. Among juveniles, for example, some are dominant and win territories or status at the expense of other juveniles which may suffer higher mortality as a result (page 231).

In many cases, the effects of agonistic behaviour are density-dependent; for example, interference between feeding animals, cannibalism, repression of sexual activity and territorial exclusion (pages 233, 234). So they can act to control the size of the population, being more influential at high population densities and slowing down the rate of population increase. The size of the population can be stabilized to an equilibrium level. Despotic territory owners, for example, gain adequate resources when the population density is high, while non-territorial individuals may not get enough of a resource and so die. If all individuals in a population were equal, each one might obtain only an inadequate share of the resource so mortality would be greater and would cause the population to crash. The despotic owners also reduce the carrying capacity of an area because of this unequal division of resources.

As a result of these population changes there are several consequences for the ecological community. Niches within a species are broadened if dominants and subordinates use different habitats or prey species (page 240) to avoid agonistic encounters. Niches become narrower, however, when the conflict is between individuals of different species. There are further repercussions: interactions between predator and prey populations and the structure of communities are both affected by the agonistic behaviour of their individual members.

(c) Evolutionary consequences

These changes not surprisingly affect the characteristics of a species, both by determining who mates with whom and by selecting for characters that are involved with the expression of agonistic behaviour such as large size or the development of weapons. Ultimately conflicts between species can change their characteristics and can contribute to their evolutionary demise.

9.6.2 General principles

Mechanisms
Territorial behaviour, dominance, and fighting, among other factors, determine who in a population is going to become sexually mature, who is going to reproduce and how successfully, who is going to leave the population and who is going to die. In this way, agonistic behaviour influences (1) the processes of population growth, stability and distribution; (2) the range of resources and habitats used by the population; (3) which genes will be passed on to the next generation; and (4) which genes will die out. The mechanisms may be direct (for example, the use of pheromones to suppress sexual activity in rodents, or the killing of a rival in a fight) or indirect (for example, the exclusion of an individual from any suitable breeding site effectively prevents breeding; and exclusion from food can lead to an animal dying).

Agonistic behaviour has strong influences on ecological and evolutionary processes and this can include the expression of agonistic behaviour itself. An animal's behaviour should maximize its lifetime fitness so there is likely to be selection for ways of overcoming the damaging effects of agonistic behaviour on an individual's chances of surviving and reproducing. This adaptive aspect of behaviour is discussed in Chapter 11.

10

Evolutionary history

10.1 INTRODUCTION

Although competition for valuable resources is universal, not all species gain access to these by fighting (page 6). The absence of intraspecific aggression is almost certainly a primitive trait, although this behaviour may be secondarily lost during subsequent evolution if its costs always outweigh its advantages (page 277). The form of agonistic behaviour patterns, the readiness with which they are used and the social context in which fighting occurs vary considerably among living animals and the topic of the present chapter is the evolutionary history of this diversity. We try to answer three questions.

1. How did the capacity for aggression evolve in the first place and what were the evolutionary precursors of the structures and movements used during fights by today's animals?
2. Once evolved, how did agonistic movements and the way these are used change to produce the diversity we observe in living animals?
3. How have weapons and the way these are used changed during evolution?

10.1.1 The fossil record

Fossils are traces, usually in rocks, of the animals in existence at the time that the rock was formed. If fossils of a particular group of organisms are present in rocks of different ages, we can reconstruct the changes that occurred during the evolution of that group. Although animal behaviour does not fossilize, palaeontology can still provide indirect information about what extinct animals did. For example, if we know what living animals do with their bodies we can make deductions about the behaviour of fossil forms of similar structure. Fossil termites from rocks that are about 100 million years old have heads and mouthparts like those of living termite soldiers and probably used these in colony defence. Fossil male dragonflies from as much as 400 million years ago have claspers on their abdomens, suggesting that, like males of some modern forms they guarded females after mating.

10.1.2 Behavioural comparisons in living species

Where fossil evidence is not available, the only way we can reconstruct the past is to work backwards from living forms. This evidence is indirect and can only provide plausible hypotheses about evolutionary events, not proven facts. Two main sources of evidence are available.

1. Comparison of the form of behaviour patterns shown in different contexts by the same animals can suggest what the evolutionary origin of their agonistic behaviour might have been. For example, claw movements during agonistic encounters between crabs (*Petrolisthes* spp.) look very like those used during exploration of the substrate. So, for example, the chela probe display is thought to have evolved by a minor change in the orientation of claw movements with which stones are picked up and examined (Molenock, 1976).

2. According to evolutionary theory, any group of related animals had a common ancestor at some time in its past and the more closely related the living animals, the more recent their common ancestor. The distribution of behavioural traits within a group of related animals can be used to make deductions about how their ancestors acted, provided independent evidence is available about the phylogenetic relationships of the various species. For example, among ants of the genus *Harpagoxenus* some species (for example, *H. americanus*) are obligatory slave makers and others (for example, *H. canadensis*) show this behaviour only under certain circumstances (they are facultative slave makers). A related kind of ant (*Leptothorax muscorum*), belonging to the genus from which the slave makers are

thought to have evolved, engages in fierce territorial battles against colonies of the same species and the broods of the defeated colonies are often carried off by the victors. The behaviour patterns used during slave raids by *Harpagoxenus* are strikingly similar in a number of respects to the territorial atacks of *Leptothorax*. This suggests that slave raiding evolved from territoriality, first producing facultative and then obligate forms (Stuart and Alloway, 1983).

10.2 EVOLUTIONARY ORIGINS

10.2.1 First appearance

In plants and in many simple animals, removal of competitors takes the form of non-directed responses. For example, bacteria release toxins into the environment in a broadcast manner, killing any susceptible individual that happens to be in the vicinity (page 14). Conjugating ciliate protists of certain strains inject poisons into their mating partners along with genetic material, so as soon as the oppor-tunity for injuring particular individuals arose, it seems to have been exploited. Apart from this last bizarre example, the sea anemone is probably the simplest animal in which directed aggression aimed at particular individuals and making use of special behaviour patterns is observed (page 14). Direct aggression as a way of resolving conflicts requires a minimum of three things.

1. A mechanism for recognizing foreign organisms (such as the primitive immune system of coelenterates and sense organs in more advanced animals).
2. A set of effectors which can, potentially, damage or remove an opponent (clusters of stinging cells in anemones and various weapons in other species).
3. A system for producing co-ordinated and directed movements (scattered muscle cells and a diffuse nerve net in anemones, special-ized muscles and nerves in more advanced groups).

10.2.2 The origins of weapons

The essential feature of a weapon is that, when deployed forcibly, it damages the tissue of another animal, causing pain (which is likely to elicit immediate withdrawal) or injury (which incapacitates rivals over a longer time period). So weapons are likely to have their evolutionary origin in structures which are used by animals to interact strongly with their environment and which are hard and sharp. On the basis of similarities in structure and position and of the movements by which they are deployed, weapons used against conspecifics are thought to have evolved in a number of different ways.

(a) Burrowing

During fights, ragworms (page 16) employ structures (the proboscis) and movements (mouth to mouth contact, proboscis extension and pushing) which are also used during burrowing. These lend themselves to removing conspecifics just as well as to removing sand; Cram and Evans, 1980). The horns of some stag beetles, which are used in certain modern forms to lift and push rivals, were also originally employed in burrowing (Otte and Stayman, 1979).

(b) Egg laying

The formidable sting of the bees and wasps evolved from the ovipositor used by their ancestors to pierce the skin of the insects in which their eggs develop. The ovipositor is sharp, strong and equipped with a pumping mechanism and egg laying involves thrusting movements; all these features were taken over when the sting became a weapon, delivering poison instead of eggs (Kasparyan, 1981).

(c) Feeding and predation

Structures and behaviour patterns used to break up tissue while feeding (by both herbivores and carnivores) represent an important evolutionary source of weaponry.

- The mouthparts of vegetarian animals have evolved into weapons in species as diverse as termites and hippopotami.
- The raptorial appendages of stomatopods (page 23) are used both for killing prey and for attacking rivals. Species that feed on soft-bodied prey have long, thin appendages, do not fight very fiercely and have not developed much in the way of protective armour. In contrast, species feeding on hard-bodied prey have heavy, spined raptorial appendages, are highly aggressive and have developed strong armour to ward off blows (Dingle, 1983).
- In ants, the primitive weapon is an elongated abdominal sting (derived originally from the ovipositor) which is used to immobilize large prey but has been taken over for use in colony defence (Kugler, 1979).
- Many carnivorous mammals use their claws to rake the skin of both conspecifics and prey.

(d) Anti-predator defence

Various structures used to defend animals against predators (them-selves often derived from feeding movements) have evolved into weapons used in fights against conspecifics.

- In decapod crustacea, claws used for counter attack against potential predators are commonly used against conspecifics as well.
- The hooves of ungulates evolved to allow fast escape from pre-dators over open country but, once evolved, they made effective weapons for slashing at conspecifics (Fig. 10.1).

(e) New structures

In other cases, weapons have developed from completely new struc-tures. Horns in ceratopsian dinosaurs first evolved as bony out-growths of the skull which probably protected the eyes and nose during head ramming fights between males; at an early stage they evolved into one nasal and two frontal horns (Spassov, 1979). The fossil record suggests a very similar origin for the horns of ungulates (Stanley, 1974). The earliest ruminants were medium sized and horn-less; horns have appeared in several lineages (cattle, deer, giraffes and antelopes) although these represent a minority of all living and fossil ungulates. In each lineage, horns appeared initially in males and in species above a critical size; this size corresponds roughly to that which, among living ungulate species, marks a change from mono-gamy to polygyny. Thus the selection pressure responsible for the evolution of horns seems to have been the need for males to compete for females. In several lineages females evolved horns as well, after a shift to open and more dangerous country where horns provided a useful defence against predators (Janis, 1982).

10.2.3 The origins of agonistic displays

Attack with a weapon can remove a rival either directly (by manhand-ling it away from a disputed resource) or indirectly (through a deep-seated response of withdrawal following pain and injury). But why should a rival withdraw in response to nothing more than a display? The form of many threat displays is such that they provide an abrupt increase in stimulation; folds of skin are raised, weapons suddenly displayed, unexpected loud noises produced, and so on. Many species have a built-in avoidance response to large looming objects and also readily associate a painful experience with a visual or auditory stimulus that preceded it. These two behavioural predispo-

Figure 10.1 Use of hooves during fights between zebras.

sitions probably facilitated the evolution of certain types of behaviour into threat displays. Comparative studies of movements used by animals in different contexts suggest that agonistic displays have diverse evolutionary origins and several of these are discussed below. To avoid repetition, in this chapter statements about evolutionary origins are not always qualified, but it should be remembered that the comparative method produces plausible hypotheses rather than proven fact.

(a) Locomotion

Animals often respond to environmental change (including the appearance of a rival) with some sort of movement and this is a common source of agonistic displays.

- Several leg displays used by spider crabs in agonistic postures have evolved from general walking movements; during the change from locomotion to display, leg movements have become more stereotyped in form (Hazlett, 1972).
- Several agonistic displays in the crab *Petrolisthes* (for example, chela hunch which involves pushing the body up and forward with the posterior walking legs) have evolved from slow following of a retreating rival (Molenock, 1976).
- Damselfly (*Xanthocnemis zealandia*) larvae confronting an opponent make stereotyped swinging movements of their abdomen. These are thought to be derived from swimming, with the addition of a retaining grip on the substrate and loss of the thoracic movements shown by swimming larvae (Rowe, 1985).

(b) Exploration

The evolutionary origin of other agonistic displays lies in another common response to any change in environmental stimulation, namely exploration.

- The visual system of most lizards is highly specialized for acute perception of objects situated to the side of the observing animal, so good vision requires lateral orientation. Many conspicuous visual displays in lizards have evolved from this sideways, scanning stance (Jenssen, 1977).
- Spiders often explore their environment by regularly tapping the substrate with legs and palps or, in web building species, by plucking at the web to set up vibrations. In a number of cases these have evolved into threat displays, by means of relatively minor changes in the form and emphasis of the movements (Buskirk, 1975).
- The upright, scanning stance assumed by many mammals in a strange environment has evolved into the upright threat posture which is a common feature of mammalian fights, often enhanced by a light coloured belly (Eisenberg, 1981).

(c) Physiological arousal

The presence of an opponent triggers physiological changes such as an increase in respiration rate, alterations in colour, defaecation and urination; these too are often common sources of agonistic displays. During the course of a fight, animals get hot and fur and feathers get ruffled, and the resulting grooming and preening movements can also evolve into displays.

- Urine and faeces are used as signals both in territorial marking and during fights in many mammals.
- Sounds are often produced as the byproduct of respiration. When frightened, many mammals produce high pitched squeaks while unharmonic low pitched sounds are produced by aggressive animals. In various groups of mammals, these have evolved into, respectively, high pitched submissive signals (perhaps effective because they resemble calls of juveniles) and low pitched threats (perhaps indicating a large body size; Eisenberg, 1981).
- Agonistic palp waving displays in spiders (Krafft, 1983) and head shaking displays in birds (Jouventin, 1982) are derived from movements used to clean the eyes.
- The abdomen arching display of damselfly larvae (*Xanthocnemis zealandia*) consists of sharp, downward bending of the abdomen and occurs just before retreating. This may be derived from movements used to groom the tail appendages (Rowe, 1985).
- The marking actions of territorial gerbils are derived from sand-

bathing used to clean the fur (Eisenberg, 1981).

(d) Protective movements

Both participants in a fight are liable to be injured, and many agonistic displays are derived from responses to potential danger.

- When threatened, many spiders raise their front legs and open their fangs in generalized defensive movements, some of which have evolved into agonistic displays (Weygoldt, 1977).
- The escape jump given when disturbed by a predator has evolved into a threat display in which the legs, bearing sharp spines, are swung forward over the head in the weta (an Australian cricket-like insect). This impressive change in behaviour is the result of a minor shift in the balance of the neurohormones octopamine and serotonin (page 138) which control activity of the femoral muscles in this insect (Hoyle, 1985).
- In mammals, immobility and freezing, common responses to danger, have given rise to the lateral postures assumed by many mammals during fights (Eisenberg, 1981).
- Several facial expressions in primates, especially those that involve wrinkling the eyes and nose and retracting the ears, are thought to have evolved from movements that protect these vulnerable organs against injury during fights (Andrew, 1972).

(e) Attack–flight conflicts

The fact that animals may get injured during a fight means that opponents arouse fear, so full aggression may be inhibited. Equally, even when animals are frightened during an agonistic encounter, flight is tempered by motivation to attack. In a number of species, inhibited attack and escape movements have evolved into agonistic displays.

- The horizontal threat and head flagging displays of gulls (Fig. 10.2) are probably derived from inhibited attack and flight movements, respectively (Tinbergen, 1959; Moynihan, 1961).
- The threat displays of sharks (with the back arched, fins lowered and head moved from side to side) may be derived from an inhibited form of the tearing movements used during attacks on prey and territorial intruders (Barlow, 1974).
- Head nodding movements in lizards may have evolved from oscillations between aggression (producing head up movements) and fear (generating head lowering; Jenssen, 1977).

Figure 10.2 Horizontal (above) and head-flagging displays (below) in black-headed gulls (reproduced with permission from Tinbergen, 1959).

10.3 EVOLUTIONARY DIVERSIFICATION

When two animals compete over important resources or outcomes, selection favours those individuals that are most efficient at bringing the encounter to a favourable conclusion with the least risk (page 277). This selection pressure is reflected in the evolutionary history of all aspects of agonistic behaviour, from the form of fighting movements to the frequency with which they are used and the social context in which fights occur. In this section we look at the evolutionary diversification of the form of agonistic behaviour, of its frequency and of territoriality and dominance. Here again, we are using comparative evidence and so can only present plausible hypotheses rather than proven fact.

10.3.1 Conservatism of agonistic behaviour patterns

The form of the motor patterns used during fights (as opposed to higher level manifestations, such as overall aggressiveness and territoriality) tend to be remarkably similar within a particular taxonomic group, in animals as diverse as flies (Spieth, 1982), fish (Ewing, 1975) and monkeys (Van Hoof, 1967, 1972). This is in marked contrast to the behaviour patterns used in courtship where rich variety is the rule. There are a number of possible reasons why agonistic movements show this relative conservatism.

(a) No selection for diversification

The sort of selective force that has caused diversification of courtship behaviour, mainly the need to avoid interspecific matings, might not have been active during the evolution of agonistic behaviour.

(b) Selection for similarity

Where closely related species are sympatric and compete for resources, a degree of communality in agonistic signals may be advantageous and so selection may have favoured evolutionary conservatism. This sort of explanation has been put forward to explain the extreme similarity of the agonistic repertoires of related species of crabs (Molenock, 1976) and for the existence of some common elements in the agonistic repertoire of so many lizard species (Jenssen, 1977).

(c) Selection against behavioural variants

The nature of aggression, with its knife-edge jostling for advantage over opponents and with the severe penalties of injury, may mean that animals that deviate even slightly from the usual pattern may be at great risk. Major changes may thus produce less effective fighting behaviour, making rapid evolutionary change difficult.

(d) Morphological constraints on the potential for change

The fact that fish have few appendages and need to retain buoyancy

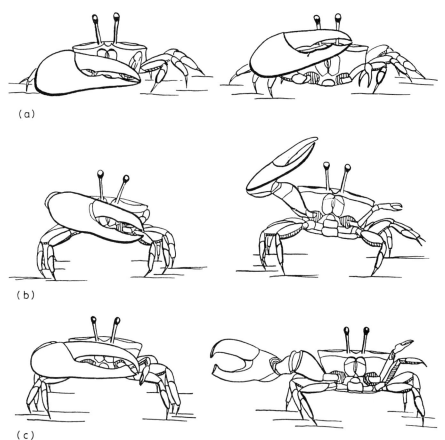

(a)

(b)

(c)

Figure 10.3 Claw waving displays of three species of fiddler crab, showing the resting position on the left-hand side and the position of maximum extension of the claw-bearing limb on the right-hand side. (a) *Uca rhizophorae*, a supposedly primitive form, (b) *Uca zamboangana*, a vertical waver and (c) *Uca lactea*, a horizontal waver. (From Crane, 1957 with permission of the New York Zoological Society).

may limit the range of fighting techniques and displays that is potentially available to them. As far as their overall body form is concerned, penguins are like fish and subject to the same constraints; there is a heavy emphasis on head movements and postures in their agonistic repertoire, because there is not much else they could use. The prevalence of lateral displays in lizards reflects both their specialized visual system (page 257) and the fact that movements in a vertical plane (head bobs, dewlap extension) are relatively easy to produce.

10.3.2 Microevolution of agonistic displays

Although agonistic displays are conservative, they do not remain completely constant over evolutionary time. In general, the changes that do occur involve minor alterations; the movements increase in amplitude and the emphasis placed on different components shifts. The form of a display may be dictated by conspicuous colour patterns already in existence, but more often bright colouration appears during the course of evolution of increasingly conspicuous agonistic displays. Complex displays arise by combining existing movements and postures and by repeating and modifying simpler displays. Minor shifts take place in the rules that relate motivational state to the performance of agonistic displays, but the broad motivational context in which these are shown usually remains conservative. These points will be illustrated in a brief systematic survey of some of the better documented studies of the evolution of aggression, in various groups of animals.

(a) Fiddler crabs

Among the fiddler crabs, waving movements of the large claw are used to attract mates and to repel rivals; these initially involved simple vertical waves, with the body close to the ground, as in *Uca rhizophorae* (see Fig. 10.3(a)). During the subsequent evolution of this display, the movement became more conspicuous by adding colour to the claw, by holding the body in an upright posture and by changing the form of the wave, enlarging the amplitude and making it more jerky. The vertical and lateral aspects were emphasized in two separate lineages (Fig. 10.3(b) and (c); Crane, 1957, 1966).

(b) Fish

Old world rivulin fish have a rich repertoire of agonistic behaviour patterns. In eighteen species (whose phylogenetic relationships are known from their biochemistry), studied in detail in the laboratory the form of these patterns is very similar. Fights also follow the same broad escalating trend of sigmoid posture giving way to fin clamp, nuzzle, tail beat and then attack. During the evolutionary history of this group, no new acts have appeared in the repertoire, the limited variation that does occur having arisen from minor changes in the form of existing patterns and the frequency with which these are performed. Figure 10.4 summarizes the sequence of agonistic actions in some of these species. Here there is no trace of conservatism; in particular, the dyadic relationships of the ancestral sequence have

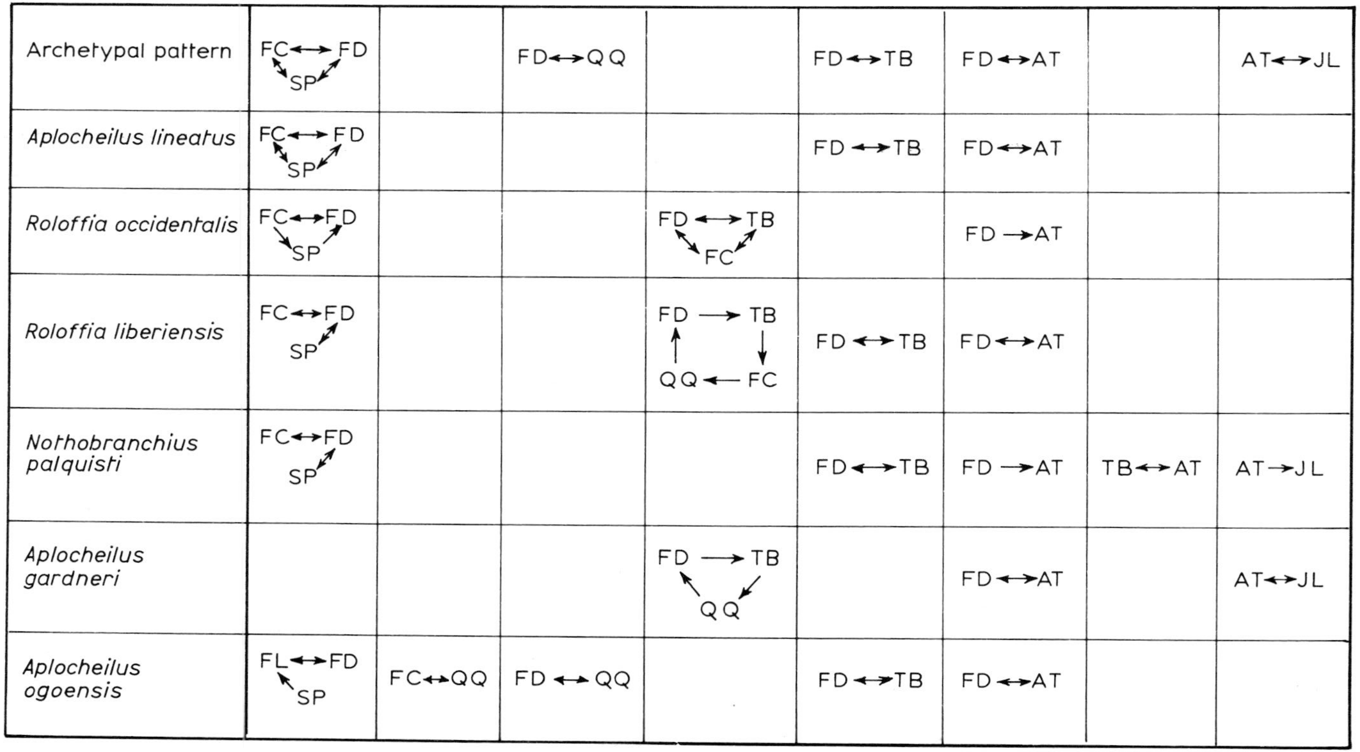

Figure 10.4 Sequential relationships between behaviour patterns used by Old World rivulin fish during fights (after Ewing, 1975). FC, fin clamp; FD, full display; SP, sigmoid posture; QQ, quiver; TB, tail beat; AT, attack; JL, jaw lock; →, one-way transitions between actions; ↔, two-way transitions between actions. Actions are arranged from left to right in order of their maximum expression during an encounter.

Species								
Archetypal pattern	FC↔FD, SP		FD↔QQ		FD↔TB	FD↔AT		AT↔JL
Aplocheilus lineatus	FC↔FD, SP				FD↔TB	FD↔AT		
Roloffia occidentalis	FC↔FD, SP (FC)			FD→TB ↘ FC		FD→AT		
Roloffia liberiensis	FC↔FD, SP			FD→TB; QQ→FC	FD↔TB	FD↔AT		
Nothobranchius palguishi	FC↔FD, SP				FD↔TB	FD→AT	TB↔AT	AT→JL
Aplocheilus gardneri				FD→TB ↘ QQ	FD↔TB	FD↔AT		AT↔JL
Aplocheilus ogoensis	FL↔FD, SP	FC↔QQ	FD↔QQ		FD↔TB	FD↔AT		

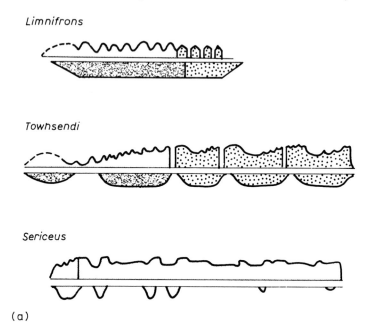

(a)

Figure 10.5 Microevolution of displays in lizards. (a) Signature displays of *Anolis limnifrons, A. townsendi* and *A. sericeus*. (b) Agonistic displays of *A. limnifrons*. From the signature display type E evolves by loss of dewlap extension, type B by delayed dewlap extension and type C by addition of head bobs. Type D has been 'welded' together from separate displays, as indicated by flexible head bob bouts (marked with arrows; after Jenssen, 1977). Dewlap position is represented below the line, head position is represented above the line, black areas denote invariant components and shaded areas denote variable features.

given way to higher order regularities involving more than two actions. For example, in *Roloffia occidentalis* repeated bouts of frontal display, tail beating and fin clamping occur and in several species, quiver and tail beat become locked together into a single, stereotyped unit (Ewing, 1975).

(c) Lizards

The agonistic repertoire of most lizards includes some form of head bobbing; the details of the movements, their timing and sequential patterning, are variable but tend to be broadly similar in groups of related species. In iguanids, smaller forms have more complex displays, and this is thought to represent a more recent evolutionary development (Carpenter, 1977). Complex displays have evolved by

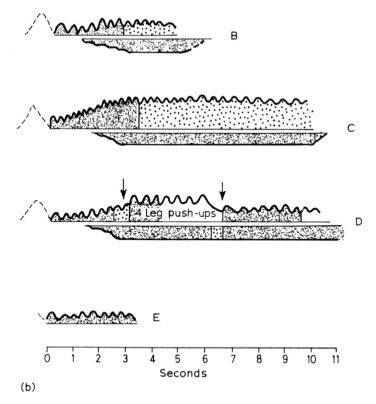

(b)

repetition and elaboration of existing displays and by joining together
simple displays. Figure 10.5(a) shows how the relatively unspecialized
signature display of *Anolis limnifrons* (shown in a variety of social
contexts) may have given rise to the more complex signal displays of
two other species (*A. townsendi* and *A. sericeus*). Figure 10.5(b) shows
how the signal display of *A. limnifrons* may have given rise to a
repertoire of four more agonistic displays (after Jenssen, 1977).

(d) Birds

The long call is a simple movement in primitive gull-like birds such as
the great skua, but it has evolved into a variety of forms by accentuating different parts of the accompanying display (Fig. 10.6) and has
been lost in terns. In most species, the call retains its aggressive
motivation but a variant (the mew call), in which the aggressive
motivation is replaced by a sexual one, evolved independently in
advanced gulls and terns (Tinbergen, 1959; Moynihan, 1961).

(a)

(b)

A

B

C

D

E

(c)

Figure 10.6 Long call/oblique postures in gull-like birds. (a) The primitive form of the display in the Chilean great skua; (b) the upward component stressed in the grey gull; (c) the downward component stressed in the brown noddy (reproduced with permission from Moynihan, 1961).

During hostile encounters, male Pacific eiders perform cooing movements in which the head is brought in contact with the water, suddenly erected and tossed back with an open bill, the breast being inflated meanwhile 'as if swallowing an egg'. In the European eider, a derivative of the Pacific eider in which the black neck colouration has been lost, the cooing movement involves a horizontal rather than a vertical component and a downward position of the bill; in the Pacific eider, this movement would obscure the black colouration. Different versions of the cooing display have evolved by repetition of the original movement. In both subspecies the display is performed in a largely hostile context, but its motivational basis is slightly different;

the display reflects a weaker attack tendency in the Pacific eider than in the European form (McKinney, 1961).

(e) Carnivores and primates

There is considerable similarity in the facial expressions used during agonistic interactions in social canids and in primates, both in terms of their form and in the context in which they are shown. For example, in both groups, vertical retraction of the lips occurs in animals that are likely to attack, while some sort of tooth chattering display is associated with a tendency to flee. Either these two groups of mammals have inherited a basic set of expressions from a common ancestor, or similar expressions in similar contexts have evolved independently in the two groups. A broad evolutionary trend becomes apparent when the facial expressions of prosimians are compared with those of monkeys and apes. The same general categories of agonistic facial expression can be recognized in all three groups but these become more complex, particularly in the apes. There is more flexibility in the mouth position which allows subtle variations in the form of a given expression (Van Hoof, 1967, 1972). This has been accompanied (and caused) by an increase in the complexity of the facial muscles so that different parts of the face can be moved independently (Chevalier-Skolnikoff, 1973). This tremendous increase in the complexity of facial expressions in monkeys and apes allows more detailed information about emotional state to be communicated.

(f) Causal and functional considerations

As far as their causes are concerned, these microevolutionary changes in the form of agonistic displays can be explained in terms of quite minor shifts, for example in the thresholds determining when different movements are performed (Manning, 1979). These could account for changes in form and emphasis within a given pattern and, perhaps and up to a point, for changes in their sequential relationships. In functional terms, many of these microevolutionary changes make the displays better at removing or discouraging a rival with minimum effort and risk.

The overall conspicuousness of the animal concerned (colour) and/or in its apparent size (high body posture, wide amplitude movements, extended flaps, sideways postures) increase. This makes them more likely to generate exactly those kinds of stimuli that elicit avoidance or fear in a wide range of species, cashing in, as it were, on existing behavioural mechanisms. In some cases, microevolution of agonistic behaviour has involved an increase in stereotypy both in the form of the movement and in their sequential relationships (fish). This may make it easier for fighting animals to jostle for advantage without betraying their intentions. In other cases, exactly the reverse is seen; in more advanced primates facial expressions have evolved in such a way that they provide detailed information about the emotional state of the animals concerned. In a complex, permanent social group, it is probably to the advantage of both sender and receiver that accurate information be provided about emotional state and agonistic intentions.

10.3.3 Frequency of fighting

In various groups of animals (ragworms, Cram and Evans, 1980; Hawaiian fruit flies, Spieth 1982; amblyopsid fish, Belcher, 1983; and penguins, Jouventin, 1982) the readiness with which aggressive responses are shown varies greatly between closely related species, even though when they do fight they use similar actions. Such between-species differences in aggressiveness, though sometimes very dramatic, could be the result of a fairly small change in the mechanisms that control agonistic behaviour, such as a change in the thresholds at which behavioural switches occur (page 85). They can generally be related to different environmental conditions, and particularly to those that dictate the costs and benefits of fighting.

Table 10.1   Colour patterns and distribution of some agonistic behaviour patterns in four species of gourami (from Miller and Robinson, 1974).

	Species and colour			
Agonistic behaviour	Snakeskin* (Series of ocelli in lateral bands)	Blue* (Two large lateral ocelli)	Pearly* (Faint, narrow lateral band and candal spot)	Moonlight† (No colour band, silvery body)
Lateral display	Common	Common	Common	Common
Tailbeating	Common	Common	Common	Common
Opercular spread	Occasional	Occasional	Rare	Absent
Mouth fight	Common	Rare	Rare	Absent

* Day time spawning habit
† Night time spawning habit

The four (partially sympatric) species of anabantid fish shown in Table 10.1 represent an evolutionary series. The distribution of traits in the four species suggests that their common ancestor was a medium sized, slightly elongated fish with a row of dark spots along the body. It probably spawned during the day and fought fiercely using a number of displays. The blue gourami and the snakeskin gourami are most similar in behaviour and morphology to this supposed ancestral form. The pearly gourami is slightly less aggressive and has a smaller repertoire of displays. The moonlight gourami is the most specialized species in terms of body form and of spawning habits (nocturnal). It has reduced overall levels of aggression and has lost the two most intense elements of the agonistic repertoire, namely opercular spread and mouth fighting (Miller and Robinson, 1974).

10.3.4 Territoriality

The occurrence of territoriality also varies among closely related species, again reflecting differences in the costs and benefits of maintaining an exclusive area (page 295). Examples can be seen in sand wasps (Evans, 1966), fruit flies (Spieth, 1982), ragworms (Cram and Evans, 1980), various species of fish (Kortmulder, 1982; Belcher 1983), penguins (Jouventin, 1982) and rodents (Eisenberg, 1981). A fairly simple change in the rules that relate agonistic decisions to environmental stimuli (page 79) could be responsible for these differences.

Some comparative studies show how territoriality appears in an evolutionary lineage. In dragonflies (Johnson, 1964) and darters (fish, Winn, 1954), defence of fixed breeding territories is seen in more specialized species. Thus primitive darters such as lake dwelling *Percina caprodes* show low levels of aggression and are non-territorial. *Hadropterus maculatus* males are also relatively unaggressive but drive other males away from the immediate vicinity of adult females. Males of two more advanced species (*Etheostoma caeruleum* and *E. saxitile*) defend individual females, but tend to stay in the vicinity of a large rock or in a patch of shallow water near the shore where females may spawn. Finally, *E. blennoides* males use high level agonistic actions to establish fixed territories which they continue to defend after the female has spawned on them (Winn, 1954).

10.3.5 Dominance

Comparative studies suggest how one rather special pattern of dominance may have evolved. In primitive members of the social hymenoptera, dominance is established and continually maintained by direct physical aggression; this suppresses egg development in the subordinates so that the dominant female lays all or most of the eggs tended by the colony. In the bumble bee and in some primitive wasps, the dominant female (who is usually larger than her companions) establishes her status by fierce and overt aggression when the colony initially forms, but subordinates are able to learn her smell and to avoid costly fights. In more advanced social insects such as the honeybee, the queen prevents the workers from breeding, not by overt aggression but by a specialized chemical produced by the mandibular glands which prevents egg production and promotes brood care (Hölldobler, 1984).

10.3.6 Weapons of offence and defence

The evolutionary history of weaponry, which is well documented in some fossil forms, reflects the need to coerce and injure with the minimum of risk. In the two best documented examples (the social insects and the ungulates) weapons evolved into larger, sharper and more effective forms. At the same time other structures evolved which protect the receiver against injury, such as thick skin, thick skulls, horns designed to parry attacks, and necks and backbones adapted to withstand strong impacts. This progressive specialization of offensive weapons and defensive armour is inextricably bound up with changes in movements by which these structures are used. The fact that fights between conspecifics, and particularly among ungulates, sometimes look like harmless wrestling matches is not an example of self-restraint among animals as was once thought (page 42). Instead it represents fierce jostling for an advantageous position from which incapacitating attacks can be delivered without putting the animal itself at risk.

(a) Social insects

In ants, the primitive weapon is an elongated abdominal sting which is used in immobilizing prey and in defence. Although sharp enough to pierce the cuticle of small insects, this primitive structure was ineffective against larger opponents and was also dangerous to use. In addition, more advanced ants changed to a largely herbivorous diet. This resulted in a gradual reduction in the size of the sting, whose function in defence and offence was replaced by biting weapons. In many lines, evolution of these biting weapons was associated with the development of poisons, initially injected with a bite but sometimes squirted from a distance, as with the sticky secretions of the Dolidioderinae and the formic acid of the Formicinae (Dumpert, 1978; Kugler, 1979).

Poisons which enhance the effectiveness of biting attack, and which can be used at a distance, evolved in the termites as well. Worker termites have two salivary glands opening on the lower lip and serving a primarily digestive purpose. The secretion of these glands has taken on an agonistic function in soldiers of species such as *Macrotermes bellicosus* which inject neurotoxins with their bites. In other species, a specialized gland (the frontal or soldier gland) evolved for the production of defensive chemicals. Initially, the oily secretion was exuded from a pore into wounds made with the mandibles whose subsequent healing they impaired. In more advanced species, the pore gradually moved forward and the tip of the head became elongated to produce the pointed weapons of specialized (nasute) soldiers, which are unable to feed and which squirt sticky poisons at intruders from a distance (Deligne, Quennedey and Blum, 1981; Prestwich, 1983).

(b) Ungulates

In various lineages of ungulates (both fossil and living forms), as well as in the ceratopsian dinosaurs (Spassov, 1979), a very consistent series of evolutionary changes has taken place. For example, titanotheres, fossil ungulates which lived about 100 million years ago, increased in size and their horns became much larger and more complex, probably diffusing the force of head-to-head blows. At the same time, the spine and skull became stronger, providing additional protection against impact injuries, and projections evolved which served to block attacks. Development of the horns on the forehead altered the direction of impact during fights and protected against neck injury (Stanley, 1974).

The weapons and fighting techniques of the species in several groups of modern ungulates can be arranged in a series which may reflect the phylogenetic history of these characteristics (Fig. 10.7; Geist, 1966, 1978). Weapons probably evolved as short, pointed horns or tusks which served to enhance the impact of blows to the body of an opponent. Initially, these would have been delivered to the flanks from a lateral position with the two opponents standing parallel, a very common feature of mammalian fights. This method of fighting and degree of weapon development is seen in the pigs and giraffes among modern ungulates. In defence against such attacks, thick matts of hair or dermal shields have developed along the sides of the body, but the skull remains thin. Wild boars have lateral shields of connective tissue up to 6 cm thick which are augmented by guard hairs, tree gum and bits of stick. The weapons have become incorporated into a variety of displays and, in many cases, overt fighting is relatively rare (as in the goat antelope), because of the risk of injury to both

Figure 10.7 Postulated sequence for the evolution of weapons and fighting in bovids. Bottom row, the American goat antelope (*Oreamnos*). Middle row cattle (*Bos*). Top row, left-hand side, sheep (*Ovis*); right-hand side, antelope (*Antelope*). (Reproduced with permission from Geist, 1966).

participants.

An alternative method of defence against weapon attack to the flanks is to catch the blows with the head; in such cases, horns became enlarged and serve a double purpose of delivering and parrying blows to the sides of the body.

- Fighting boars catch an opponent's blows with their curved upper tusks while slashing at each other's sides with the lower tusks.
- Cattle jostle with heads together for a position from which blows to the flanks can be delivered.

- Muntjac males wrestle with each other using their small horns, struggling for an advantageous position from which to deliver attacks on the opponent's flanks with their sabre-like canines (Barette, 1977).

This form of fighting has given rise to two different patterns of front-to-front combat. In one line, horns (or teeth) became increasingly ornamented to ensure contact during twisting, wrestling fighting in a frontal position, as in antelopes; the ultimate target for attack is still the side of the body. In the other line, the weapons became massive but compact and are used to ram the opponent at high speed (as in muskox and mountain sheep). In this case, the head is the main target of attacks and thick skull bones evolved to protect both the striker and the stricken against the impact of such attacks (Geist, 1966).

10.4 OVERVIEW

10.4.1 Have our questions been answered?

(a) Origins

All kinds of structures which are well placed for making injurious contact with an opponent have evolved as weapons. Injury from these weapons elicits deep-seated protective responses, causing the rival to withdraw. Equipment used during predatory attack or counter attacks by prey provide the evolutionary origin of weapons in many species (stomatopods and horses for example, but digging (in beetles and ragworms and egg-laying structures (ants) have also evolved into weapons in some animals (page 254).

Agonistic displays have different evolutionary origins. Many are the byproduct of the context in which fights occur, arising from the locomotion and exploration that an opponent triggers (in crabs, for example, page 256), from symptoms of the physiological arousal that accompanies a fight (vocalizations in mammals, page 257), from protective responses (facial expressions in primates and canids, page 258) or from inhibited attack and escape (in gulls and lizards, page 258).

(b) Diversification of agonistic behaviour

Once agonistic displays have evolved, they tend to remain broadly constant over quite long periods of evolutionary time. The changes that do occur are small, and probably represent relatively minor changes in the causal mechanisms for fighting. They make the display more conspicuous and more effective at influencing the behaviour of a rival; the changes are often such as to conceal information about motivational state and intentions (in crabs and rivulin fish, for example, pages 262 and 263) but sometimes they have the opposite effect (in primates, page 268). Higher level manifestations of aggression are much more labile; the evolutionary history of overall aggressiveness, of territoriality and of dominance has been one of fairly rapid change in response to altered environmental circumstances (page 270).

(c) Diversification of weapons

During their subsequent evolution, weapons have become increasingly effective (sometimes larger, sometimes sharper, sometimes backed up by poisonous chemicals), as have the movements used to deploy them. In the social insects, the need to attack effectively without being injured has led to the evolution of poisons which can be used at a distance. In ungulates, the continual need for animals to protect themselves both against the impact of their own blows and against retaliation by their opponent led to the evolution of morphological protective structures (antlers to catch blows, shields to parry them and strong skulls to prevent impact injuries). Sophisticated fighting techniques evolved for jostling for a position from which blows can be safely and effectively delivered (page 271).

10.4.2 General principles

(a) Pre-adaptations and constraints in the evolution of agonistic behaviour

Some kinds of structures and behaviour patterns are, by their nature, quite readily converted into equipment used in fights. Thus hard structures on the ends of extremities and violent movements are a rich source of agonistic actions. Those that generate strong, conspicuous stimuli are also important, since they can draw on pre-existing behavioural dispositions to avoid certain stimuli. Pre-adaptation may also lie in the context in which the original behaviour patterns are shown. The unfamiliar aspect presented by a potential opponent induces exploratory behaviour; the stress of encountering a rival stimulates physiological responses; risk of injury generates protective responses; fear of counter-attack produces inhibited attacks and escape. So these kinds of behavioural response have often given rise to the agonistic postures which complement overt fighting. On the other hand, there are certain aspects of behaviour which have not served as an evolutionary source of agonistic displays. Behaviour patterns shown during parent–offspring encounters and during nest building commonly give rise to courtship displays (where the need is to elicit approach and not withdrawal) but have rarely evolved into agonistic displays.

Within different groups of animals certain categories of response are more likely than others to evolve as agonistic displays. The sensory capacities of the species concerned and the environment in which encounters take place impose fairly obvious constraints (page 39). The visual system of lizards predisposed them to evolve lateral displays (page 257). Fish rarely have specialized weapons, so colour patterns and fin postures that increase apparent size rather than weapon displays are the rule, and gulls with dark heads have evolved head postures in various social contexts (though which came first is not clear). Finally, feathers and fur which can be raised for the purpose of thermoregulation in birds and mammals also increase apparent size and so make good candidates for threat displays. The same fur and feathers get ruffled during fights, so grooming and preening movements are also common sources of displays.

(b) Selection pressures

The way that agonistic behaviour has evolved reflects the strength of the selection pressures acting on animals to come out on top in conflicts. We see this in the great diversity of structures which have been put to use in the evolution of weapons and in the fact that as soon as animals had the physical equipment to use force during competition they evolved the behavioural mechanisms to do so. Anemones, with no more than the most rudimentary nervous and muscular system, deploy their formidable weapons effectively to remove conspecific competitors.

The need to bring conflicts to a satisfactory conclusion, with as little risk as possible, has favoured animals that can settle disputes without overt fighting, and hence the multiple evolutionary origin of threat displays. Once fights are under way, the pressure is to coerce, immobilize or injure the opponent without being injured oneself; hence the spectacular evolution of offensive weapons and defensive structures in ungulates, and of the behaviour patterns by which these are deployed. This knife-edge aspect of agonistic behaviour probably puts behavioural variants at a disadvantage and so may explain the relative conservatism of agonistic behaviour patterns. The selective advantage of adapting levels of aggressiveness, territoriality and dominance to the immediate environment of a species and the relative ease with which such changes might be effected explains the evolutionary lability of these traits.

11
The behavioural ecology of animal conflict

11.1 INTRODUCTION

Animals invest time and energy in agonistic behaviour and can run serious risks of injury or even death from fighting. Evolutionary theory suggests that this behaviour should help to maximize the fitness of animals – to improve their chances of producing as many viable offspring as possible. Otherwise it should not persist. So the occurrence of agonistic behaviour raises the following questions.

1. Why do animals show agonistic behaviour?
2. In what circumstances is agonistic behaviour worthwhile?
3. Can animals win a conflict without engaging in potentially lethal fighting?

To answer these questions some biologists have used concepts from the theory of economics and studied agonistic behaviour in terms of cost/benefit analyses. A behaviour may be worthwhile if the benefits gained (such as acquiring mates) outweigh the costs (such as risking serious injury).

11.1.1 Costs and benefits of conflict

Animals often grow structures such as antlers or horns for use as weapons, and develop protective structures such as manes or fat deposits (pages 272). That this is costly is suggested by the fact that such structures are sometimes lost when they are no longer needed, for example the 'mane' of male bison and the antlers of deer (Lott, 1979). On a day-to-day basis the cost of maintaining a territory or defending a resource can also be high: 3–20% of the energy used each day by nectar-feeding territorial birds is spent in territorial defence (Gill, 1978) and male fruit bats defending harems increase their daily energy requirements by almost 50% compared with non-harem males (Morrison and Morrison, 1981).

Injury and death are obvious risks of fighting but displays and fights can also expose an animal to predators. The call of the tungara frog, for example, attracts frog-eating bats (Tuttle and Ryan, 1981). These frogs also burn up a lot of energy when they call (Bucher, Ryan and Bartholomew, 1981). When alone, they make a simple call that does not require much energy and that is difficult for the bat to detect, but when competing with other males they use a call that is more complex, takes more energy to produce and is more detectable, because the females find this type of call more attractive (Rand and Ryan, 1981). Apart from the risk of attracting predators, males on lekking grounds (see page 33) run the risk of losing body condition or even starving because of the need to stay on the territory and keep displaying. If they leave in order to feed they will miss out on attracting females (Vehrencamp and Bradbury, 1984).

As well as the costs, however, there are also substantial benefits to being aggressive. Individuals can thereby gain exclusive use of a resource such as a food source, or may win exclusive mating rights. Losers of fights gain either nothing or only partial access. Males on a lek, for example, compete amongst themselves for the best positions (Vehrencamp and Bradbury, 1984); males are not excluded, but those occupying inferior positions or who do not perform well fail to attract females (for example, indigo birds, Payne and Payne, 1977, frogs, Ryan, 1980).

The more aggressive an animal is, the more benefits it may gain (such as extra food). But if an animal is too aggressive it might face unacceptably high costs (such as serious injury) so the animal must weigh up the relative costs and benefits of its action and choose an optimum level of aggression (i.e. maximize the net benefits, Fig. 11.1). If the costs are too high and the benefits too low, avoiding a fight may be preferable to competing. In other cases it may be worth fighting vigorously for a valuable resource.

But there are problems with adopting a cost/benefit approach to animal behaviour. Most importantly, short-term benefits and costs may be relatively easy to measure but it is lifetime costs and benefits

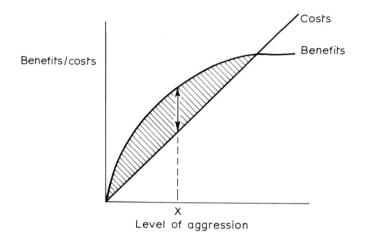

Figure 11.1 The costs and benefits of aggression. When an animal uses more aggression to get its own way, the benefits of doing so may increase, but probably not indefinitely. At the same time the costs of aggression also increase and may continue to do so. It is worth being aggressive only when the benefits outweigh the costs (shaded portion) and the optimal level of aggression to use is that which maximizes the differences between the benefits and costs (i.e. the net benefit, X).

that we need to know if we are to have an accurate picture of the function of agonistic behaviour. In addition, some costs and benefits are difficult to evaluate. For example, an animal may spend very little time and energy defending its offspring against a predator, but the possible costs in terms of injury or even death are very high.

11.2 STRATEGIES FOR RESOLVING CONFLICTS

When animals compete with one another the behaviour each adopts depends on what other individuals in the population do. The techniques of game theory, originally applied to human conflicts, have been used to investigate which behaviour is best for an individual to use in relation to what others are doing. Game theory treats evolution as a game in which the players use different patterns of behaviour. These behaviours are termed strategies but conscious decision-making is not necessarily implied. Many games try to answer the question of how aggressive an animal should be when fighting over a disputed resource. For example, in the hawk–dove game a hawk always fights fiercely either until the opponent retreats or until one or other combatant is seriously injured. A dove, however, tries to settle the dispute amicably by displaying rather than fighting and will retreat if attacked. Hawks always beat doves but if they fight another hawk they stand an equal chance of winning or being injured. If two doves fight, each has an equal probability of winning; neither is injured but they both spend a lot of time displaying (Table 11.1).

The best strategy to use depends on the resulting costs and benefits and the frequency of its use in a population. Thus if the benefits (of gaining the resource) exceed the costs (of injury), a hawk in a population of hawks will do better than a dove because the dove never reaps the benefits of winning. But if the costs are greater than the benefits, a dove in a population of hawks does better than a hawk because it never bears the costs of escalated fighting.

In the first case, where benefits exceed costs, the strategy of playing hawk is termed an evolutionarily stable strategy (ESS). Such a strategy cannot be outcompeted by any other strategy defined in the model that might invade the population (dove in this case). Where costs exceed benefits, however, neither hawk nor dove is an ESS. The evolutionarily stable strategy here is termed a mixed one: play hawk with probability p and dove with probability $p-1$. At this evolutionarily stable ratio the pay-offs in terms of costs and benefits of being a dove or a hawk are equal.

The first animal models of game theory were developed by Maynard Smith (1974 et seq.) whose ideas have spawned a great deal of literature, mainly in the form of more theoretical models. The early models were unrealistic in assuming that opponents in a fight were equal, that individuals reproduced asexually and that encounters between individuals were random. More recent models have tried to

Table 11.1 The hawk–dove game (after Maynard Smith, 1974).

Payoffs:	V to the winner of the game
	O to the loser of the game
Costs:	I is the cost of injury
	T is the cost of spending time displaying

Payoff to:	When fighting:	
	Hawk	Dove
Hawk	a. $0.5(V-I)$	b. V
Dove	c. 0	d. $0.5V-T$

a. Wins half the time, loses half the time.
b. Hawk always wins.
c. Dove never wins.
d. Always displays, and wins half the time.

deal with these problems (for example, Grafen, 1979; Fagen, 1980; Maynard Smith, 1981; to name a few). Much of the literature deals with the strategies used during fights but the techniques have been applied to other situations such as parent–offspring conflicts and mate choice strategies. Game theory as applied to the evolution of animal behaviour has been extensively reviewed recently (for example, Maynard Smith, 1982, 1984; Parker, 1984) and only a brief account of its contribution to our understanding of the adaptive significance of animal conflict can be given here.

11.2.1 Asymmetries in animal fights

Early game theory models assumed that the combatants in a fight were equal in all respects. Clearly this is not so and more recent models consider contests in which individuals differ. The most studied questions have been whether animals use these differences (asymmetries) to settle fights and to decide how much to escalate, and whether the way they fight conveys information about the asymmetries. The asymmetries between the combatants are of three possible types (Table 11.2; Maynard Smith and Parker, 1976).

1. Resource-holding potential. One combatant may be better able to fight and defend a resource than the other. A large, obviously superior opponent might quickly win a dispute without any escalation to physical contact being necessary.
2. Resource value. One combatant may value a resource more highly than does its opponent. Food may be more valuable to a hungry animal than to a satiated one, for example.
3. A third type of asymmetry is not related to resource-holding potential or to a payoff, being quite arbitrary. A convention such as 'if the opponents are otherwise equally matched, the owner of a territory wins, the intruder retreats' is of this type.

There are examples of each type of asymmetry being used to settle a dispute (Table 11.2). But in cases where the ownership convention is used, some sort of asymmetry in resource-holding potential or resource value is likely to be involved to some extent. For example, the owner may value the resource more than an intruder would. Thus among common shrews, territory owners win against weak opponents and when food is scarce. If both shrews regard themselves as residents the one experiencing the lowest food density is most likely to win contests (Barnard and Brown, 1984). Other asymmetries may often be involved in contests between territory owners and intruders. Thus owners are likely to be more familiar with the site and have reduced defence costs because they know the neighbours (Krebs,

Table 11.2   Asymmetries between combatants used to resolve fights.

Asymmetries	Animal	Cue or resource	Reference
In fighting ability	Beetles	*Size	Eberhard (1979)
	Toads	*Pitch of croak	Davies and Halliday (1978)
	Red deer	*Roaring rate	Clutton-Brock and Albon (1979)
	Buffalo	*Strength	Sinclair (1977)
In resource value	Hummingbirds	†Food abundance	Ewald (1985)
	Spider mites	†Possession of females	Potter (1981)
	Pygmy sunfish	†Hunger	Parker and Rubenstein (1981)
	Wagtails	†Feeding rate	Davies and Houston (1981)
	Shrews	†Hunger	Barnard and Brown (1984)
Non-correlated (owner wins)	Zebra spiders		Jacques and Dill (1980)
	Speckled wood butterflies		Davies (1978)
	Dark-eyed juncos		Yasukawa and Bick (1983)
	Funnel web spiders		Riechert (1984)
	Hamadryas baboons		Kummer et al. (1974)
	Fiddler crabs		Hyatt and Salmon (1978)

* Cue used to assess asymmetry
† Resource asymmetry

1982). In the long term, owners may be stronger than intruders because there may have been repeated contests over the resource in question with the strongest individual always winning and becoming the new owner (Leimar and Enquist, 1984).

Although asymmetries are usually involved in settling fights, if they are very weak and difficult to detect then the ownership convention alone may be used. In the funnel web spider *Agelenopsis aperta*, large individuals win against smaller ones if there is more than 10% difference in their size. If the difference is less than this, the owner of the disputed web is the victor (Maynard Smith and Riechert, 1984; Fig. 11.2). Unnecessary and potentially harmful escalation is thus avoided.

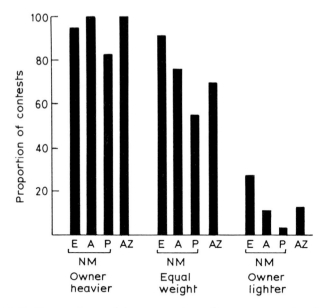

Figure 11.2 Comparison of the proportion of contests won by the territory owner in spider, *Agelenopsis aperta*, interactions. NM: New Mexico grassland; AZ: Arizona riparian; E: excellent quality site; A: average quality site; P: poor quality site. The outcome of the contests depends on the relative weights of the territory owner and the intruder, and on the quality of the site. (After Maynard Smith and Riechert, 1984).

11.2.2 Assessing asymmetries

An asymmetry in resource-holding potential such as size can sometimes be clearly assessed. Only large toads have low-pitched croaks, for example, so the pitch of a croak is a good indicator of a toad's size (Davies and Halliday, 1978). The value of a resource may also be obvious. A funnel web spider intruding on the web of another, can gain information about the quality of the web apparently without any overt behavioural cues being given by the web owner (Maynard Smith and Riechert, 1984).

An animal's ability to fight or the value of a defended resource may, however, be difficult to perceive. In such circumstances it might derive some benefit by signalling that it has a high resource-holding potential and so is likely to win a fight. Unnecessary escalation can then be avoided. But do individuals give false information, pretending they have a higher resource-holding potential or a more valuable resource than they really do? It would seem advantageous to lie, but there can be costs to doing so; and the more obvious the asymmetry involved, the harder and more costly (in terms of energy and time or risk of injury) it is likely to be to fake it. Conversely, cues will be reliable only if they cannot be faked (Zahavi, 1979). Where assessment is by a trial of strength, faking may be too costly: roaring contests in red deer stags, for example, are physically exhausting (Clutton-Brock and Albon, 1979).

11.2.3 Assessing intentions

We would expect evenly matched opponents to conceal their intentions (to attack or to retreat) during an agonistic encounter – there is no point in saying you are going to give up right at the beginning of the fight because your opponent might also be prepared to give up later on. Whether or not displays do provide information about an animal's intentions to attack or retreat has been much discussed recently (for example, Hinde, 1981; Caryl, 1982; Barlow and Rowell, 1984; Enquist, 1985).

Intentions are often concealed. For example, when male cichlids, *Nannacara anomala*, fight the patterns of behaviour, and their frequency, shown by the eventual winner and loser are identical except towards the end of the contest (Jakobsson *et al.*, 1979). However, there are a few examples (both fish: mouthbrooders and Siamese fighting fish) of the loser reliably signalling early on in the fight that he is the one who will lose (Bronstein, 1985a, b; Turner and Huntingford, 1986).

Whether or not an animal uses a display that might reveal its intentions about persisting or giving up the fight may be determined by the relative costs and benefits of that display. Displays can be time- or energy-consuming and an individual using them can face varying degrees of risk of retaliation so a combatant can afford to use high cost or high risk displays only if they are effective. In fulmars, for example, the effectiveness of a display is positively correlated with its cost in terms of risk of retaliation (Enquist, Plane and Röed, 1985; Fig. 11.3 and Table 11.3). Complex signals may be particularly effective (the varied songs of territorial birds may be better at deterring intruders than a simple song, Krebs, 1977) but complex, conspicuous displays also bring extra costs such as a higher predation risk (for example, calling frogs attract frog-eating bats, page 278).

Faking intentions may be possible within certain constraints (Maynard Smith, 1984), but the cost of lying may be high. A commitment to escalate may lead to serious injury if the cheat is physically unable to continue fighting and cheating will thus be limited. There are few intraspecific examples of animals faking signals. Mantis shrimps when they are moulting and unable to fight still

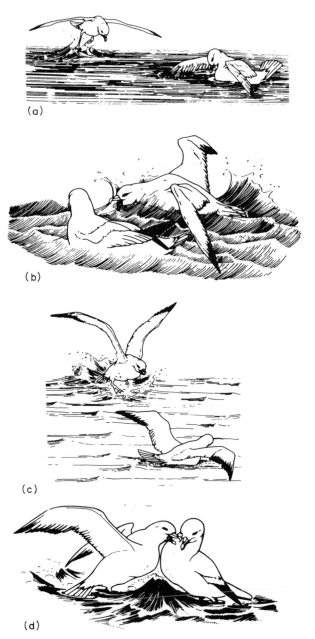

Figure 11.3 Agonistic displays of fulmars. (a) Challenge from the air (left bird) and wing raising combined with bend (right bird); (b) bill pointing (upper bird); (c) rushing (left bird) and wing display (right bird); (d) breast-to-breast; (a) to (d) represents a series of increasing cost (reproduced with permission from Enquist, Plane and Röed, 1985).

Table 11.3 Effectiveness and cost of fulmar's response to challengers that rush in an attempt to steal food (from Enquist, Plane and Röed, 1985).

Behaviour	N	Opponent giving up	Winning	Next action being dangerous	Present action being dangerous
			Probability of		
Rushing	55	0.818	0.891	0.200	0.200
Bill pointing	30	0.700	0.733	0.133	0.167
Bend	19	0.632	0.684	0.158	0.211
Leave	65	0	0.308	0	0.138
Significance (p)		0.001	0.001	0.001	N.S.

N.S. = not significant

use a display that suggests a readiness to fight (Steger and Caldwell, 1983).

11.2.4 Badges of status

While information about the opponent is acquired during a contest, there are a few examples, mainly in wintering bird flocks, of this information being clearly given prior to any fight (Fig. 11.4). Dominant Harris' and white-crowned sparrows have distinctive plumage markings and so status (probably age-related in these cases) may be signalled (Watt, 1986). Fugle *et al.* (1984) demonstrated that immatures and adult females painted to mimic dominant adult males were able to win contests with normal birds of their own age and sex and Rohwer (1985) produced cheats in Harris' sparrows by dying young birds to resemble adults.

Status signals may be advantageous because they obviate the need for agonistic encounters. But if it is so advantageous, why do not all individuals signal dominance whatever their true status? There are several possible reasons why status signals can exist without cheats invading and taking over the population (Balph, Balph and Romesberg, 1979; Maynard Smith, 1982; Rohwer, 1982; Fugle *et al.*, 1984; Studd and Robertson, 1985). In situations where individuals frequently encounter and recognize one another, cheats may be discovered. In

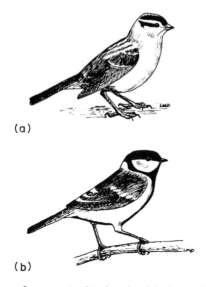

(a)

(b)

Figure 11.4 Badges of status in birds; the black head stripes in white-crowned sparrows (a) and the black chest stripe in great tits (b) are more marked in dominant birds.

addition, it may not pay some individuals to pretend to be dominant if (because of factors such as small size) they are not capable of living up to the demands of high status. Where encounters are most frequent between individuals of similar rank a cheat would often be involved in a contest with superior birds (Maynard Smith and Parker, 1976). Cheats may then be discovered or they may suffer costs as a result that outweigh the benefits of signalling dominance.

Finally, status may be signalled because both dominants and subordinates benefit from doing so. In a study of flock-feeding birds, Rohwer and Ewald (1981) suggested that high- and low-ranking Harris' sparrows coexist because the high-ranking individuals defend spaces in which subordinates feed and use the subordinates as food finders. The subordinates benefit because they suffer fewer indirect supplants when feeding close to dominants than when they feed next to other subordinates.

11.3 COSTS AND BENEFITS OF BEING DOMINANT

Within a group of animals there is usually some form of dominance hierarchy (page 46), with associated threats, supplanting and sometimes fights. The prevalence of these dominant–subordinate relationships suggests there are substantial benefits to having high rank when one lives in a group.

The question of whether high status bestows benefits has been extensively investigated in the past few decades but unfortunately the research has taken many different forms so that studies are not often directly comparable. The most frequent subjects have been primates and the most frequent possible benefit studied potential reproductive success. But reproductive success has been measured in several ways, such as the number of mounts, copulations or ejaculations achieved by a male. Dewsbury (1982) has recently reviewed many of these studies and has outlined several problems in interpreting them.

The main difficulty is that studies have tended to concentrate on short-term benefits of being dominant, often dealing with conflicts over one resource. Yet high status may increase an animal's fitness only in the long term, not as a result of every agonistic encounter. Consideration of lifetime benefits, and possible costs, of being dominant and, equally important, of being subordinate are necessary for a full understanding of the adaptive significance of dominance hierarchies.

11.3.1 The benefits for primates

The many studies of this group show that in some primates, in some circumstances, dominant animals derive some benefit from their high status. Most studies dealing specifically with the relationship between rank and reproductive success do reveal a positive effect (see reviews by Robinson, 1982; Chapais, 1983).

In a captive troop of bonnet macaques, for example, high-ranking males were the main companions of fertile females and the three top males obtained nearly three quarters of all the observed copulations with them (Samuels, Silk and Rodman, 1984). There were some changes of rank during this three-year study: males rising in rank increased the amount of time they spent exclusively with fertile high-ranking females and copulated more frequently than before. The reverse was seen in males dropping in rank (Fig. 11.5).

In baboons, male rank is correlated with mating success (Hausfater, 1975; Packer, 1979) but in the yellow baboon status accounts for only about 50% of the males' variance in mating success; in the olive baboon other factors such as experience and the ability to form coalitions also have an effect. Dominant male Japanese macaques have priority of access to females in oestrus, but this does not necessarily lead to a successful mating since females have a say in the matter and can reject even a high-ranking male (Takahata, 1982). Other studies have found no effect of rank on measures such as copulation frequency or ejaculation frequency in rhesus or Japanese macaques (for example, Wilson, 1981; Johnson, Modahl and Epton, 1982).

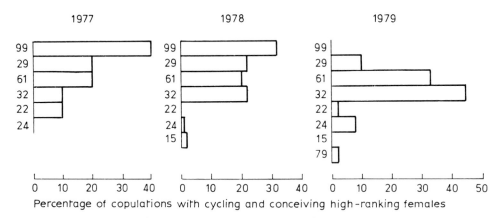

Figure 11.5 In bonnet macaques, access to fertile, high-ranking females depends on a male's rank, but an individual's rank can change. These graphs show the percentage of copulations with fertile (cycling and conceiving) high-ranking females. Each bar represents the percentage of all copulations with those females for which a single male was responsible. Males are listed in 1977 (descending) rank order. During 1978 the top two males were challenged and defeated and several changes in rank occurred. (Reproduced with permission from Samuels *et al.*, 1984).

Measuring reproductive success of females is less difficult and studies of females tend to show a positive effect of rank. High-ranking female baboons produce more surviving offspring, possibly because of less interference and harassment from other females (Dunbar and Dunbar, 1977). Survival of the offspring of high-ranking rhesus and bonnet macaques is also better than that of offspring of low-ranking mothers (Drickamer, 1974; Silk *et al.*, 1981). High-ranking females often also mate with high-ranking males and the offspring may benefit both genetically and through increased protection (Seyfarth, 1978). Not all studies have shown a positive advantage to having a high rank: for example, Gouzoules *et al.* (1982) found that there was no effect of rank on fecundity, survival of infants to one year or age at first parturition in female Japanese macaques.

Several studies have demonstrated benefits to high status other than reproductive success. During a water shortage, for example, high-ranking vervet females had priority of access to the preferred tree holes where water was still available and survived better than did low-ranking females (Wrangham, 1981).

Results of studies on feeding benefits have been equivocal. High-ranking chacma baboon males are the most likely to be the first consumers of any vertebrate prey caught, but subordinate males that have possession of such a food item can retain it (Hamilton and Busse,

1982). Among yellow baboons, high-ranking males can feed for longer and suffer less interference than can subordinates but there is no difference in the proportion of the time each spends feeding (Post, Hausfater and McCluskey, 1980). In toque monkeys, subordinates are likely to be displaced from food, and have less access than do dominants to sleeping sites and refuges from predators; growth and mortality are also related to dominance status in this species (Dittus, 1977, 1979). Dominant vervets also have improved access to preferred food items (Wrangham and Waterman, 1981). Body fat has been related to rank in vervets and the offspring of high-ranking Japanese macaques are heavier than those of low status females (Mori, 1979).

11.3.2 The benefits for other animals

(a) Rodents

Dominant male rodents sometimes benefit by siring most, sometimes all, litters or offspring produced within a group (for example, mice, Singleton and Hay, 1983; rats, Gärtner, Wnakel and Gaudszuhn, 1981; deer mice, Dewsbury, 1981), or obtaining a majority of the copulations or matings (mice, Oakeshott, 1974). But not all studies have demonstrated a clear relationship between dominance and reproductive success.

(b) Carnivores

Among those carnivores studied, dominance rank is generally related to mating success within a group (Dewsbury, 1982). Dwarf mongoose packs, for example, consist of a dominant pair, the dominant female producing nearly all the offspring. The dominant male has sole access to her at the start of her oestrus period (Rood, 1980). Survival of the offspring of high status individuals is also sometimes better than for subordinate offspring. In wild dogs, for example, dominant females kill the pups of others (Van Lawick, 1974).

Dominant carnivores also benefit when feeding. Male lions dine before the lionesses do (Eaton, 1974) and dominant spotted hyaenas in the Namib desert can exclude subordinates from small carcasses (Tilson and Hamilton, 1984).

(c) Ungulates

There is also good evidence from this group that high status is associated with certain benefits, the best data coming from the study of red deer on Rhum (an island off the west coast of Scotland,

Clutton-Brock *et al.*, 1982). Among red deer stags, winter dominance is correlated with success during the rut in getting and keeping a large harem. During the winter, high-ranking individuals get exclusive use of the best feeding areas which may contribute to their general fitness and reproductive success (Appleby, 1980). High ranking hinds benefit by producing more successful sons (but not daughters) than subordinate hinds, and also produce a higher ratio of sons to daughters than do subordinates (Clutton-Brock, Albon and Guinness, 1986).

(d) Birds

There are several studies of the benefits of dominance in birds. In

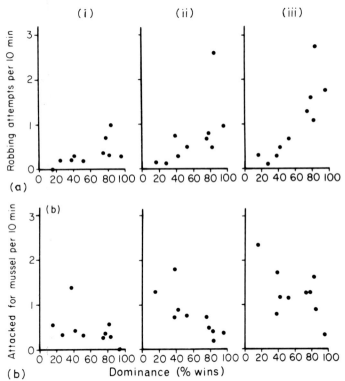

Figure 11.6 Dominant oystercatchers (a) attempt to steal more food and (b) are attacked less than are low-ranking birds. But the amount of this fighting also depends on the density of the birds on the mussel beds. (i) low density, less than 72 birds per hectare; (ii) intermediate density, 72–104 birds per hectare; (iii) high density, more than 104 birds per hectare. Boundaries were chosen such that an equal number of observations pertained to each class of bird density. (From Ens and Goss-Custard, 1984, reproduced with permission of the British Ecological Society.)

feeding flocks, dominant individuals often benefit by increasing their food intake (dark-eyed juncos, Baker *et al.*, 1981), sometimes by stealing (oystercatchers, Ens and Goss-Custard, 1984, Fig. 11.6), sometimes by occupying the best feeding areas to the exclusion of subordinates (oystercatchers, Goss-Custard *et al.*, 1984, herring gulls, Monaghan, 1980), and sometimes by occupying the safest areas in a flock (woodpigeons, Murton, Isaacson and Westwood, 1971) or nearest to cover (Schneider, 1984). The dominant individuals can also afford to take fewer risks, either because they are less hungry or because they are more capable of getting their food in a short space of time. The most dominant blue tits are the last to return to a feeding site after a predator has been seen (Hegner, 1985).

Dominance during the breeding season is often related to getting the best territories (page 296). But in some species, such as dunnocks, groups of birds can be seen on a territory with a dominance hierarchy among them. The most dominant male dunnock benefits by doing most, if not all, of the mating (Houston and Davies, 1985).

During the breeding season, a group of birds often consists of one breeding pair with one or more helpers (such as Florida scrub jays) or several pairs breeding communally (such as groove-billed anis). Among the latter the dominant pair produces most of the offspring that are reared (Vehrencamp, 1978).

(e) Other vertebrates

There have been very few studies of dominance in other vertebrates. Fish, and especially amphibians and reptiles, are often territorial or else they fight over resources rather than live in stable groups with dominance hierarchies. But where such hierarchies are evident then benefits can be seen. In Chapter 2 we saw how, among anemone fish, only the dominant male mates with the female in a group, the testicular development of subordinates being retarded (Fricke, 1979). Similarly, in schooling mollies dominant males mature earlier and do most of the mating compared with subordinates (Parzefal, 1969).

Laboratory studies also suggest that dominants benefit during feeding. In an experiment with juvenile firemouth cichlids, the dominant individual of a pair was able to feed on the most profitable patch of food, the subordinate had to feed elsewhere. Dominance status in these fish affects survival rates and growth as well (Hodapp and Frey, 1982).

(f) Invertebrates

Although the costs and benefits of winning fights, guarding resources

and being a superior competitor have been investigated for invertebrates, there are few studies of group dominance hierarchies reflecting
their scarcity in this group.

Dominant male cockroaches (*Nauphoeta cinerea*) seem to mate more
often than would be expected if mating were random. This is probably due to the female choosing a dominant to a subordinate partner,
which she can distinguish by smell, rather than a dominant getting
priority of access over a female (Breed *et al.*, 1980).

Among ants, the slave-making *Harpagoxenus americanus* forms
linear dominance hierarchies (Franks and Scovell, 1983). All the
workers are fed by slaves but dominant workers are fed more frequently and they also steal food by interrupting slaves feeding subordinate individuals. This extra feeding seems to affect reproduction
because ovarian development is correlated with rank. Dominant
workers also stay at home while subordinates undertake the dangerous task of scouting for slaves. The dominant workers do not have it
all their own way, however. The queen solicits food predominantly
from these high-ranking individuals, perhaps in an attempt to stop
them producing eggs.

11.3.3 When is high rank advantageous?

It is clear from past work, especially that on primates, that rank is only
one of many characteristics of an individual and of its social and
ecological environment that determine its overall fitness.

(a) Individual characteristics

Success in reproduction or gaining access to other resources may
depend on such factors as age, experience, physical fitness and intelligence. An ability to share, co-operate and form alliances and coalitions
can also be important, although they may be related to an individual's
rank anyway.

(b) Group characteristics

Acquiring a resource does not depend only on what the individual
does. Dominance hierarchies are about relations between individuals.
Males, for example, sometimes have to take into account the wishes of
females and females do not always choose a mate on the basis of
dominance alone. Female chimpanzees prefer males who are willing
to share food (Tutin, 1979).

Group size and composition also affect decision making. A dominant male primate may find it difficult to monitor and control mating

behaviour in a large troop and if other members of the group are his relatives, aggressive interactions may be undesirable. Even if a subordinate is not related, the costs of confrontation may outweigh the benefits of the action, for example if the dispute is over a minor item of food (Hamilton and Busse, 1982).

(c) Environment

Perhaps the most important factor determining an individual's success in life is the environment it finds itself in and the abundance and predictability of resources in that environment. An interesting study from this point of view is Whitten's (1983) research on female vervets in Kenya. Rank-related differences in feeding time and diet, timing of breeding and number of offspring produced were related to the distribution of food plants in the home ranges of two troops, high-ranking females being more successful only when their food was clumped, not when it was randomly distributed. Clumped food is defensible; if a food source is spread over a wide area, low-ranking animals can still take it without suffering interference from dominant individuals.

So high rank does not automatically bestow benefits on an individual. It may do so in the right social and ecological circumstances, which may occur frequently (such as clumped food sources) or only very rarely (such as a severe drought). Either way, having a high rank is an insurance when such conditions do occur; it is likely to be easier to deal with them if one is already dominant than if one has to try to become dominant in such situations.

Subordinates may enjoy fewer benefits but they may also suffer lower costs than dominants. For example, a dominant great tit or pied flycatcher has a higher resting metabolic rate and so needs relatively more food than a subordinate one (Røskaft et al., 1986). Dominant individuals may also lose body condition through defending a resource over long periods, and thus may drop in rank. The dominance status of yellow-rumped caciques (a South American bird) declines with age probably because they spend much of their time consorting with females instead of feeding (Robinson, 1986).

As we saw earlier, in certain circumstances being subordinate or showing some submissive behaviour is not necessarily a poor strategy for an individual. Dominance status may also be related to age so each individual may get an opportunity to become dominant at some time in its life. Subordinates may do as well as dominants or, in some cases, they may be making the best of a bad job. Low status may be environmentally or phenotypically determined and may be frequency-dependent. A small male's best strategy may be to avoid fights, a large male's best strategy may be to initiate them.

It is perhaps not surprising then that studies on rank-related success have come to different conclusions. Such success may not always be apparent. In particular, studies on laboratory or confined groups may be missing out on the right conditions for rank-related differences in behaviour to appear. The study of Japanese macaques by Gouzoules *et al.* (1982), for example, showed no effect of rank on female reproductive success, but this study was done on a captive, food-provisioned troop in which even subordinates might have obtained sufficient resources to breed successfully.

11.4 COSTS AND BENEFITS OF TERRITORIAL BEHAVIOUR

Many animals have a territory, which they defend against intruders. As a consequence, animals tend to be spaced out and have the exclusive use of resources. The resource being defended can be anything from a year-round food supply/living area to just enough physical space for a pair of adults and a nest or for a male to mate with a female. But any territory takes time and energy to defend so why do some animals own them when others seem to do without?

Why a particular species tends to be territorial and why it defends an area of a certain size and type depends very much on its lifestyle, especially its feeding ecology, which determines how easily or economically the animal can defend a resource (Brown, 1964).

11.4.1 Benefits of having a territory

Many studies have looked at the benefits of having a territory, although relatively few consider how well non-territorial individuals fare and whether they are at a disadvantage compared with territory owners. The main benefits of having a territory relate to an improved access to resources, which in turn may increase survival or reproductive output.

(a) Survival

One benefit of having a territory is indirect in that individuals or nests are well spaced out so it may be more difficult for predators to find them. Predators tend to concentrate their searches in areas where they have already found prey, so could easily find nests that are clumped. For instance, clay-coloured sparrow nests that are close together suffer more predation than those that are further apart (Knapton, 1979). Individuals, not just nests, benefit. Juvenile Atlantic salmon that have territories, for example, suffer three times less predation by brook trout than do non-territorial individuals (Symons, 1974). Spacing out may also decrease the likelihood of a disease spreading.

(b) Access to resources

One main type of territory is used for feeding and can be beneficial in several ways.

1. The food lasts longer if only one or a few animals are eating it. Some trees, such as holly and hawthorn, can bear fruit all winter when each is defended by a thrush; undefended trees are quickly stripped (Snow and Snow, 1984).
2. Secondly, if the food is renewable – nectar in flowers, for example – levels will be higher because it is depleted at a lower rate. For example, territorial golden-winged sunbirds lose 8% of their nectar each day to intruders but non-territorial individuals lose 36–46% (Gill and Wolf, 1979). So the amount or availability of food can be higher on than off territories. Territorial red-backed salamanders thus have a higher net rate of energy gain than do non-territorial individuals (Jaeger, Nishikawa and Barnard, 1983).
3. Thirdly, the abundance of food may fluctuate less within a territory than outside it. For example, pied wagtails that have a territory also have a less variable feeding rate than do flock-feeding wagtails, whose feeding rate can fall dangerously low (Davies and Houston, 1983). Their territories are large enough to ensure their survival on most days during the winter (Houston, McCleery and Davies, 1985). Territories can thus provide a long-term insurance against poor feeding conditions, which may be unpredictable in time. Wagtail feeding territories are most important during cold weather; salamander territories are needed most during dry weather when the salamanders' leaf-litter prey become scarce.

The second main type of territory is used to get and fertilize mates. At its most extreme, the lek, no resource is defended and males compete among themselves to attract females, with the best quality males attracting the most (Payne and Payne, 1977; Ryan, 1980).

Some territories are used for rearing offspring and a good quality one may both attract mates and enhance the success of the breeding attempt. The largest male bullfrogs have the best territories for eggs: these are sites with warm water for fast egg growth (but not too hot as this leads to developmental abnormalities) and have sparse submerged vegetation so that predation by leeches is low. The female bullfrogs prefer this sort of site (Howard 1978). For glaucous-winged gulls, the most important characteristic of a territory is its size. Because neighbours often cannibalize chicks, the larger the territory when food is scarce and chicks are most likely to wander, the better is the chicks' survival (Hunt and Hunt, 1976). Among insects, female aphids, *Pemphigus betae*, produce over 50% more progeny on territories on the basal parts of leaves than on those on distal parts (Whitham, 1986).

11.4.2 Costs of having a territory

Territory owners may get lots of benefits from having a territory but they have to pay some costs as well, costs that a non-territorial individual would not suffer.

1. First of all, an animal has to acquire a territory, which may involve a fight with a current owner. Once the animal has the territory it has to proclaim that this patch is now occupied and keep out intruders who have not yet realized this. For an animal that uses scent, this may mean frequent marking at the territory boundary. For birds it may mean more displays. Black-chinned and Anna's hummingbirds spend 17% of their time defending a territory when they first acquire one, whereas established birds spend only 7% of their time in defence (Copenhaver and Ewald, 1980).

2. Once the territory owner has become established some time and energy must still be put into displays, patrolling, and seeing off intruders. So less time and energy are available for other activities such as feeding and breeding. If the territory is not defended, intruders may deplete the owner's resources. Intruders can also interfere with the owner's mating or feeding. Red-backed salamanders lose some feeding time and also take smaller prey than usual when intruders are present, so overall the amount of food they can catch drops (Jaeger *et al.*, 1983).

3. Defending a territory may result in the owner being injured or even killed. Fights between red-backed salamanders often result in damage to chemosensory organs and so reduce the injured animal's ability to get food. However, fights are infrequent; intruders are usually put off by threats and displays (Jaeger *et al.*, 1983).

4. A territory that is just used for mating may not have enough food, shelter or other resources for the owner to stay there all the time without losing condition. Some Antarctic fur seals, for example, have landlocked territories and so have to leave them when they need to enter water to cool themselves down (McCann, 1980) and male impala that are defending females on their territories are seldom able to feed and so lose condition; another male may then take over the territory (Jarman, 1979).

11.4.3 Cost/benefit analyses

There are clearly several costs as well as benefits to owning a territory. Analysis of these costs and benefits helps us to understand many aspects of territoriality, such as why only some individuals own them, why individuals sometimes give them up and what determines the size of territories.

There are two main reasons why not all individuals have a territory.

1. When suitable areas are limited some individuals may be forced to use suboptimal areas, but the benefits of keeping a territory in such areas may be low compared with the costs so a particular individual may fare better without one. In some cases, no space for territories may be available and non-territorial individuals may have no choice in the matter.

2. Some individuals may be unable to defend a territory because the environment is unfavourable or because they are phenotypically unsuited to territory ownership. Again, the costs outweigh any possible benefits. Males of the rufous hummingbird are more manoeuverable, hence better suited to defending a territory than are the females, which are often non-territorial (Feinsinger and Chaplin, 1975; Kodric-Brown and Brown, 1978).

Although individuals of a given species are often usually either territorial or non-territorial, the factors determining costs and benefits of territories vary so at times the benefits will not be sufficient or the costs will be too high for a territory to be worth maintaining. Thus an owner may give up a territory if it does not contain enough food to meet the energy costs of both its daily activities and defence of the site (for example, Gill and Wolf, 1975; Carpenter and Macmillen, 1976; Davies and Houston, 1984). The owner may also give up its territory if the abundance of food or intruder pressure becomes too variable, fluctuating from very high to very low (Lima, 1984). Sometimes a territory is given up when it contains abundant resources, either because the territory owner no longer derives any benefits from keeping out competitors or because defence costs increase too much. For example, water striders cease to be territorial if supplied with abundant food (Wilcox and Ruckdeschel, 1982) and flocks of birds can force individual thrushes to give up their defence of fruit trees (Snow and Snow, 1984). Where defence costs are too high, a system of exclusive territories can change to a dominance hierarchy with priority of access (Craig and Douglas, 1986).

The same costs and benefits that determine whether or not an animal owns a territory also determine how large that territory will be. It must be large enough to provide sufficient food or other resources, but must not be so large that it is too costly to defend. Recent theoretical models (for references and a review see Schoener, 1983b) predict a certain optimal size that provides the maximum benefits less costs. However, the optimal size predicted depends on the initial assumptions of the models and they do not have a general applicability; their value is mainly restricted to the animals and circumstances for which the models were made. Experimental studies investigating the relationship between resource abundance, the costs of defence and territory size are perhaps more fruitful, especially where investment in defence is measured as a continuum rather than as a simple presence or absence.

Food is a particularly important resource for an animal. If food (particularly high quality or preferred types of food) is scarce an animal may need a large territory to find enough for its needs (see for example, Jones and Norman 1986). If food is abundant a small area may provide sufficient food without the animal having to spend time and energy maintaining a large territory.

Many studies, both experimental and observational, have looked to see if territory size is related to food abundance and if it can be changed

Table 11.4   Territory size and food resources. 0 indicates no relationship, − a negative relationship, and + a positive one.

Species	Relationship	Reference
Limpets	−	Stimpson (1973)
Salmon	0	Symons (1971)
Salmon	−	Dill et al. (1981)
Brook char	+	McNicol and Noakes (1984)
Trout	−	Slaney and Northcote (1974)
Reef fish female	+	Ebersole (1980)
Reef fish male	0	Ebersole (1980)
Lizards	−	Simon (1975)
Red grouse	0, −	Miller et al. (1970)
Great tit	0	Krebs (1971)
Blue tit	−	Krebs (1971)
Towhee	0	Franzblau and Collins (1980)
Nuthatch	−	Enoksson and Nilsson (1983)
Longspurs	−	Seastedt and Maclean (1979)
Hummingbirds	−	Gass et al. (1976)
Belted kingfisher	− (non breeding)	Davis (1982)
Belted kingfisher	0 (breeding)	Davis (1982)
Squirrels	−	Smith (1968)

by manipulating the food supply. In a variety of animals the size of territories is inversely related to natural food abundance or decreases if the food supply is supplemented. Territory sizes do not always change when extra food is provided or the natural food supply is abundant, however (Table 11.4). There are other important factors in deciding the best size for a territory.

1. If a territory is used just for feeding, then its size is likely to depend on food abundance. But if it is used for breeding then the territory owners may not be able to spare the time and energy to defend a larger territory if food is scarce. Conversely, food may be more abundant during the nestling-rearing period than when the territory was established but a large territory may still be needed to maintain a high internest distance which protects against predation from other species and from cannibalistic conspecifics (Burger, 1981).

2. Extra food may be irrelevant to the animal's needs. Some animals can make use of all the food they can get (for example if they are still growing). Others need only a fixed amount of food and optimize feeding time rather than food gain. The former may benefit from a larger territory with more food, the latter may not (Schoener, 1983b).

3. Food abundance may be important to individuals establishing territories, but not to those well established. Salmon do not change their territory size when food is increased (Symons, 1974) except when they are setting up the territory (Dill et al., 1981).

4. Resource abundance tends to fluctuate over time and space. So a territory may need to be large enough to ensure that sufficient food is available in the very long term (Maclean and Seastedt, 1979; Getty, 1979; Rothstein, 1979) or to minimize the probability of the owner starving (Lima, 1984). Such a territory may be unaffected by minor short-term changes in food abundance.

5. If females choose males that defend large areas, then, over many generations, there will be positive selection for larger and larger territories (Rothstein, 1979; Hixon, 1980).

6. Although it is generally considered that a territory should be of such a size as to provide just sufficient resources for an animals particular needs (e.g. Brown, 1964; Wilson, 1975), Verner (1977) suggested that an animal might defend, out of spite, a territory larger than was necessary for its own requirements. But this is unlikely to happen, except in special circumstances such as small and disjunct populations, since the spiteful territory owner will bear all the costs of defending this large territory but does not gain anything directly. The neighbouring territory owners, on the other hand, have no extra costs but may gain from the reduction in competition. Spiteful behaviour has not been demonstrated and it seems to be unnecessary as an

explanation for animals apparently having too large a territory. Other explanations, such as bet-hedging in a variable environment and sexual selection are probably adequate (Davies and Houston, 1984, and inclusive references).

There are two reasons why an animal might decrease the size of its territory when food is abundant (Myers, Connors and Pitelka, 1979). Firstly, the territory owner may be able to assess food abundance directly and 'decide' that a smaller area will be sufficient for its needs (for example, nectarivores, Gass et al., 1976). It can then have the benefit of plenty of food without having the extra costs of patrolling and defending a large area. Secondly, an abundant source of food may attract competitors who might want to set up their own territories in the area or to sneak in and steal some of the food. If the number of competitors increases, the cost of defending a territory of a given size will also increase and so it may be worth while giving up part of the territory and keeping a more economically defendable portion; or the territory owner may be unable to keep up defence of its territory (for example, tits, Krebs, 1971).

Sanderling in winter set up feeding territories, the size of which varies inversely with prey density (Myers et al., 1979). But intruders, non-territorial birds which feed in flocks most of the time, are attracted to the areas with abundant food; if the interaction between prey density and intruder density is controlled statistically, territory size is no longer related to prey density. So here territory size is determined by intruder pressure. In western gulls, too, the rate of intrusion is the main cause of variation in territory size (Ewald, Hunt and Warner, 1980). A decrease in territory size related to an increase in the number of competitors has also been seen in fish (juvenile brook char, McNicol and Noakes, 1984).

We have seen in this section that a territory owner is in conflict with other individuals wanting to use the territory and its resources and so finds that ownership brings costs as well as benefits. Experimental studies investigating the relationship between food abundance, intruder pressure and territory size suggest a compromise is made between the costs and benefits of having a territory. Rufous hummingbirds adjust the size of the territory daily to one which provides the fastest rate of weight gain attained at any stable territory size, suggesting the birds are striving to maximize their energy intake and providing an example of what the territory owner considers to be a territory of optimal size (Carpenter, Paton and Hixon, 1983; Fig. 11.7).

Other field studies also suggest a compromise is made. If western gulls have too small a territory, cannibalism of their chicks increases,

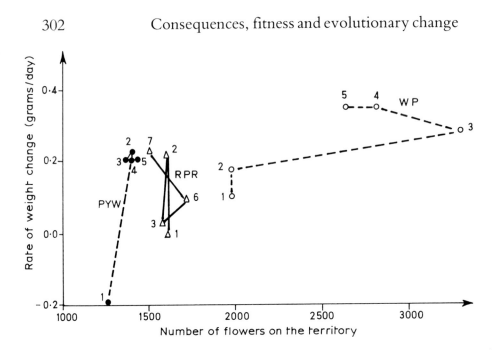

Figure 11.7 Rufous hummingbirds adjust the size of their feeding territories from day to day and thus the rate at which they put on weight prior to migration. The graph shows the daily weight changes of hummingbirds in relation to the number of flowers on their territories. Numerals by data points indicate successive days on the territory, beginning with the first day of intensive observations and ending on the evening before the birds migrated. Capital letters refer to the individual colour rings of the three birds. (After Carpenter *et al.*, 1983).

but too large a territory might make them more vulnerable to costs such as predation (Hunt and Hunt, 1976). Hummingbirds may compromise between flower density and costs of defence (Kodric-Brown and Brown, 1978). Travel costs may influence the optimal territory size of animals having to return to a particular site such as a nest (Andersson, 1978). A pied wagtail's territory is large enough to provide enough food most of the time; a larger one would need more defending but would not increase the number of days when feeding is good (Houston *et al.*, 1985). Thus animals appear to adjust their territorial behaviour according to both costs and benefits.

11.5 CONFLICT BETWEEN THE SEXES

11.5.1 Mating systems

There are several social contexts in which mating and rearing offspring occur. These mating systems are both the end product of competition between rivals and the arena in which conflict within a pair of breeding animals takes place. Some species are monogamous (one male mates with one female); others are polygamous (an individual of one sex (usually the male) mating with more than one individual of the opposite sex); yet other species are basically monogamous but individuals will mate with others promiscuously if the opportunity presents itself. These mating systems are variable both within and between species and the way in which they have evolved probably also varies between groups.

Males and females may or may not agree about the mating system they use. Conflicts between the sexes arise over three main questions. How many individuals should an animal mate with? What type of individual should he or she mate with? How much care should be given to the offspring? The answers to these questions are not necessarily the same for a male and a female mating with each other. The conflict arises because of the differential production of gametes between the sexes (Bateman, 1948; Trivers, 1972). Females produce relatively few, but large, eggs. The male produces many small spermatozoa and potentially could fertilize the eggs of many females. Females, however, would not usually benefit from mating with more than one male. Because males are therefore inclined to be polygamous or promiscuous it is less advantageous for them to commit themselves to mating exclusively with and rearing the offspring of a single female.

11.5.2 Conflict over the number of mates

Does it matter to a female if her mate copulates with other females? If the male is only occasionally promiscuous and still invests his full share in the care of the offspring, or if the female can rear the offspring by herself such behaviour may not harm the female's interests but it may do so if she produces fewer offspring without the full help of the male. This problem has been studied in pied flycatchers (Alatalo, Lundberg and Stählbrandt, 1982). Some males are monogamous, others are polygamous and have a primary and a secondary female. Secondary females get less help in chick feeding than do primary females and, consequently, they raise fewer young than if they were paired with a monogamous male, and these chicks are also smaller and lighter in weight. So there is no advantage in a female mating with an already paired male whereas the male does increase his reproductive output by being polygamous. Alatalo *et al.* (1982) have suggested that the polygamous males deceive females into becoming secondary mates.

Females mating with more than one male seem to be less frequent than male polygamy, presumably because females can generally produce fewer eggs than males can sperm. But there are situations when multiple matings by normally monogamous females do occur, either because another male forces them to copulate or because they actively solicit copulations from other males. Multiple matings may benefit a female if (1) there is a possibility that her previous mate was completely or almost infertile; (2) mixed paternity of the offspring could promote their survival and increase their fitness, especially in unpredictable environments; (3) her previous mate was inferior in quality to another prospective mate.

There is some evidence of paired females producing offspring not sired just by their own partners and also some of females actually soliciting copulations from other males (rooks, Røskaft, 1983; swallows, Møller, 1985). Among dunnocks, when there are two males on a territory, the female encourages both males to mate and benefits by having both males help feed the nestlings. If only the dominant male mates with her, the subordinate male not only does not feed the brood but may actually destroy the eggs. Here multiple matings allow the female to produce more offspring than would otherwise be possible, but the number of offspring produced by the dominant male is reduced so he tries to prevent such copulations from occurring (see Houston and Davies, 1985; Davies, 1986, Davies and Houston, 1986, and references therein).

11.5.3 Mate guarding

Male promiscuity and mate guarding by males (to prevent the female copulating with another male, whether or not the female is a willing partner) have been well documented for birds (for example, bank swallows, Beecher and Beecher, 1979; magpies, Birkhead, 1982; mountain bluebirds, Power and Doner, 1980; starlings, Power, Litovich and Lombardo, 1981; rooks, Røskaft, 1983; dunnocks, Fig. 11.8(a), (b), to name a few). The guarding frequently takes the form of males keeping close to, and following, their mates especially during the few days when the female is fertile. Females can, however, escape the attentions of the male, as in the case of the dunnock. The dominant male tries to get round this by removing sperm from the female's cloaca and copulating again himself (Davies, 1983).

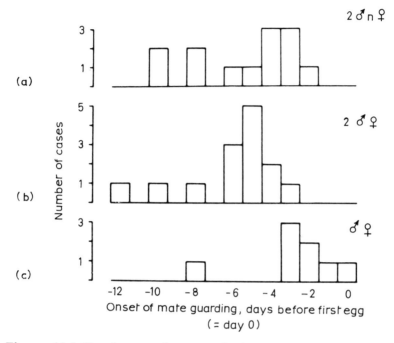

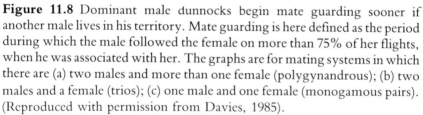

Onset of mate guarding, days before first egg
(= day 0)

Figure 11.8 Dominant male dunnocks begin mate guarding sooner if another male lives in his territory. Mate guarding is here defined as the period during which the male followed the female on more than 75% of her flights, when he was associated with her. The graphs are for mating systems in which there are (a) two males and more than one female (polygynandrous); (b) two males and a female (trios); (c) one male and one female (monogamous pairs). (Reproduced with permission from Davies, 1985).

In other groups of animals, guarding of females by males can take other forms, such as the keeping of harems in mammals. In some insects, guarding occurs after copulation, the male often keeping hold of the female or just staying close to her to reduce the chances of the female remating. In some crustacea such as *Gammarus pulex*, males guard moulting females before copulating to ensure they are the first to mate with her (Grafen and Ridley, 1983).

However, mate guarding is by no means always a result of females desiring other mates and thus coming into conflict with the male. Mate guarding can be beneficial to both male and female by reducing unwanted interference from other males trying to court or force copulations onto the female. In the dragonfly *Pachydiplax longipennis*, for example, females that are guarded can lay eggs without being disturbed for four times as long as can unguarded females (Sherman, 1983).

11.5.4 Conflict over the quality of mates

An individual may have genes that make it a successful competitor but, usually, not all of those genes can be passed on to its offspring. The best that can be done is for it to choose a high quality mate to complement its own genes. An unsuccessful animal may improve its reproductive success by choosing a mate of superior quality (Hamilton, 1964). However, a poorer quality mate than oneself is likely to be undesirable, unless one is a male who can mate several times and on average can mate with good quality females. This leads to another area of conflict between the sexes. The female that a male chooses may not see him as a desirable mate, and vice versa.

Females can exercise some choice in the matter, preferring males with high status, males who share food, and so on. Female choice of males may sometimes be superfluous in that males who are capable of keeping them in a harem or otherwise guarding them are most likely to be the best quality males around and therefore the best mates anyway, and in some cases the females may choose a good territory and incidentally get a good male. For instance, bullfrog females mate with large males which have the best territories for ensuring the survival of the eggs (page 296).

Males sometimes force females to copulate with them, thereby acquiring a new, additional or better partner for a breeding attempt, or, by being promiscuous, fertilizing eggs without bearing the full costs of parental care. In green-winged teal, forced copulations seem to be associated with attempts by males to make new pair bonds, for example if their own mate has died. But females that are already paired resist the advances of these males (McKinney and Stolen, 1982). Forced copulations also occur among other waterfowl such as mallard, especially when the paired female is at her most fertile time (Cheng, Burns and McKinney, 1982). Sometimes, however, they result in the female being drowned by large numbers of males.

The other group in which forced copulations are well documented are the scorpion flies (Thornhill, 1981). Males of the genus *Panorpa* mate in three different ways. Males can attract females to a salivary mass or to a dead arthropod which the females eat, but some males just grasp passing females and forcibly copulate with them without offering any food. Males fight over the prey that they use as nuptial offerings and it is the males that fail to get any prey that try to force matings on passing females. But females never initiate interactions with these males and interact less with them than with males who have food. Consequently, their copulation success is much lower than that of other males. The nuptial offering is probably important in increasing fecundity and survival of the female, hence it is in her interest to choose males who can provide some sort of food.

11.5.5 Conflict over parental care

Parents come into conflict with each other and with their offspring. Which parent should look after the children? How much investment should be made in them?

Each parent could increase its reproductive fitness by deserting and leaving its mate to look after one lot of offspring, then finding another mate to rear some more (Trivers, 1972; Dawkins, 1976; Parker, 1979). But why is it sometimes the female, sometimes the male, sometimes both parents and sometimes neither that care for their young? Research on these questions is still largely at the theoretical modelling stage (Knowlton, 1982; Vehrencamp and Bradbury, 1984), but some general comments can be made.

When fertilization is external, it is often the male that shows parental care; when fertilization is internal the female generally does. Dawkins and Carlisle (1976) suggested that the sex that deserts is the one that is first free to do so, for example the male if the female retains the eggs inside her body. Males may also have more confidence in their paternity of the offspring during external fertilization, so may be more prepared to look after them. Among mammals, however, only the female can feed the young offspring, so her mate rarely shows parental care. Other factors are likely to be important in determining whether or not a parent deserts. There may be no point in deserting, for example, if the deserter cannot get another mate or if the deserted partner also deserts in order to re-mate (Ridley, 1978; Maynard Smith, 1982).

Sometimes a parent will not desert because both parents are needed to rear the offspring successfully. In birds, the eggs require a lot of investment in time and energy to keep them warm so both parents often incubate whereas in fish, guarding involves relatively little investment and one parent can do the job. Similarly, altricial bird nestlings need more care and are more likely to have both parents caring for them than are precocial nestlings. The energy cost of egg production may preclude the female from rearing the offspring as well so both parents may be needed as is often the case in birds. The environment is also important: parental care is more likely to be needed in harsh or unpredictable conditions (Bayliss, 1981).

Once the parents have decided who is going to look after the offspring there is still the problem of how much time and energy to invest and what risks are worth taking during their care. Even if both parents share parental duties, they can still disagree about how much each should do, male birds may do less incubation or nestling feeding, for example (Hails and Bryant, 1979) but more defence (Shields, 1984).

The amount of time and energy invested, and the level of risk taken, in any one reproductive effort is thought to depend on the value of the current breeding attempt and on the future reproductive value of the parent (Williams, 1966; Trivers, 1974; Dawkins, 1976). A brood increases in value as it ages because it will have an increasing probability of surviving to reproductive age and so is worth a larger investment. Large broods are also more valuable than small ones. There are now many studies, especially of songbirds, that demonstrate an increased level of parental care, usually defence against predators/ conspecifics, with increasing value (such as size and age) of a brood (Ricklefs, 1977; Andersson, Wicklund and Rundgren, 1980; Grieg-Smith, 1980; Shields, 1984). Also, in fish, the readiness of a male three-spined stickleback to defend its nest also depends on the number and age of the eggs (Pressley, 1981).

As the parent ages the total number of offspring that it can produce in the future declines so it is better for an old parent to invest in a current brood than to save for the future. The reproductive values of the brood and the parent are closely linked. As animals get older they may produce larger broods, so current investment should increase for both reasons. A parent that invests heavily in a current breeding attempt may also thereby decrease its ability to reproduce in the future (for example, double-brooded female house martins have poorer chances of survival than do single-brooded females, Bryant, 1979). But the level of care and defence of the brood should also depend on how effective these are in enhancing the growth and survival of the brood (Sibly and Calow, 1983).

Few studies deal with the effect of the future reproductive value of the parent on its level of parental care. Californian gulls aged 11–18 years invest more time in collecting food for their chicks and in defending them from conspecific predators than do younger birds (Pugesek, 1983). Also, female lace bugs, *Gargaphia solani*, defend their young more aggressively as both they and their offspring grow older (Tallamy, 1982).

When two or more individuals co-operate to rear the brood, several factors such as the value of the brood and what other members of the group are doing may determine how much work each one does. In dunnocks, for example (Houston and Davies, 1985), monogamous males and females work equally hard at feeding the brood, but when the female is part of a trio with an alpha (dominant) male and a beta (subordinate) male, the alpha male feeds the nestlings less than he would if he were monogamous and the beta male provisions less than the alpha male. This reflects the shared paternity of the brood which is thus less valuable to the males, and particularly to the beta male, who would have fathered fewer chicks than the alpha male.

11.6 CONFLICT WITHIN THE FAMILY

11.6.1 Parents and offspring

Young animals try to get as much individual attention from their parents as they can. This brings them into conflict with their parents and with their siblings. Following Trivers (1974) model of parent–offspring interactions, there has been a series of papers concerned with the theoretical aspect of this conflict (see Parker, 1985, and Lazarus and Inglis, 1986, and references therein) but there are few non–theoretical studies. Nestling birds may try to get more food by begging more frequently but this may not improve the parents' reproductive success (Stamps *et al.*, 1985). Conflict may be particularly intense when a young animal becomes independent of its parent(s) or when a younger sibling is born and the parents consequently lower or terminate their investment in it. For example, among some primates the parents threaten and attack their mature offspring, leading to its emigration from the group (Kleiman, 1979, Lyon, Goldman and Hoage, 1985). Aggression toward offspring may also help socialize young animals by punishing unacceptable behaviour (Bernstein and Ehardt 1986).

11.6.2 Sibling rivalry

Siblings often compete for resources among themselves and this can involve them fighting and even killing each other. Nestling birds sometimes attack and kill younger siblings (page 34); in some species this is most common when food is scarce. An individual nestling may thereby increase its own inclusive fitness, for example if there is insufficient food for the whole brood (O'Connor, 1978). Some young animals, in such diverse groups as beetles and sharks, not only kill but also eat siblings, thus both removing competitors and gaining a meal (Table 11.5).

11.7 INFANTICIDE AND CANNIBALISM

Infanticide is a well-documented phenomenon among a range of animal groups (page 43; Table 11.5). Although such killings are sometimes the result of non-adaptive pathological conditions (Troisi and D'Amato, 1984), there are several adaptive explanations for the

Table 11.5 The adaptive significance of infanticide and cannibalism.

1. Sexual selection – killing to gain a reproductive advantage	Wasps (*Polistes*) – queens invade combs of rivals and eat their broods.	Kasuya *et al.* (1980)
	Arrow poison frogs – males eat eggs being looked after by potential mates.	Weygoldt (1980)
	Barn swallows – males destroy chicks of potential mates.	Crook and Shields (1985)
	Lions – males kill cubs when they take over a pride.	Bertram (1975)
	Redtail monkeys – male kills infants when he takes over a group.	Struhsaker (1977)
	Mice – males kill litters of potential mates.	vom Saal and Howard (1982)
	Collared lemmings – females remove litters of competitors for nest sites.	Mallory and Brooks (1978)
2. Exploitation – using conspecifics as food	Scorpions – eat immatures.	Polis (1980)
	Red land crabs – eat immatures when protein is scarce.	
	Ants – eat war casualties.	Gerard *et al.* (in press)
	Beetles – eat their relatives; the clutch may thereby increase their average inclusive fitness.	Eickwort (1973)
	Tadpoles – feed on other tadpoles.	Crump (1983)
	Perch – feed on younger age classes.	Kipling (1983)
	Herring gulls – adults eat chicks.	Parsons (1971)
	Belding's ground squirrels – males use pups as food.	Sherman (1981)
3. Resource competition – killing to remove competitors	Tenebrionid beetles – eat competitors.	Tschinkel (1978)
	Parasitic wasps – larvae kill other larvae present in the same host.	Fisher (1970)
	Sharks – eat their siblings in the uterus.	Wourms (1977)

Table 11.5   (*contd.*)

	Tadpoles – eat competitors.	Crump (1983)
	Buzzards – siblings are killed when food is scarce.	Mebs (1964)
	Boobies – alpha-sib pushes subordinate one out of nest where it is left to die when food is scarce.	Newton (1978)
	Wild dogs – dominant female kills pups of subordinate thus removing competition for food for her own pups.	Van Lawick (1974)
4. Parental manipulation – killing offspring	Social insects eat their own brood when food is scarce.	Wilson (1971)
	Coccinelid beetles – females provide eggs as a food for their larvae.	Kawai (1978)
	Arrow poison frogs – female provides eggs for larvae.	Weygoldt (1980)
	Sticklebacks – males use their eggs as food.	Van den Assem (1967)
	Black eagle – two eggs are laid as an insurance against one being infertile; if both hatch one chick is killed by its sib.	Gargett (1978)
	Wood storks – desert offspring when hunting is too difficult.	Kahl (1964)
	Lions – mothers abandon litters of one.	Rudnai (1973)

occurrence of infanticide. There is considerable controversy, however, about the significance of infanticide in non-human primates (Hrdy, 1979; Sherman, 1981; Hausfater and Hrdy, 1984).

1. Sexual selection. Takeovers of groups such as prides of lions and breeding groups of primates sometimes involve the killing of infants by the incoming male(s). The new male benefits because the females come into oestrous sooner than they would otherwise have done (Bertram, 1975).

2. Exploitation. Infants are often treated as food and can thus help increase the fitness of the cannibalistic individual. For example, the red

land crab becomes cannibalistic on the young stages when the available food is poor in protein (Wolcott and Wolcott, 1984).

3. Competition for resources. Competition between young animals can end with some of them being killed and even eaten. The survivors can increase their fitness as a result. Cannibalism in tenebrionid beetles results in better larval growth rates and higher fecundity (Tschinkel, 1978).

4. Parental manipulation. Parents sometimes kill their own offspring, particularly if conditions for rearing them are poor. Thus wood storks desert their chicks during sustained rain, which prevents them hunting (Kahl, 1964).

11.8 ALTERNATIVE STRATEGIES

Throughout this chapter and in Chapter 2, we have seen that agonistic behaviour can bring benefits to the victor of contests but also that there are other ways of being successful. In wagtails there are territory owners and territory satellites, in Harris' sparrows dominant 'shepherds' and subordinate food finders ('sheep'), in frogs and crickets calling males and silent males, in bluegill sunfish some individuals look after their offspring while others desert them, in ungulates there are males who defend females and males who sneak copulations, in scorpion flies there are males who court females and males who force females to mate.

Different strategies may be played in two ways. An individual may use more than one strategy, either opportunistically or sequentially during its life. For example, young males may be sneaks until they are old and large enough to own territories and females. Alternatively, the strategies may be played by different individuals. (See Davies, 1982; Dunbar, 1982; Barnard, 1984, for reviews.)

Firstly, an individual may just do the best it can in the circumstances; for example, small males cannot win fights against large males so they may as well be scroungers. In the bee *Centris pallida*, large males patrol for emerging females and will fight with other males for them, the larger male winning. Small males hover around nearby vegetation and wait for females that did not mate on emergence. The payoffs may be unequal, but the strategy of waiting may be the best for the individual in those circumstances (Alcock, 1979).

Secondly, each of two alternative strategies may be as good as the other. Dominant yellow warblers have good territories but they suffer high territorial costs, whereas low-ranking individuals have poorer territories and lower costs. Yet in terms of nestling production they do just as well as the dominant birds (Studd and Robertson, 1985). Alternatively, one strategy may be best in a particular situation. For example, in some species of fig wasps the males are dimorphic, being either winged or wingless (Hamilton, 1979). When male density is low, winged males find females easily without engaging in costly fights. But when males are abundant and so most females are mated, wingless males which mate with all the females on the home fig are at an advantage.

Which strategy is used can depend on the environment, both physical and social, the animal's development or its genes, or on learning, (for a discussion of how alternative tactics can arise, see Caro and Bateson, 1986).

1. The environment may limit the strategy used in either the long or the short term. Speckled wood butterflies, for example, patrol for females in cloudy conditions but guard patches of sunlight when these are available (Davies, 1978) and cactus flies change from being territorial to searching for females when it is windy and females seek refuges (Mangan, 1979).
2. The strategy used in a social group may depend on its frequency and on the size of the group. Scroungers, who usurp a resource obtained by others, the producers, need to have some producers in the group. But if scroungers are outnumbered by producers they may do less well (Barnard and Sibly, 1981). In house sparrow flocks, scroungers seem to be less efficient foragers than are the producers, and they also tend to do poorly in agonistic encounters.
3. Phenotypic variations in strategies are common, small males scrounging on large males. Whether a scorpion fly courts females with a gift of food or forces her to copulate depends on his size. The strategies do not have equal payoffs, the large, courting males gaining more copulations and suffering less predation from spiders (Thornhill, 1981). Such phenotypic variations are sometimes a result of conditions during development. In the spider *Nephila clavipes*, food intake affects size and development time: large males fight over females ready to mate, small males stay longer with females, perhaps waiting for them to mature (Vollrath, 1980). Some phenotypic variation is genetic in origin. For example, ruffs (wading birds) can have dark or light plumage, the dark individuals being dominant on the leks with the light males being satellites to the dark males (van Rhijn, 1973). Some variation, however, is simply due to animals changing in size and appearance as they mature.
4. Learning can also influence the choice of strategy. The distribution of animals in a population depends on the behaviour of other individuals (page 230). Subordinate ducks distribute themselves between patches of food according to the relative profitability of the patches and so they do as well as each other, although they do not fare as well

as the dominants who take most of the food at each site (Harper, 1982).

11.9 CONFLICT IN SOCIAL GROUPS

We have seen so far in this chapter that there are many advantages to animals in winning fights. But despite the basically selfish behaviour of individuals trying to increase their own fitness, co-operation between individuals is common throughout the animal kingdom. Primates help each other to win fights (page 215); many animals warn each other of approaching danger; Cape hunting dogs share their food; and there are many birds and mammals in which some individuals help or share in the breeding (Emlen, 1984).

The individuals doing the helping apparently contribute to someone else's fitness but not to their own. They may even put themselves at risk, for example of predation if they give an alarm call and thereby attract the attention of a predator. But in reality such an 'altruistic' individual can benefit from his or her action.

1. When animals recognize and frequently meet each other, an altruistic act by one can be repaid by the one who is helped. Vampire bats, for example, give food to each other; any individual who refused to share food would not be given any (Wilkinson, 1984).
2. The altruistic individual may help its relatives who carry some of the same genes so its overall genetic contribution to the next generation (its inclusive fitness) may still be increased (Hamilton, 1964). Primates prefer to help their relatives rather than non–relatives during fights (for example, pigtail macaques, Massey, 1977).

11.9.1 Co-operative breeding

When members of a group help to rear offspring of another pair, both the helper and the helped may gain. The breeding pair may gain from the helper feeding and defending their offspring, the helper may gain from the experience and from the possibility of inheriting a territory or mate when these are scarce commodities and/or breeding alone is too risky. In addition, helpers are often related to those they help (for example, superb blue and splendid wrens, Rowley, 1965, 1981), so their inclusive fitness may be improved. But the benefits to each vary and the dominant and subordinate individuals can come into conflict in two main areas (Emlen, 1984).

1. The subordinate may interfere with the dominant's sexual behaviour, compete for resources, be incompetent in its help or make the

group more conspicuous to predators (Zahavi, 1975, 1976; Gaston, 1978).

2. Conversely, the dominant may gain but the subordinate may lose out from helping. The subordinate could increase the benefits of helping by breeding and the dominant may try to disrupt such attempts. But if this is not possible the dominant could share parentage, or at least copulations, with the subordinate. As we have seen (page 228), suppression of breeding is common.

In either case, the dominant individual or pair tries to skew the benefits of the group in their favour without doing so to such an extent that the subordinate will benefit by leaving the group (Vehrencamp, 1983). In anis (a bird), for example, females lay in a communal nest but they have a dominance hierarchy that results in considerable differences in the number of eggs laid and in the amount of parental care shown. The fitness of the most subordinate female is the same as she would have if she bred alone (Vehrencamp, 1978).

Similar conflicts can be seen in social insects (Brockmann, 1984) between workers and between workers and queens. Queens manipulate workers by the use of pheromones or direct aggression and by eating any eggs laid by workers. The group remains together despite these conflicts, probably because each individual still fares better than if it tried to breed alone. In *Polistes metricus*, for example, some colonies have two, related, foundresses and these have greater fitnesses than do single foundress nests. The more subordinate of the two foundresses has a higher inclusive fitness than she would have as a solitary foundress (Metcalf and Whitt, 1977; Vehrencamp, 1983).

11.10 OVERVIEW

11.10.1 Have our questions been answered?

(a) Why show agonistic behaviour?

By winning fights, having a high rank or owning a territory, an animal can gain exclusive use of, or access to, resources such as mates, food, water and shelter from predators. Thus dominant individuals get better feeding sites (page 292), more food (page 290), and more mates (page 288) than subordinates. Territory owners get more food and have a less variable or longer lasting food supply (page 290) than non-territory owners. Winners of conflicts between the sexes may get more or better mates (page 303). The winners of any of these conflicts may in the long run reproduce more successfully and have better chances of survival than losers.

(b) In what circumstances is it worthwhile?

We have seen in this chapter that animals can derive many benefits from agonistic behaviour, but that they also incur costs. Decisions about whether and how to fight are highly tuned to these probable costs and benefits and to the probability of winning. Fighting is likely to be advantageous when the benefits of winning outweigh the costs. Whether or not they do so depends on several factors.

Firstly, characteristics of the individual (or its resource-holding potential, page 281) are important. A large, mature and experienced animal is most likely to win a dispute. Secondly, the outcome of a dispute can depend on the value of a resource (page 281). Thus a small amount of food may not be worth fighting over, especially if an animal is not hungry anyway. Finally, relative benefits and costs can depend on the environment, both social and non-social, where the dispute takes place. Thus a territory in a poor-quality habitat may not be worth defending and scattered sources of food may be difficult to keep exclusive control over (page 295).

Clearly, there are circumstances when it is not worth pursuing a dispute; the costs outweigh the benefits. When this is the case the animal may use another method of getting access to a resource. Sometimes alternative strategies are as good as each other, sometimes an animal must do the best it can (page 312).

(c) Avoiding lethal combat

Although it can be worth fighting over a resource, animals can settle disputes without escalating to the point of fatally injuring themselves. Combatants often avoid escalation by assessing the asymmetries between themselves. One of them is likely to be a better fighter because of its superior size, for example, or is likely to value the resource more highly than the other. Sometimes an owner of a resource wins by convention without having to fight. Such asymmetries are often clear to both combatants and are difficult to fake.

11.10.2 General principles

(a) The extent of conflict

The examples in this chapter illustrate the many diverse circumstances in which animals have conflicts of interest. Competition is common, both between and within the sexes, within families and groups, and the competitors are unlikely to be equal. There is a great deal of variation between individuals in how well they can fight and win resources, in the value they place on resources and in the environmental circumstances in which they find themselves. Because of this variation, disputes are sometimes settled before the fighting becomes too fierce. However, injuries are well documented phenomena. In some circumstances even killing others is adaptive, allowing an animal to gain food or mates. Conflict is not universal and animals sometimes co-operate with each other. But even in these cases they usually benefit by doing so. And even co-operative systems involve some conflicts of interest.

(b) Behavioural ecology

There is a vast literature on the function of animal behaviour, but interest in cost/benefit analyses is comparatively recent with much of the work being done in the past decade. Current cost/benefit analyses, however, have several weaknesses.

- In most respects, theory has far outstripped the data base and has usually been too simplified, depending heavily on assumptions and sometimes leading to contradictory predictions (as can been seen with the many models of territoriality). Short-term costs and benefits have generally been investigated, whereas we need analyses over lifetimes to understand the function of behaviour.
- Studies have generally focused on energy costs, copulation frequency or some other small part of the whole gamut of an animal's behavioural repertoire and its ecological and social environment.
- As well as simplifications, current explanations of behaviour do not always consider the possibility that a particular trait is non-adaptive.

Despite these criticisms, some detailed and elegant studies have been done, for example of wagtails, dunnocks and red deer, which have made substantial contributions to our understanding of behaviour. And theories, although simplified, have led to many tests and explorations of behaviour. The search for spiteful territory owners, for example, although unfruitful, has helped us to understand other, real, functions of territories.

In general, our understanding has been greatest where our knowledge of an animal and of its ecology is good. The integration of behavioural and ecological studies has developed mainly over the past decade with the appearance of the science of behavioural ecology, but a continued awareness of ecology in behavioural studies is needed.

12

The biology of human aggression

12.1 INTRODUCTION

12.1.1 Is there a biology of human aggression?

The title of this chapter and its inclusion in this book imply that the rules determining how and why animals fight may be relevant to the behaviour of our own species. We are part of the animal world and so may have inherited behavioural traits from our primate ancestors. There is ample evidence that this is the case at the level of morphology, physiology, and biochemistry.

However, just because we are primates does not necessarily mean that we will behave in the same way as other primates do. There are three main complications:

1. *Species diverge during evolution.* Although in some respects humans share some characteristics with other primates, inherited from a common ancestor, in other respects we have diverged from our closest relatives during the evolutionary process. We know from studies of animals (page 269) that while the movements and postures

used during fights are conservative across animal groups, levels of aggression and the social context in which animals fight are very labile, generally reflecting recent adaptation to local ecological conditions rather than the evolutionary heritage of the species concerned. Any information we may have about the presence or absence of territoriality, for example, in our nearest relatives may be completely irrelevant to patterns of space-related aggression in our own species.

2. *Learned behaviour patterns cannot evolve.* While the forms of agonistic motor patterns are relatively resistant to environmental modification during development in most species (page 211), when and how often these are used depends very much on experience. This is particularly the case for long-lived species such as primates and especially humans.

3. *Unique characteristics of humans.* Our species differs markedly from other primates in having a well-developed culture in which language plays a vital role. In humans, cultural influences are all-pervasive; from birth onwards, children are exposed to a complex set of beliefs, traditions and customs, all of which will mould the way their behaviour develops. These cultural processes make it hard to compare human behaviour directly with anything other primates do.

Language is a unique component of human behaviour. We use it to persuade and coerce others, and, through religion, ethics and the political and legal system, to justify aggressive actions and, in some circumstances, to make them seem acceptable and even beneficial to society. Other primates have the behavioural mechanisms that are needed to make complex decisions and to transmit information between generations (so they have a rudimentary culture) but these are immeasurably better developed in humans. In particular, human language greatly increases the power of culture to influence human behaviour and lifts our decision making into a new non-biological domain.

Because of our mobility and communications our sphere of influence is not restricted to a small group of both related and personally known individuals as it is in other primates, especially in technologically advanced societies. Our weapons also contribute to this picture of a very distinctive animal. They are more sophisticated and vastly more destructive than anything any other animal has come up with (page 42). Only humans use long-range weapons and participate in large-scale warfare. Such cultural characteristics might have biological roots, but in modern humans they have little in common with those of other primates.

12.1.2 How can we identify the biological roots of human behaviour?

If the complex and pervasive influence of culture means that there really are no aspects of human aggression that depend on behavioural capacities inherited from our ancestors, then biologists would have nothing other than some hypotheses to offer students of human behaviour. How can we find out if this is right?

(a) The fossil and archaeological record

Perhaps we can identify such inherent behavioural capacities by studying the remains of the earliest human beings, arguing that they had less time than modern humans to diverge from their ape-like ancestors. The earliest creatures with the skeletal features characteristic of the human species were the australopithecines (dating from around four million years ago) and the more advanced *Homo erectus* (1½ to ⅓ of a million years ago). The australopithecines were markedly sexually dimorphic, males being larger than females; the degree of dimorphism is less marked in more advanced hominids and relatively slight (compared to other primates) in modern humans. According to current theory on the functions of sexual dimorphism, this could mean that australopithecine males fought over females but that this competition has become less intense or that weapons have become more important than body size in this process. However, there are other explanations for size differences between males and females, including different economic roles for the two sexes. So the implications of the existence of sexual dimorphism in australopithecines and its reduction in modern forms is not clear (Frayer and Wolpoff, 1985).

Some fossil skulls of australopithecines, *Homo erectus* and of Neanderthal man have depressions and fractures which suggest a violent death, although it is impossible to tell whether these were the result of accidents, of attack by predators or intraspecies fighting. However, Palaeolithic cave paintings (dating from at least four thousand years BC) show human figures pierced by spears and arrows (Fig. 12.1), suggesting that homicide has a long history. Archaeological data suggest that the appearance of warfare between larger groups corresponded to the development of permanent agricultural settlements (Roper, 1975).

(b) Comparisons with other primates

Are there any behavioural traits characteristic of primates in general

(a)

Figure 12.1 Possible evidence of homicide in early humans. (a) A group of archers fighting over cows. North African rock painting *c.* 4000 BC from *The First Artists*, D. Samachson, and J. Samachson, 1970, Windmill Press, Tadworth, Surrey. (b) Group of archers fighting. Mesolithic rock painting. Morella la Vella, eastern Spain. (c) A dying warrior. Mesolithic rock painting from Cueva Saltadore, eastern Spain. (b) and (c) are from *The Art of the Stone Age*, Bandi *et al.*, 1961, Methuen, London.

(b)

(c)

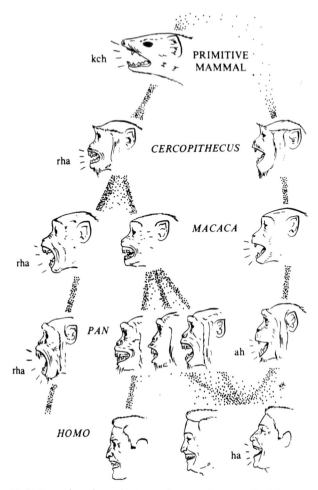

Figure 12.2 Postulated evolution of some human facial expressions from those of non-human primates (reproduced with permission from Van Hoof, 1972).

that are typical of humans as well? The facial expressions used during fights are very conservative among primates, and humans are no exception (Fig. 12.2) so we may well have inherited them from the common ancestor we share with other primates. Rough and tumble play and the ability to form dominance relationships are also widespread among primates (page 199) so we may have also inherited our propensities for such behaviours (Van Hoof, 1972). The occurrence of territoriality, on the other hand, is spasmodic among primates, relating rather to the presence of economically defensible resources (Chapter 11). In this case, then, comparative data suggest not simply that humans will or will not show territorial behaviour but that, should economically important resources exist, they have the behavioural capacities to do so.

(c) Cultural comparisons

It has been argued that because the earliest humans were almost certainly hunter gatherers, the behaviour of modern people who pursue this way of life is most likely to reflect our biological heritage, pared of the complexities imposed by our modern Western culture. However, this ignores the fact that present day hunter gatherers are not necessarily primitive at all.

Alternatively, we can compare modern cultures to see if there are universal and hence possibly biologically based characteristics. For example, studies of a variety of different cultures have demonstrated considerable uniformity both in the facial expressions used during social encounters and in the response of children to strangers (Eibl-Eibesfeldt, 1980). However, even this extensive study does not tap the full range of cultural variability; we should be aware that children growing up in different cultures may well share many childhood experiences.

A particularly detailed study of cultural similarities and differences was the six-culture study of a New England Baptist community, a North Indian caste group, a Philippine barrio, a village in Okinawa, an Indian village in Mexico and a rural group in Kenya (Lambert, 1978). In all the cultures about 10% of acts by children out-of-doors were aggressive, most frequent being insults and hitting. However, cultural differences are also seen reflecting different child–care practices. Nuclear families have sociable, intimate children; non-nuclear families have more aggressive, authoritarian ones. The use of physical aggression also differs between cultures. In the USA boys are encouraged to fight but in France verbal rather than physical aggression is encouraged. In Haiti the dominance hierarchy depends on age and an elder child of any sex can punish a younger sibling, but in Japan the hierarchy is based primarily on gender, so a younger brother can be dominant over an elder sister (Bourguignon, 1979).

(d) The behaviour of very young children

It might be that the behaviour of very young children, especially before they can talk, is close to that of our primate ancestors. However, by the time they are old enough to show any behaviour at all, children have been exposed to a whole complex of influences from their pre- and post-natal environment. Even so aggressive responses of one sort or another, together with appropriate facial expressions, are shown by very young children, and from an early age these are structured along dominance lines and are often commoner in boys than in girls (Strayer and Trudel, 1984).

12.1.3 A disclaimer

Attempts to apply facts about the behaviour of animals directly to humans are very likely to result in mistaken conclusions, unless carried out with great caution, a quality notably lacking in much writing on this topic. Rather than indulge in this exercise of extrapolating from animals to humans, in the remainder of this chapter, we take each of the aspects of the biology of conflict covered in the previous chapters, see what work on these topics has been done directly on humans and compare the results with what we know about animals. Our aim is to see what human conflict looks like from a biological perspective and to identify any special attributes of humans which move their behaviour out of the realm of biology.

As biologists, we are not competent to undertake a comprehensive review of the psychological, anthropological, sociological and criminological literature on human aggression, and we do not wish to pose as experts in those fields. We have tried simply to identify the behavioural phenomena which are special to people, to spell out their implications for the biology of human aggression and, where possible, to refer to up-to-date reviews or representative references for those wishing to pursue these topics further. Neither do we make any moral statements about the behaviour of humans. Even if human aggression does have biological causes and functions, this does not imply that our aggression is either justifiable or incapable of being modified.

12.2 WHAT IS HUMAN AGGRESSION?

12.2.1 Defining and measuring human aggression

All the problems about defining aggression in animals are exacerbated where human aggression is concerned. In particular, questions about how both participants interpret their opponent's behaviour become acute; if an act which causes injury to another is seen as justified, we may not consider it to be aggression at all (Nettleship and Esser, 1976).

The actual form that human aggression can take is extremely variable (Fig. 12.3), and its students have included in their definitions such diverse forms as overt attack, verbal abuse, teasing, hostile

Figure 12.3 Some manifestations of aggression in humans.

thoughts, vindictive dreams, and so on. Where overt aggression is concerned, it is particularly important to bear in mind the distinction between hostile aggression (the attack of an angered person on another) and instrumental aggression (where the attacker is using aggression as a means to an end and may not even be angry). Aggressive acts can occur at many different levels of organization, between individuals within a group, between small groups and in the form of warfare between nations. But it is by no means clear whether the processes responsible for fighting between individuals are also involved in warfare between nations (Scott, 1976).

All these things make human aggression particularly difficult to define, recognize and measure, and yet these things must be done if it is to be studied rigorously. Students of human aggression have used various methods to measure human aggression (Edmunds and Kendrick, 1980; Bertilson, 1983); prominent among these are the use of crime statistics, questionnaires and psychologists' reports. Aggression is often measured in laboratory experiments as the amount of punishment (such as an electric shock) that a subject delivers (or thinks she or he is delivering) to a victim, supposedly to facilitate learning; this may seem rather far removed from real aggression, but the level of 'aggression' shown in such a set-up correlates rather well with other measures of aggression. However, the samples used are frequently biased, the subjects often being psychology students. Finally, some scientists identify and measure human aggression by direct observation of behaviour; to a certain extent this is a legacy from biology.

12.3 ISSUES AND CONCEPTS

Before we can attempt to understand human aggression from a biological perspective we need to ask how comparable it is to the behaviour of other animals. Do humans show the same types and patterns of behaviour or are they distinct in some way? Such a comparison is complicated by the fact that biologists have often borrowed words such as dominance hierarchies from human behaviour to describe the behaviour of other animals. So similar terms may be used for behaviours that have different causes and functions.

12.3.1 Displays and escalated aggression

The social contexts in which agonistic behaviour occurs are similar in humans and other animals. Both fight over access to resources, especially mates, and in both there is conflict within family and other groups.

Humans, like non-human animals (page 40), use threats including facial expressions and fist shaking. Humans also use verbal threats and, owing to the complexity of human language and our reasoning powers, it may be that in some situations verbal threats are perceived by the opponent as being just as aggressive as actual physical aggression. Verbal as well as physical aggression may thus escalate in humans.

Deaths as a result of aggression occur throughout the animal kingdom, including our closest relatives (page 45) and ourselves. Such deaths in humans result from a variety of acts such as premeditated assassinations, crimes of passion involving jealous lovers, and infanticide – either as a direct killing or as abuse or neglect leading to the death of the child. In addition, some cannibalism may occur (or may have in the past) although the evidence for this is disputed (Arens, 1979).

12.3.2 Territoriality

There is a great deal of confusion in the literature about the occurrence of territoriality in humans because of a misuse of this concept. As we described in Chapter 3, a territory is a fixed area, usually containing one or more resources, which is defended against intruders by the owner. Whether or not an animal has a territory, it will defend its own body from attack and thus has a 'personal space'. In the human literature, this personal space is often quite mistakenly thought of as a territory. The functions and stimuli involved in defence of a personal space or territory are quite different. Although in some animals a territory is used only for mating, among the higher primates it is a living area, usually occupied by a family or group throughout the year.

It is clear that humans, just like other animals, will defend themselves. This has been observed and studied in several groups, especially children (even toddlers), mentally retarded and psychiatric patients (review in Malmberg, 1980). There is also evidence that individuals showing high levels of aggression prefer more personal space around them than do normal individuals (Cavallin and Houston, 1980).

If we consider a true territory in the primate sense of an area permanently lived in, it is apparent that this is seen in humans as well. A further misconception is that a territorial animal must always have a well defined and rigorously defended territory. We have seen, however, that what is defended, and when, depends on the economics of defence at that time (page 298). Hence one cannot say that because humans do not always defend rigid territorial boundaries they are not territorial. For example, present day hunter gatherers forage over a fixed area, but defence may be centred on the most important and economically defensible resource, such as a waterhole, within that area (Peterson, 1975).

The basic territorial unit in humans seems to be the tribe, in which the members are often closely related; for example, the Ituri pygmies in Africa and aborigines in Australia (Malmberg, 1980). In modern industrialized societies the concept of territory is not restricted to the family or close-knit tribe. Our culture, in particular our pervasive oral, written and electronic communication system, permits an extension of these units. So a family may own a house but also be part of a neighbourhood in which some members will be related and reciprocal altruism may be common. Neighbourhood ties can be strong and gangs may defend the area as their own territory. But in addition, humans are perhaps unique in being part of a larger society which has elements of the territoriality seen in other animals. Families can be part of a nation and perhaps a federation. Humans can create a common identity such as a nationality within a defensible area despite the lack of kinship or of a one-to-one reciprocal altruistic relationship, and can maintain that identity especially when it is threatened from outside.

Intergroup fights over territory and resources occur in both humans and non-human animals. The use of sophisticated weapons, particularly long-range ones, makes human warfare quite different in destructive potential from the fights of other animals. It is often seen as an evil or a sickness unique to humans (for example, Stoessinger, 1982) but there is no need to deny its biological roots in order to eliminate it. Primitive cultures are sometimes said to be peaceful, but aggression of some sort is a universal characteristic. Hunter gatherers do fight, although to a lesser extent than pastoralists or agriculturalists (Wright, 1942, cited in Malmberg, 1980). Groups organized into states rather than clans are also more warlike. Nevertheless, it is important to distinguish between warfare on the tribal and on the national scale. The political machinations that send soldiers into large-scale international wars have little to do with the biological causes of tribal fights over mates or territory.

12.3.3 Dominance

Our knowledge of possible dominance hierarchies in humans is very biased and not helpful for comparing them with hierarchies found in other animals, since good quality human studies concentrate on children.

From a functional point of view, dominance relationships are more important among adults, but may differ from those found in children. Adult status probably depends on many things such as the possession of wealth, family ties, intelligence, political coalitions, and so on, that do not appear in childhood relationships.

Many studies of children have concentrated on nursery school groups playing freely together. In this situation dominant–subordinate relationships based on who wins a fight or who is verbally or physically aggressive against whom are commonly observed. The children in the group can be ranked into stable hierarchies with few reversals of rank (Fig. 12.4; McGrew, 1972; Sluckin and Smith, 1977; Strayer and Trudel, 1984).

Some studies have also considered how authority and attention giving affect dominance relationships. Children look at high-ranking peers more often than by chance, for example (Vaughn and Waters, 1980), and children can be ranked by who initiates and who accepts group decisions (Barner-Barry, 1980). However, attention and authority should not be seen as equivalent to dominance, since other factors such as the popularity of the child play a part in determining who gets attention and who is listened to.

Teenagers also form stable dominance hierarchies based on verbal and physical aggression. The stability of the hierarchy depends on the situation, being most stable during organized activities. Assertion of dominance is usually verbal; physical displacements of, and threats to, another child are the least likely form of coercion to be used. Differences between children of high and low rank are clearly seen, dominance being related to athletic ability, pubertal maturity and group leadership (Savin-Williams, 1977, 1979).

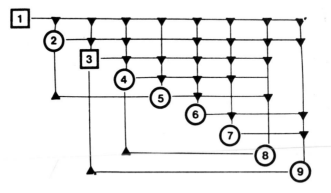

Figure 12.4 Network of dominance interactions in 18-month-old children. Open circles represent females, while squares represent males. Numbered individuals are arranged diagonally, according to their position in the hierarchy. Arrows represent the observed outcome of dyadic relationships (reproduced with permission from Strayer and Trudel, 1984).

Differences in children's dominance relationships can be seen between individuals both of different ages and of different sexes. Toddlers tend to play in mixed-sex groups and form a single hierarchy, whereas older children associate with peers of their own sex and each group has its own dominance hierarchy (Sluckin and Smith, 1977). The structure of girls' groups is less stable and depends more on the situation than that of boys, and girls use less physical means of asserting dominance (Savin-Williams, 1979). Girls tend to be more assertive in such single-sex groups than when they are in a group comprising boys and girls (Cronin *et al.* 1980).

These studies suffer from trying to deal with dominance in a situation where children who are strangers to each other, are put together and are expected to develop normal social relationships. As we have pointed out in Chapter 3, this is an unusual situation. Young animals usually learn social interactions as they grow up within a group, or else enter an established group if they disperse from their natal one. When young children are reared together in a large group, as in the Synanon community in the USA, girls can be as aggressive as boys and can occupy high-ranking positions (Missakian, 1980).

A further criticism of dominance studies in humans is that dominance is often taken to be an immutable characteristic of a child, related to its personality and physical development. It should not be forgotten, however, that one can talk about dominance only in relation to another individual (page 48). A child may be high-ranking in nursery school with a particular group of children but its status may be quite different in another situation.

12.3.4 Differences between the sexes

Many studies have considered differences in aggression between boys and girls or men and women and they show a more or less consistent tendency for males to be more aggressive than females. Maccoby and Jacklin (1980) suggest that females are inherently less aggressive. Other authors suggest females use a different type of aggression or are more inhibited in their expression of anger. Brodzinsky, Messa and Tew (1979) found that girls use less physical aggression and more indirect aggression (such as stealing or tattling) in stories that they are asked to write about a picture and that girls use more controls against aggression. That is, females seem to be more inhibited in their use of physical aggression but will use more verbal aggression (see also Crassini, Law and Wilson, 1979, on adults). In a questionnaire study, Frost and Averill (in Averill, 1982) found that males and females experience anger in similar ways, but that females express it less overtly.

Human gender differences in aggression seem to be dependent on the situation and the provocation received. Among young children playing freely, females use more verbal aggression and males more physical aggression. But in a more adult-controlled classroom situation these differences may only be seen in older children or not at all, and it is clear that there is considerable individual variation in the frequency of aggression used (Barrett, 1979; Archer and Westeman, 1981). Gender differences also depend on who is the recipient of the aggression. Boys in one study of 5–8 year-olds at a summer camp used more verbal aggression than girls when the recipient was a girl but not when the recipient was another boy (Barrett, 1979).

From the biologist's viewpoint, it would appear that females are less likely to escalate a fight and that instead they use low-intensity behaviour to resolve conflicts.

12.4 THE CAUSES OF AGGRESSION IN HUMANS

Conflicts between non-human animals are the result of various environmental circumstances, including pain and frustration (page 89). During a fight, levels of aggression and fear fluctuate as a result of both performing and observing agonistic actions (page 80). In some cases, variation in readiness to fight depends on differences in endocrine physiology (especially the reproductive and adrenal hormones; pages 104, 118) or on differences in brain physiology and biochemistry (especially in the limbic system; page 150). Similar factors may be at work during human conflict, but cognitive processes (a person's understanding and interpretation of the world she or he lives in) have a pervasive influence that is unique to humans, in degree if not in kind. For example, physiological arousal may or may not enhance aggression, depending on the causes to which that arousal is attributed. Such cognitive processes limit the importance of organic factors as determinants of human aggression. So, although (for example) certain physiological abnormalities are found in criminally violent individuals, their role as causal agents is probably small compared to that of other factors that promote aggression.

12.4.1 Organization

Observations of freely interacting school children (Blurton-Jones, 1972; Cheyne, 1978) have identified four independent categories of behaviour, very similar to those identified in non-human animals and especially in primates (page 60; Table 12.1). Taking these studies together, we see actions promoting social contact (prosocial behaviour),

Table 12.1 The organization of social behaviour in four-year-old-boys. Some results of a factor analysis of correlations between behaviour patterns. Groups of mutually correlated actions show up as positive loadings on the same factor. (From Blurton-Jones, 1972.)

	Tentative label			
Behaviour pattern	*Rough and tumble play*	*Aggression*	*Fear*	*Social*
Cry			+ +	
Pucker		+ +	+	
Red			+ +	
Fixate		+ +		
Frown		+ +		+
Hit, contact		+ +		
Grab, tug		+		+
Push		+		
Wrestle	+ +			
Hit at, no contact	+ +			
Jump	+ +			
Run	+ +			
Laugh	+ +			
Work	− −		−	
Smile				+ +
Talk				+ +
Receive				+ +
Give				+ +
Points				+ +

rough and tumble play, aggression and subordinance or fear. In adults as in children, although the manifestations of agonistic behaviour are complex, specific facial expressions are regularly associated with particular emotional states, including anger and fear, and these states are reliably recognized by people of different cultures (Izard, 1977).

The experience of emotions in adults is often, but not inevitably, accompanied by signs of physiological arousal (such as increased blood pressure and pulse rate). It has been suggested that the various emotions are associated with, and caused by, specific physiological states; the experiences of fear and anger are indeed accompanied by different patterns of physiological activity. However, both experienced emotion and overt behaviour are related to physiological state in a very complex way. For example, the same state of moderate arousal (drug induced) produces different emotions, depending on whether the subjects are in a pleasant or an unpleasant situation and on their understanding of the causes of the arousal (see Ferguson, Rule and Lindsay, 1982, for an example and Leventhal, 1980, for a review).

12.4.2 Antecedent conditions

In people as in other animals, a broad spectrum of unpleasant events can give rise to aggressive behaviour, but these can be very complex in nature. The same conditions can also give rise to escape or to some quite different response (including increased achievement), depending on the subject's past history, present circumstances and interpretation of the events (Bandura, 1983; Berkowitz, 1982; Geen, Beattly and Arkin, 1984). Physical assault and a variety of other painful experiences enhance the probability of aggression in people, but this is influenced by the subject's assessment of his/her chances of winning or avoiding retaliation (Olweus, 1978; Harrel, 1980). Painfully cold or hot temperatures (Berkowitz, Cochran and Embree, 1981), unpleasantly loud noises (Geen, 1978) and offensive smells (Rotton et al., 1979) also potentiate attack in people, both in the laboratory and in the real world. Crowding (which can be seen as an invasion of personal space) enhances physiological arousal and can produce aggression (Aiello, Nicosia and Thompson, 1979). Aggression is not the only response to crowding, the effect of which is to intensify ongoing behaviour (Freedman, 1978). Aggression can be a response to unpleasant events which are verbal as well as physical. Threats to reputation or humiliating affronts are the most reliable predictors of violence in assault-prone subjects and perceived loss of power may trigger aggressive attack (Worchel, 1978).

Short-term frustration (perhaps a special category of aversive event, Berkovitz, 1982) can activate aggression in people as it does in animals (page 90). This has been seen in children and in adults, both in standard laboratory aggression tests and in unstructured encounters (Fredericksen and Peterson, 1977; Ahmed 1982). However, frustration is not the only cause of aggression and aggression is not the only response to frustration. Whether frustrated people behave aggressively depends, among other things, on their personality, on the perceived intentions of the frustrator and on the availability of non-aggressive alternatives (Bhatia and Golin, 1978; Dodge 1980).

The factors that influence the occurrence of aggression in people are often the result of complex mental or cognitive processes and not of events that are aversive in any simple sense. Thus people will deliver stronger (simulated, see page 328) electric shocks to a victim in a laboratory setting if they believe this is justified, if the victim is violating a norm or deprived of individuality, if the aggressor is anonymous and censure unlikely or if the responsibility for any suffering can be placed elsewhere (see Mann, Newton and Innes, 1982, for an example, and Zimbardo, 1978, for a review). These kinds of influence have been identified in artifical laboratory experiments, but similar phenomena must surely influence hostile encounters in real life.

12.4.3 The dynamics of hostile encounters

So far we have discussed the factors that precipitate an aggressive encounter; we now look at the determinants of the course of the encounter, at whether it escalates and when it stops. This depends on some of the same factors that initiate an aggressive act, but to these are now added the various consequences of attacking someone. Some of these consequences depend on how effective the attack is; these include pain caused to the victim or to the attacker. Others are the result of simply performing an aggressive action, regardless of its immediate effects. Signs of pain in a victim can provoke further aggression, but only in angry subjects (Berkowitz et al., 1981; Sebastian et al., 1981). Experience of pain by the aggressor also heightens aggression, provided it is not experienced immediately after an attack (Zeichner and Phil, 1979). These two effects tend to promote escalation of fights.

The effects of the actual performance of an aggressive action regardless of its consequences are complex. The endogenous drive theory of aggression (page 86) formalizes the widespread belief that acting aggressively has a cathartic effect, reducing levels of an internally generated aggressive drive. This has important implications for control of aggression, but, although a few studies have demonstrated an aggression reducing effect, the majority show quite the opposite effect (Table 12.2; see Geen and Quantry, 1977, for a review). Physiological arousal is reduced in angry subjects as a result of performing overt aggressive acts against the person who made them angry in the first place; this effect depends on there being nothing to make the subjects anxious (such as a high status victim or feelings of guilt). A few studies suggest that overt aggression as well as arousal is reduced by performing a hostile action. For example, women show less aggression and less arousal if they are given an opportunity to express their hostile feelings (in writing) after being provoked. This looks like catharsis, but could reflect the development of alternative, non-hostile strategies for coping with anger and, anyway, the converse effect is seen for men (Frodi, 1978).

Table 12.2 Some effects of behaving aggressively. (a) Intensity of supposed electric shock (maximum = 10) given by male subjects to a male victim who had previously angered them after *either* being allowed to injure the victim *or* spending a control period without injuring him. (b) Drop in blood pressure during the injury/control period (from Geen and Quantry, 1977).

	(a) Shock levels given	(b) Drop in blood pressure
Victim shocked	6.65	−8.80
Victim not shocked	5.20	−1.30

The great majority of studies agree with this last result, and show that performing either verbal or physical aggression or engaging in competitive sports increases hostility. For example, the opportunity to retaliate verbally against a tormentor increases subsequent aggression towards that person, to an extent proportional to the amount of verbal hostility shown. School football players are more aggressive at the end of a season than at the beginning; PE students who had been equally active in a non-aggressive sport over the same period showed no change in aggressiveness (see Geen and Quantry, 1977). On a quite different level, comparative studies show that cultures which indulge in combative sports are among the most likely to be warlike (Sipes, 1973).

There are various ways in which this aggression-enhancing effect might come about, and these are not mutually exclusive.

1. There may be some direct positive feedback to aggressive arousal from the physical sensation of performing a vigorous aggressive act.
2. A person may learn that behaving aggressively reduces unpleasantly high arousal to an acceptable level.
3. The act of attacking someone might lower the socialized inhibitions that most people have against hurting someone else.

12.4.4 Filmed violence; short-term effects on aggression

If the aggression-promoting effects of indulging in aggressive behaviour come about through the weakening of inhibitions against hurting people, we might expect that watching other people behave in this way would also promote aggressive behaviour. Because of widespread concern about the pervasiveness of violence in the media and in

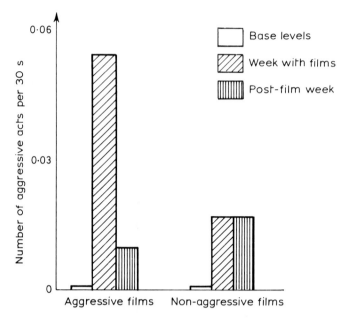

Figure 12.5 The effects of filmed violence on aggressive behaviour in a naturalistic setting. Levels of aggression (including physical and verbal threats and attacks) observed in 14–18 year-old boys in an institution for offenders before and after a week in which subjects saw *either* aggressive *or* equally acute but non-aggressive films. The figures are for boys ranked as non-aggressive; those for aggressive boys give similar results. Differences are significant at $p < 0.05$ (after Parke *et al.*, 1977).

sport, there have been many studies of effects of seeing other people behaving aggressively. In general, observing violence (and particularly successful violence) increases aggression in the viewers, whether in the laboratory or in a less structured setting (Fig. 12.5; Parke *et al.*, 1977). However, the effect depends on the subject's mood and on the way the viewed aggression is presented. Provided that they are angry, subjects deliver more painful stimuli (in a 'learning test') after seeing a violent film than if they have seen an equally arousing, non-violent film (Geen *et al.*, 1984). This effect is stronger if the violence witnessed by the angry subjects has a successful outcome (Lando and Donnerstein, 1978) and if a commentary justifying the violence accompanies the film; filmed violence preceded by a commentary condemning this behaviour reduces subsequent aggression (Berkowitz and Powers, 1979).

These effects of watching other people behave violently could come about through a direct stimulatory effect on aggressive arousal, could teach viewers novel forms of aggression and it could change their assessment of whether aggressive acts are likely to be rewarded or punished (Malamuth and Donnerstein, 1982).

On this last point, careful analysis of the crime statistics shows that homicides (over the whole United States for the period 1973 to 1978) increase by over 12% three days after a well publicized boxing fight, where injuring another person is both intentional and rewarded. The victim is particularly likely to be similar (in age and colour) to the loser of the fight. In contrast, homicide levels (weekly figures for London for the period 1858 to 1921) were significantly reduced in the week following a well publicized execution (representing severe punishment for violent behaviour). Homicide rates increased to above average levels in the second to fourth week after the execution. In other words: 'The lesson of the scaffold' is real, but only short-lived (Phillips, 1986).

Viewing pornographic films which contain an element of violence has also attracted attention, given the prevalence of 'video nasties'. In male subjects previously made angry by another man, pornographic and non-pornographic aggressive films are equally effective at enhancing aggression to a male victim. However, when faced with a female victim, aggressive pornography produces a more marked increase in aggression to the victim than either non-aggressive pornography or non-pornographic aggression; this is true whether or not the subject was angry. Quite what causes this increase is not clear, since the assessed pleasantness or aversiveness of both pornographic and non-pornographic films influence their effect on aggression (Malamuth and Donnerstein, 1982; Geen et al., 1984; Zillman, 1984).

12.4.5 Hormones

In humans as in other mammals, male foetuses produce more androgens than do female foetuses and the hormonal environment in which a baby develops influences its morphology, physiology and behaviour as an adult. Early exposure to androgens can promote aggression in humans but the effect is not a strong one; there are all kinds of problems of interpretation and so its nature is poorly understood (see Meyer-Bahlberg, 1981a; Mazur, 1983 and Brain, 1984, for reviews). Human males with a genetically determined partial insensitivity to testosterone and both males and females whose mothers were treated with an artificial progesterone during pregnancy (to prevent miscarriage) are slightly less assertive and competitive than normal. On the other hand, females with abnormally high levels of androgens during development (produced by the adrenal glands as a result of a congenital malfunction in the mother) are more physically active and very slightly more aggressive than normal females. Both females and males exposed to androgen-derived progestagens during pregnancy (again, to prevent miscarriage) show more physical (but not more verbal) aggression compared to siblings who had not been exposed to the hormone (Reinisch, 1981).

In adolescent boys, plasma testosterone levels are positively related (with a correlation of 0.44 for a sample size of 58) only to those measures of aggression which concern response to provocation or intolerance of frustration; this relationship persists when differences in body build are allowed for (Olweus, 1984). In adult men the relationship between androgens and aggression is both weak and variable (see Meyer-Bahlberg, 1981b, Mazur, 1983 and Brain, 1984, for reviews). There is no association at the individual level between testosterone levels and self-ratings of overall aggressiveness in non-criminal men or in convicted criminals. However, men with a record of violent, aggressive crimes have significantly higher levels than those convicted of non-violent crimes and non-criminal men have lower levels still (Fig. 12.6; Ehrenkrantz et al., 1974; Schiavi et al., 1984).

The relationship between androgens and aggression in human males is complicated, as it is in primates, by the fact that experience of agonistic encounters can itself alter hormone levels. Male hockey players who are consistently likely to retaliate aggressively during a game have higher plasma testosterone levels than less aggressive players (Scaramella and Brown, 1978). On the other hand, engaging in competitive sports (such as tennis or wrestling) increases testosterone secretion in both participants, but this is more marked in the winner than the loser (Mazur and Lamb, 1980; Elias, 1981). In addition, neither surgical nor chemical castration has a clearcut effect on aggressiveness in human males (Meyer-Bahlberg, 1981b), so while there does seem to be a relationship between testosterone production and some aspects of aggressiveness, it is not a strong one.

In adult females, feelings of hostility and aggression change in a predictable way with the menstrual cycle. Both negative emotions and criminal convictions are at a peak during the pre-menstrual phase; positive emotions peak and convictions drop just before ovulation when oestradiol levels are high. These results could mean that ovarian hormones have a direct effect on aggressiveness, but the hormonal shifts are accompanied by a complex collection of physical changes, any of which could be responsible for the behavioural changes (d'Orban and Dalton, 1980; Sanders et al., 1983).

There are few data on the effect of the pituitary–adrenal hormones on aggression in humans, although cortisol levels are increased by fights and by aggressive sports training (Elias, 1981; Carli et al., 1982).

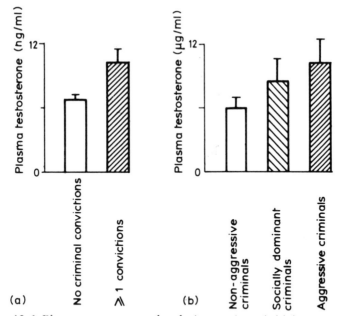

Figure 12.6 Plasma testosterone levels (mean ± s.e.) (a) in men with and without criminal convictions, (after Schiavi *et al.*, 1984) and (b) in different categories of convicted criminals (after Ehrenkrantz *et al.*, 1974).

Higher ratios of noradrenaline to adrenaline have been found in the urine and plasma of certain male criminals who have records of extreme personal violence. These men also show markedly reduced behavioural and endocrinological responses to stress (Woodman, 1979; Hinton, 1981a). These physiological differences may influence susceptibility to the negatively reinforcing consequences of aggression.

12.4.6 Neurobiology

The human nervous system is very similar in structure to that of other primates. There is no particular reason to suppose that it works in a radically different way, so the neurobiological factors that influence agonistic responsiveness in other mammals may be important in our own species as well. However, the fact that the cerebral cortex is relatively larger means that the effects of social and experiential factors are likely to be even more marked. Most of the available information on the relationship between brain activity and aggression in humans comes from studies of people who show abnormal patterns of behaviour; the two major areas of interest are prisoners with careers of violent crime and people with medically defined brain misfunctions, such as epilepsy. Such studies suffer from a number of deficiencies including small and highly selective samples, poor assessment procedures and lack of controls (Hinton, 1981b). Violent criminals differ from non-violent criminals and from the non-criminal population in all kinds of ways and it is usually not clear what part any special neurophysiology might play in causing their deviant behaviour (Spellacy, 1978).

Violent offenders tend to differ from non-violent offenders (who themselves differ from control subjects) in their EEG patterns; these indicate chronically high levels of cortical reactivity, which may impair attention (Mandelzys, Lane and Marceau, 1981). The EEG patterns resemble those produced by children, leading to the suggestion that maturation of the nervous system is in some way retarded (Surwillo, 1980). However, the significance of EEG patterns is poorly understood and the association between EEG pattern and criminal violence is by no means clearcut (Monroe, 1981).

Average levels (in the cerebrospinal fluid) of certain metabolites of the neurotransmitter serotonin (page 153) are significantly lower in criminals convicted for murder of a sexual partner compared to those convicted for other categories of murder and a control, non-criminal group. However, there is considerable overlap between the groups, with some of the lowest scores of all being found in the control population (Fig. 12.7). Scores on a standard aggression test are negatively related to cerebrospinal fluid levels of a serotonin metabolite (correlation coefficient = -0.78; see Goodwin and Post, 1983, for a review), while oral tryptophan (which enhances 5HT levels) reduces aggressiveness in violent schizophrenics (though for a sample size of three, Morand, Young and Ervin, 1983). These observations suggest that some aspects of aggression in humans may be related to low levels of the inhibitory neurotransmitter serotonin. This is in agreement with the data on other mammals, although these data also suggest that the effects of serotonin are unlikely to be specific to aggressive behaviour.

Serious attacks in response to minor provocation are sometimes given the name of episodic discontrol. The appearance of episodic discontrol in a number of clinically-presenting cases occurred after some sort of head injury or in people with brain tumours. Other cases were found to have temporal lobe epilepsy, miscellaneous neurological disorders (sic) or the (vaguely defined) condition of minimal brain dysfunction (Elliot, 1983). These observations could mean that violent behaviour is caused by abnormal brain activity, but to assess the results properly we need to know the frequency of these various conditions in people who do not suffer from episodic discontrol. Violent outbursts are sometimes seen in association with an attack of

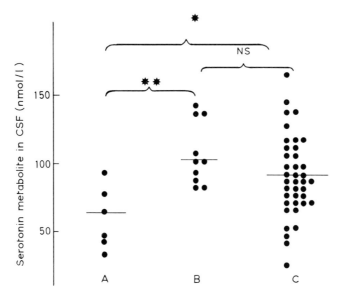

Figure 12.7 Levels of a serotonin metabolite (5-hydroxyindole acetic acid) in the cerebrospinal fluid (CSF) of violent criminals and non-violent controls. Group A, murderers whose victim was a sexual partner; B, murderers of other victims; C, healthy, non-criminal control subjects. The data are complicated by the fact that some individuals in groups A and B were alcoholics. *, p<0.05; **, p<0.01; NS, no significant difference. (After Asberg *et al.* in Goodwin and Post, 1983).

temporal lobe epilepsy in patients who do not normally behave in this way; these outbursts of aggression can sometimes be suppressed by drugs that reduce seizure thresholds (Devinski and Bear, 1984). This suggests that there might be a link between temporal lobe dysfunction and aggressive behaviour, although most seizures are not accompanied by violence and the majority of unpredictable violent episodes in humans are not of this nature (Monroe, 1981). Electrical stimulation through electrodes placed in the human amygdala (which lies in the temporal lobe, see Fig. 6.5) can elicit feelings of anger and hostile behaviour (Brain, 1984).

The evidence for a link between localized brain abnormality and aggression has been thought sufficiently compelling to justify surgical interference with healthy brains in people who show pathologically high levels of violence. The hypothalamus (see Fig. 6.5) and the amygdala have both been implicated in the control of aggression in other mammals (though their roles are complex and poorly understood and are among the brain regions destroyed in such psychosurgical operations. Besides their medical objectives, such operations might possibly help us to understand the role of these structures in the control of human aggression. However, because these are medical operations and not scientific experiments, the subjects are by no means a uniform sample and it is hard to find adequate controls (for example, for placebo effects or spontaneous remission). To the extent that the results can be assessed at all, they do not demonstrate a convincing reduction in aggression following surgery (Carroll and O'Callaghan, 1981).

12.4.7 Alcohol and violence

Alcohol has wide-ranging and complex effects on a huge variety of bodily processes, including some which have been implicated in the control of aggression. There has been a great deal of research into the relationship between alcohol consumption and aggression, but the results are conflicting (depending, among other things, on the dose levels and the time scale over which the drug is administered) and the conclusions unclear (see Brain, 1986, for a review).

The consumption of alcohol has been linked with many types of violence ranging from petty crime, family quarrels and hooliganism to child abuse, murder and rape. There are, however, several problems in interpreting the data on this subject. For example, studies often do not use adequate controls and there may be biases in the reporting of alcohol consumption. Also, a lot of alcohol is drunk without apparently leading to violence in the consumer so there is no inevitable causal link. Other factors are probably involved in linking alcohol consumption and violence. For example, alcohol is often drunk in places and circumstances where violence is not unusual anyway, so it would not be surprising to find a correlation between the two.

Where an effect of alcohol on aggressive behaviour has been demonstrated, this may come about in a variety of ways. Alcohol may influence the expression of aggression directly, but it may also impair a person's judgement of what is acceptable behaviour and lower inhibitions against being violent. Even the expectation of the presumed effects of alcohol can increase a person's aggressiveness, so once again, human cognitive processes seem to be more important than physiological factors. Finally, chronic alcoholics may suffer damage to the brain which may enhance aggressiveness.

It thus seems that alcohol is just one of many possible causes of aggression in humans. It is linked to violence in only a small proportion of the many circumstances in which it is drunk and so should be seen as a contributory but not a major factor in making people aggressive (Brain, 1986).

12.5 GENES, ENVIRONMENT AND THE DEVELOPMENT OF AGONISTIC BEHAVIOUR

In humans as in other animals (page 171) some differences in aggressive behaviour can be ascribed to genetic differences between the people concerned. However, their role is small compared to that of the environment. A baby's first aggressive action may well be a deep seated, unlearned (and adaptive) response to aversive stimuli, but from this first act onwards, the behaviour of young human beings (like that of young monkeys, for example) brings a complex pattern of positive and negative consequences. They also see aggressive and non-aggressive behaviour in other people bring rewards or punishment. As a result they acquire a set of expectations about the probable consequences of aggression and a set of values about how justifiable such behaviour is. These environmental influences (interacting with genetic predispositions) produce adults with a certain tendency to show aggression rather than some other response to conflict.

12.5.1 Genetic mechanisms

Two main kinds of evidence have been used to find out whether differences in aggressiveness between individual human beings are inherited. In the first place we can look for patterns in the occurrence of aggression among relatives in the normal population. Secondly, aggressive behaviour in people with known genetic abnormalities can be compared with that of the rest of the population. As in so many studies of human aggression, it is usually extreme, anti-social manifestations of aggressiveness which have been studied.

Monozygotic (or identical) twins have exactly the same set of genes while dizygotic (or fraternal) twins have the average 50% genetic relatedness typical of any siblings. A greater resemblance in aggressive behaviour among identical twins than fraternal twins would indicate that genetic factors contribute to individual variation in this trait (provided identical twins do not experience a more similar environment than fraternal twins). Twin studies have identified inherited differences in a number of personality traits (emotionality, activity, fearfulness, irritability) which may influence the expression of aggression, but the identified genetic effects are weak (see Hay, 1985, for a review). Levels of concordance (both twins having the same condition) are low and variable for the occurrence of criminal violence, but in general they are higher for identical twins (Christiansen, 1977a,b). So genetic factors make a small but real contribution to the occurrence of criminal behaviour.

There is a tendency for adopted children to resemble their biological rather than their adoptive relatives in certain personality traits, including some which might influence aggressiveness (see Hay, 1985, for a review). The frequency of psychopathy is higher in biological relatives of identified cases (14%) than it is in their adoptive relatives (7%), so that heredity seems to be a contributing factor for this condition (Schulsinger, 1977). Adoptees resemble their biological parents for convictions on property crimes more than they do their adoptive relative. This result holds when matching the background of adopted parents and their adoptees is allowed for, and when adoptive parents have no knowledge of the criminal history of their children's biological parents. However, this effect is not found for crimes of violence (Mednik, Gabrielli and Hutchings, 1984).

A number of rare conditions which are known to depend on a mutant allele at a single locus have behavioural as well as medical effects, including abnormal aggression. For example, episodes of serious aggression are observed in patients with Lesch-Nyhan's disease (inherited as an X linked recessive) (Palmour, 1983). Such observations show that allele substitutions at a single locus can have potentially dramatic effects on aggression in humans as they can in non-human animals (page 172). However, it is unlikely that such drastic differences contribute much to variation in aggressiveness in the normal population.

Certain men have an extra Y chromosome, as a result of a mistake in the process of gamete formation (Hay, 1985). Notwithstanding the sensationalism, biased sampling and lack of appropriate controls of earlier reports, such men are slightly over-represented among inmates of prisons and mental institutions; 42% of XYY males in one study had been convicted of a criminal offence, compared to 9% of (equally tall) XY males (Schiavi et al., 1984). This observation (taken together with the literature suggesting an aggression-promoting effect of the Y chromosome in some strains of mice, page 183) has been interpreted as showing that XYY men are genetically predisposed to behave violently, but various additional facts show that this is not the case.

For example, men with one additional X chromosome (XXY men; 19% convicted, Schiavi et al., 1984) as well as women with an extra X chromosome (XXX) are also over-represented in prison populations, so we may be looking at a non-specific effect of an imbalanced genome, rather than a specific effect of the extra Y chromosome. Anyway, the crimes for which XYY are convicted are not of a particularly violent nature. This is supported by the results of careful psychological studies of uninstitutionalized men; compared to XY men, XYY men (who have slightly higher testosterone levels, see Fig. 12.8) are slightly more likely to find themselves faced with a conflict, but are no more likely to behave aggressively if this occurs. XXY

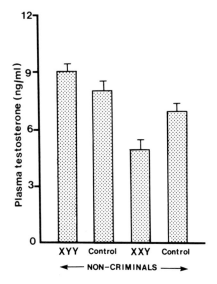

Figure 12.8 Plasma testosterone levels (means and s.e.) in men with sex chromosome anomalies and fully matched controls (after Schiavi *et al.*, 1984).

males do not differ in aggressiveness from controls; however, they tend to be 'anti-aggressive', responding to aggression with anxiety and submission (Schiavi *et al.*, 1984). In general, any effects of an extra Y chromosome on aggression in human males are weak and probably indirect.

12.5.2 Development

Aggressive behaviour can be seen in very young children (Fig. 12.9): babies of less than twelve weeks old give what have been described as rage reactions to painful stimuli, and even at this age individuals differ in their readiness to show this behaviour (Parens, 1979). Eighteen-month-old children of various cultures have quite an extensive repertoire of well organized agonistic behaviour patterns which are often expressed in the context of dominance–subordinance relationships (Blurton-Jones and Konner, 1973; Cheyne, 1978; Strayer and Trudel, 1984). Play fighting in very young children is much like that seen in other young primates and, again as in other primates, boys tend to indulge in this activity more than do girls (but see page 333).

As described for so many other species (page 203), physical aggression is progressively replaced by threats, facial expressions and gestures which are not overtly violent. In humans these non-violent alternatives are particularly complex, often taking the form of verbal

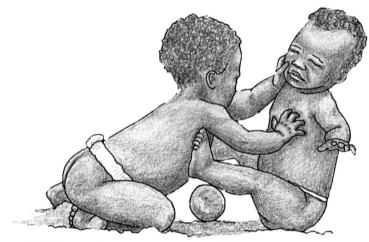

Figure 12.9 Aggression in very young children.

aggression, symbolic aggressive actions, rejection or prosocial responses (Cheyne, 1978; Perron *et al.*, 1980; Milavsky *et al.*, 1982).
Dominance hierarchies are more structured in older children, with
roles more clearly differentiated and challenges less common (Strayer
and Trudel, 1984). Older children are less aggressive towards their kin
and show less aggression when in the presence of their friends or an
adult; the converse is true for younger children (Lambert, 1980).

Differences in aggressiveness appear to represent a rather stable
feature of personality which is manifest from an early age, although
between-individual variability in style and level of aggression become
more marked with age (Milavsky *et al.*, 1982). The relative aggressiveness of children in kindergartens correlates well with their behaviour in primary school and as adolescents. This in turn can be used
to predict anti-social behaviour some years later (Olweus, 1984). Of
course, aggressive children may prefer activities and situations which
themselves promote conflict, and they elicit negative responses from
their peers and from adults, so that initial (possibly chance) behavioural differences may be exaggerated (Bullock and Merrill, 1980).

Although various aspects of the environment in which children
grow up can affect their aggression as adults (undernourished children
are less aggressive, Smart, 1981), by far the most important influences
come from interactions with social companions (peers, parents, other
adults) which offer many opportunities for learning. The study of
learning in humans is a huge discipline in itself; here we only touch on
some of the processes that can influence aggressive behaviour. Our
aim is simply to stress the fact that from an early age children

experience a complex mixture of negative and positive reinforcement
for performing (and watching) aggressive behaviour and that this
experience modifies the way they behave.

Aggressive behaviour often has negative consequences; from 3 to
60% of aggressive actions between young children result in retaliation
by the victim or punishment by an adult caretaker. On the other hand,
the consequences can be positive; depending on circumstances, anything from 24 to 97% of aggressive actions in children have positive
consequences, as when the opponent responds by giving up a disputed
object (see Bjerke, 1981). Parents often encourage sons to behave
aggressively and this can be an important determinant of adult social
behaviour; aggressiveness in teenagers correlates positively with the
degree of encouragement for aggressive behaviour given by parents
(Neapolitan, 1981).

Children who are subjected to harsh or frequent physical punishment are often unusually aggressive as children (George and Main,
1979; Eron, 1982), as teenagers and as adults. Violent delinquents
often come from homes where harsh punishment is common (Farrington, 1979) and people who were themselves beaten when young
tend to resort to violence towards their own children. These effects
could come about by a process of habituation of fear (people who were
beaten as children are described as more daring) or through a process
of modelling, in which beaten or abused children learn to accept
violence as an appropriate form of behaviour. Beating by fathers has
the most effect on sons, especially if the sons identify strongly with
them (Neapolitan, 1981).

12.5.3 Violent films and aggression in children

Children might learn the lesson that violence is an appropriate,
rewarding way to behave from watching such events in films or on
television. In the short term, watching a violent film (especially if the
person seen behaving violently is praised) makes children more
aggressive and more likely to consider violence an appropriate response (Puleo, 1978; Belsen, 1978; Singer and Singer, 1981). However, if filmed violence is portrayed in negative terms this inhibits
subsequent aggressive behaviour in children.

It is not clear whether these short-term effects accumulate to
produce a major influence on the development of aggression, since the
results of long-term studies are conflicting. Levels of aggression in
children increased following the introduction of television into an area
lacking this facility (see Roberts and Bachen, 1981, for a review).
Positive correlations have been found between children's preference
for violent TV programmes and their levels of aggression and anti-

social behaviour several years later (Eron, 1982). On the other hand, Milavsky *et al.* (1982) failed to find any long-term correlations (after three years) between patterns of TV viewing and subsequent aggressive or anti-social behaviour. Where correlations between viewing television violence and aggressive behaviour persist over several years, this may still reflect a short-term rather than a long-term effect if viewing patterns remain stable in the intervening period. Whichever is the case, unsupervised exposure to violent films is likely to promote aggressive behaviour in children.

12.6 THE BEHAVIOURAL ECOLOGY OF HUMAN CONFLICT

Does human aggression have a function(s) or has our culture destroyed any function that it once had? In other animals aggression is not usually abnormal or deviant, rather it is highly adaptive (Chapter 11) and we might expect the same to be true for people as well. However, although a functional view of aggression may tell us about the selection pressures acting on behaviour in the past, it does not tell us about any selection pressures that might be relevant to our present day societies.

12.6.1 Costs and benefits of aggression

As in other animals (page 277), humans (as individuals and as members of groups, Durham, 1976) seem to take into account the costs and benefits of aggressive behaviour. Benefits can be possession of or access to an important resource such as land, food or mates (Meggitt, 1977; Table 12.3). Costs can be time, energy or risks of injury or punishment; rape is most frequent when women are vulnerable and the likelihood of punishment is low (Shields and Shields, 1983). In fights between individuals, a low cost form of aggression such as verbal threats is used most often since injury to the instigator is less

Table 12.3 The benefits of warfare: the outcome of wars over land among the Mae Enga tribe of Papua New Guinea. (from Meggitt, 1977)

	Losers wholly evicted from their land	Victors secure some land	No land gained	Outcome unknown
Number	6	19	9	7
Percentage	14.6	46.3	21.9	17.1

than if physical aggression is used. Similarly, children use verbal rather than physical aggression towards adults and, in cultures where families are non-nuclear, physical aggression between older children is inhibited when adults, and the threat of punishment, are present (Steinmetz, 1977; Bourguignon, 1979).

12.6.2 Overt fighting

In human fights there is often some sort of asymmetry (page 281) such as the value of the disputed resource to the combatants or the possession of weapons. The use of many weapons by humans on an individual basis, in contrast to that of animals, is low cost in that retaliation by the opponent is minimized if he has been shot rather than punched or stabbed. So in a sense killing an opponent is more likely to happen in human fights than in other animals.

Fighting occurs among animals to gain access to mates, food, or other resources. Among humans, fights can be seen in similar contexts as a means of gaining resources, especially sexual partners, or perhaps maintaining relative status, for example if a male's sexual reputation is questioned. Access to women is a frequent cause of fights among the Yanomamo (Chagnon and Irons, 1979). In modern Western societies, many disputes that result in murder concern either access to sexual partners and other resources or the maintenance of status (Daly and Wilson, 1982; Zillman, 1984; Wilson, and Daly, 1985) but murder is also often committed by people who are mentally ill or who are under the influence of alcohol (Petursson and Gudjonsson, 1981).

Cannibalism has probably provided a source of food during periods of starvation and has probably sometimes had a religious significance, but the importance of humans as a food source is a matter of dispute (for example, de Montellano, 1978).

Humans are by no means the only animals that kill each other, but our culture and weapons do make the circumstances in which we kill unusual, if not unique. The use of a gun may result in a killing when hand-to-hand combat would have resulted in one individual losing and escaping. Some killings, such as assassinations, are premeditated in humans, an unlikely occurrence in other animals.

12.6.3 Costs and benefits of dominance

In other animals high rank can sometimes confer priority of access to one or more important resources (page 287). In humans, too, high-ranking individuals are privileged in one way or another, gaining both popularity among peers (Deluty, 1981) and other more tangible benefits. Among children the more dominant ones gain possession of preferred toys, play areas, sleeping areas, and food (Savin-Williams, 1977, 1979).

Status is important in adult humans. In some cultures only the highest ranking males in a group have more than one or two wives (Dickemann, 1979). Low-ranking males may fail to marry. The headmen of Yanomamo groups have an average of 3.6 wives compared with 2.4 for other males, and consequently their reproductive success is higher (8.6 versus 4.2 children); some men do not get any wives, some only late in life (Chagnon, Finn and Melancon, 1979). In modern Western societies, reproductive success is considered less important, but high status can still give the bearer access to the best resources.

Females also fight for resources and have dominance hierarchies. In Zambia, for example, women within a household (in-laws, sisters and co-wives) fight over resources provided by the male head of the household. The highest ranking 'elite' females use frequent verbal aggression against middle-ranking females, but not against the lowest ranking, perhaps because competition for husbands is only between close-ranking females (e.g. high or middle ranks for high-ranking men) (Schuster, 1983).

12.6.4 Costs and benefits of territoriality

In many cultures, the territory clearly holds valuable resources which are defended by the tribal owners. Attempted theft of the resource can readily lead to fights or even battles. In Australia, for example, the Waibiri tribe fought to protect water sources and the Washo Indians of the Great Plains fiercely defended their fishing lakes and areas containing large game (King, 1976). Among the Mae Enga of Papua New Guinea, disputes over land were by far the commonest cause of warfare (Table 12.4). Defence is usually focused on the resource rather than the boundaries, perhaps because a large range suitable for hunter gatherers would be difficult and costly to defend in its entirety (Peterson, 1975).

As with other territorial animals, the presence or absence of a territory and the size of the territory depend on the economic defensibility of the resources within it. On the Great Basin of North America, for example, the Western Shoshons ate mainly widely dispersed and unpredictable seeds during the summer, living in family groups. These families came together in the winter to feed on locally abundant nuts. Such resources are not easy to defend economically and were not owned by the Indians. In contrast, the Owens Valley Paiute lived in a richer area in permanent villages and they irrigated wild seed patches. Groups of villages formed well-defined bands which defended a

Table 12.4 Reasons for intergroup warfare among the Mae Enga tribe of Papua New Guinea, c.1900–50 (from Meggitt, 1977).

Ostensible reason	Number	Percentage of known reasons
Land	41	57.7
Theft of pigs	12	16.9
Avenging homicide	8	11.2
Homicide payments (recompense for a homicide)	3	4.2
Theft of pandanus nuts	3	4.2
Title to trees	1	1.4
Garden theft	1	1.4
Rape	1	1.4
Jilted suitor	1	1.4
Unknown	11	–

territory (Dyson-Hudson and Smith, 1980)

Ecological factors thus affect the possession of a territory and the size of the group in non-industrialized societies, including probably that of our hominid ancestors. Movement patterns in industrialized societies are more fluid, so territorial behaviour is less conspicuous and its function less clear.

A consequence of humans being territorial is that, at least in hunter gatherer cultures, they tend to be spaced out a great deal. In agricultural and industrial societies a low population density is unnecessary because of the increased localization and defensibility of the resources. Some authors have invoked population regulation as a function of territoriality in humans as well as in other animals (cf Malmberg, 1980) but other functions are more likely (see Chapter 11).

12.6.5 Gang warfare

Another aspect of human inter-group conflict that may be based on territoriality is the existence of street gangs comprising adolescents and young men. The phenomenon has been around for a long time and particular gangs, as opposed to their individual members, may last many years; one gang in Los Angeles has been in existence for more than fifty years (Stumphauzer, Veloz and Aiken, 1981). Such

Table 12.5 Occupations of convicted soccer hooligans (from Harrington, 1968, in Dunning *et al.*, 1982).

School or apprentice	79
Unskilled/labourer	206
Semi-skilled	112
Skilled	50
Salesman/clerical	19
Professional/managerial	2
Not known or unemployed	29
Total	497

gangs have attracted widespread attention recently because of the hooliganism associated with them. Football hooliganism is a major concern, especially in Britain, but it occurs throughout the world and has a long history. The hooligans are members of gangs which comprise hundreds of individuals, mostly unemployed or manual workers (Table 12.5), although they are dominated by a few individuals. They identify strongly with their local community and local football club and, although there may be conflicts within segments of the group at other times, they unite to confront groups from other localities. The members of the gang gain status by fighting, something they may be unable to gain by other means because of lack of education and job opportunities. In any culture (and probably in our hominid ancestors) young people form single sex, same age groups. However, hooliganism is particularly noticeable now because of media attention, which may have attracted people who enjoy fighting. Football itself provides an area with ready-made opponents to take the place of the more usual street gang fight (Dunning, Murphy and Williams, in press).

12.6.6 Aggression within the family

From a sociobiological viewpoint we would expect to see conflicts of interest in a family setting, between parents, between parents and offspring and between siblings (page 309), and all of these certainly occur in human families.

(a) Sibling rivalry

This is frequently seen in animals including humans. It is the most frequent type of family aggression in the USA, for example (Straus *et al.*, 1980). Even infants engage in it, with an older child typically being aggressive to a younger one and the aggression can increase over time (Kendrick and Dunn, 1983). In a study in which college students were asked about their relations with siblings at junior high school, Felson (1983) found that both verbal and physical aggression were more common amongst siblings than between unrelated children. In fights between siblings there were no sex differences in the frequency of aggression, whereas boys were more likely to fight with non-kin than were girls. Most sibling fights are over property ownership and about a quarter in Felson's study were over division of labour in the household, a source of quarrels that would be expected to increase in importance as the children grow older and take on more responsibilities.

(b) Within-pair conflict

Among animals, there are often conflicts between mates (page 303); similar conflicts occur in humans but most of the causes are probably uniquely human. Wife battering is widespread. Some of it is associated with sexual jealousy but the use of alcohol is also frequently involved (Gelles, 1980; Kahn *et al.*, 1980). Alcohol probably lowers the male's inhibitions against being aggressive towards his wife. Thus wife batterers are likely to experience a low job satisfaction and are unlikely to be religious (Gelles, 1980), and society's stereotyped view of women may also be involved (Straus, 1978). Some wife-battering also involves mentally ill husbands (Walker, 1981).

(c) Child abuse

Child abuse may have, or have had, a functional explanation, but there are proximate causes and explanations. Abusive parents are likely to have been abused themselves and they show poor parenting skills (Browne, in press). Social isolation, poverty and poor education may contribute to a parent being unable to cope with a child without using coercion. The parent's perception of a child is also important; a mother may fail to form a strong bond with a handicapped or premature child.

Data available for child abuse suggest that it occurs in similar instances to those in other animals (page 309; although that does not imply that the causes, either functional or proximate, are the same or that it is inevitable. Possible functional reasons for child abuse are (1) to increase an abusive male's own reproductive success, (2) to reduce competition for one's own offspring or (3) to terminate investment in offspring that are unlikely to survive to reproductive age. Among

Consequences, fitness and evolutionary change

Table 12.6 Circumstances of alleged infanticide in society (from Daly and Wilson, 1984).

Number of societies in which circumstance is alleged

Number of societies studied	Not sired by husband	De-formed or very ill	Twins	Birth too soon or too many	No male support	Mother dead	Mother unwed	Econ-omic hardship
Africa								
9	2	6	2	1	1	1	2	0
Asia								
2	1	2	1	0	0	0	0	0
Europe								
2	1	0	0	0	0	0	1	0
Middle East								
1	0	1	0	0	0	0	1	0
North America								
7	3	3	3	2	2	1	4	0
Oceania								
7	4	2	3	4	1	2	3	2
Russia								
2	0	1	0	0	0	1	1	1
South America								
9	5	6	5	4	2	1	2	0
Number of societies								
	16	21	14	11	6	6	14	3

humans, children are at greater risk of being attacked if they have a step-parent; only the step-children, not the step-parent's own offspring, are likely to be abused. Handicapped and premature children are also more at risk from abuse than are normal infants. Men are more likely to be abusive than women – males are less certain than mothers that the child is their own. Conversely biological mothers are least likely to abuse their children. Child abuse and neglect are also likely to occur in families with limited resources (Lenington, 1981; Lightcap, Kurland and Burgess, 1982; Daly and Wilson, 1984).

Infanticide occurs for several reasons in many societies (Daly and

Wilson, 1984; Table 12.6). For example, infants are commonly killed because they are deformed in some way or because they have been born too close in time to an older sibling and the mother cannot cope with both; one member of a pair of twins is often killed in Yanomamo society (Chagnon et al., 1979). Killing of females occurs in some human societies where sons are of more value than daughters, or in stratified societies where reproductive opportunities for high status women are poor and daughters are costly because dowries must be paid when they marry (Dickemann, 1979).

12.6.7 Alternative strategies

There is sometimes more than one way in which an animal can gain a resource (page 312) and the same is true of humans. In modern industrialized societies several possible reproductive strategies for males can be seen (Thornhill and Thornhill, 1983).

1. Males can pairbond and look after their offspring.
2. They can pairbond but give little parental care.
3. They may provide some resources to a female and the offspring but without pairbonding.
4. They may copulate, deceiving the female into believing resources will be provided (Shields and Shields, 1983).
5. Copulation may occur without explicit or implicit commitment.
6. Males may favour their sisters' offspring rather than reproducing themselves.
7. Males can forcibly copulate with females.

Low status males may adopt the last two alternatives in particular.

The statistics of rape are consistent with this sociobiological view. Most victims are of reproductive age and some rapes do result in full-term pregnancies (more probably did so in the past before abortion became commonplace). Most rapists are young, subordinate and have a low self-esteem (Shoham, 1982) so may gain few other mating opportunities. As Shields and Shields (1983) suggest, men rape when costs are low and potential benefits are high. That is victims are not usually related to the rapist and are usually vulnerable (during wartime for example).

Rapists keep costs down by planning the rape and frequently carrying weapons. Costs from retribution are low because rapes in Western society are under-reported, few convictions are made and punishments are mild anyway. When sanctions against rape are high, as it is in many highly polygynous societies, rape is infrequent or absent. Even in Western society, however, many males may contemplate rape but fear the punishment.

This functional view of rape is consistent with the facts, but function should not be confused with proximate causes of a behaviour. In the case of rape, it seems to be motivated by a sense of hostility and a desire for power rather than for sex and is associated with an acceptance of interpersonal violence and sex role stereotypy (Stark-Adamec and Adamec, 1982). There is thus more than one level (proximate and ultimate) on which rape can be prevented.

12.6.8 Co-operation

Both conflict and co-operation can be seen in human societies. From studies of other animals, particularly primates, biologists would suggest that the co-operation occurs because of kinship and reciprocal altruism (page 314). We can see this in non-industrialized societies. For example, among the Yanomamo physical assistance over disputes is most likely to come from kin (Chagnon and Burgos, 1979). Among !Kung bushmen only parents and their offspring, siblings and spouses usually help each other if the help is not reciprocated, and in potlatch rituals (where gifts are given) where non-kin benefit, reciprocation is expected.

In large industrialized societies it is more difficult to prevent conflict. The basis for altruism – kinship and individual reciprocity – has been largely removed because these societies are so large, mobile and impermanent, although on a personal level co-operation is apparent; for example, reciprocity for gifts is expected (Essock-Vitale and McGuire, 1980).

12.7 PREVENTING, PREDICTING AND CONTROLLING HUMAN AGGRESSION

One of the reasons for the interest in studying human aggression is to find ways to control or prevent it. However, whether or not we want to do so depends on complex moral and ethical issues which are outside the realm of biology. People sometimes justify aggression and make it acceptable to society. Religious violence, terrorism, political violence against individuals, groups or races, and of course warfare can all be justified by one society or another. Behaviour associated with dominance relations may be acceptable at a certain level and controls may only be needed to prevent excessive amounts of aggression occurring.

Many factors are involved in the expression of aggression, some of which society can change if it so desires. Biologists can only point out possible factors contributing to violence; it is up to society to decide whether or not to do something about them. Many factors contributing to human aggression are not biological in origin, but are based in our society.

12.7.1 Prevention

Means of preventing aggressive behaviour can be placed into one of three categories. Individuals showing particularly aggressive behaviour can be treated, for example with drugs or behaviour therapy; individuals at risk of developing aggressive behaviour can be identified and the behaviour modified; or the incidence of aggression in society can be reduced by global social measures (Rappaport and Holden, 1981).

Moral standards may play a role in reducing aggression since some aggression is seen as justifiable by society and hence by the individual (Feshbach, 1978). The condoning of violence, especially against women, has been blamed for criminal acts such as rape (Walker, 1981). Religion has been invoked as important in inhibiting aggression. It may be so for the individual, but seems more likely to provoke aggression between groups.

Changes in socio-economic factors are more likely to be important in reducing unwanted aggression. For example, Dunning et al. (1986) suggest there are several contributing factors to football hooliganism which could be controlled to some extent: youth unemployment, too much force used to control the fighting, poor relations between the football clubs and their supporters, and wide media coverage. It is also often suggested that child abuse and wife battering can be prevented in the long term by improving the environment in which the offender grows up.

12.7.2 Prediction

In order to develop preventive rather than curative measures, it is necessary to be able to predict which individuals, and in which situations or environments, are likely to become aggressive.

Many studies have now identified certain personality traits that are characteristic of aggressive individuals, although they are usually biased in examining only criminals or psychiatric patients. For example, in one study, violent prisoners were also neurotic and introverted (Mefferd et al., 1981); in another study, aggressive boys attending a psychiatric clinic also resisted discipline and were egocentric, impatient, impulsive and excitable (deBlois and Stewart, 1980). A correlation between particular personality traits and aggression, however, does not imply causality and the same traits may be seen in the general public without being associated with violent behaviour.

12.7.3 Control

Firstly, it is important to remember that animals, including humans, do not have an aggressive drive that must have an outlet (page 86). We do not need to look for alternatives to aggression, such as highly competitive games, to reduce feelings of aggression, although an adult-organized and supervised activity may inhibit the expression of that aggression.

Biological methods of controlling aggression encompass the use of drugs, hormones and psychosurgery. However, these are likely to be relatively unimportant in the overall control of aggression compared to non-biological methods.

(a) Drugs and hormones

Some aggressive behaviours are associated with certain mental illnesses, but the cause of the aggression is often unknown. Drugs may only control the symptoms of the illness rather than cure it. Aggression in schizophrenics can be eliminated with major tranquillizers or neuroleptics such as chlorpromazine; in manic-depressives with mood stabilizers such as lithium; in severe depression with antidepressants such as imipramine; and in psychomotor epilepsy with anticonvulsants such as primidone. Hyperactive, aggressive children have been successfully treated with amphetamines, and aggression in autistic and retarded children responds to treatment with lithium. Some hormones and drugs have been used on criminals, for example anti-androgens and progestational agents on sex offenders, but with limited success (Liberman, Marshall and Burke, 1981; Valzelli, 1981; Campbell, Cohen and Small 1982).

(b) Psychosurgery

This has also been used to control some aggressive patients. Removal of the temporal lobes reduces fear and rage among a whole range of complex behavioural effects, and temporal lobe lesions made to relieve epilepsy have similar effects (Valzelli, 1981). However, the results of such treatments are not clear-cut, and other behavioural changes often accompany the alteration in aggression (page 343).

(c) Non-biological treatment of individuals

An alternative treatment for institutionalized patients is behaviour therapy. This involves giving positive reinforcement for desirable behaviour and negative reinforcement for undesirable ones (Liberman *et al.*, 1981). Behaviour therapy may be useful for certain categories of aggressive people such as rapists (Abel *et al.*, 1981). Where aggressive behaviour seems to be due to a lack of skill in interacting with other people the teaching of social skills can be useful (Fredericksen and Rainwater, 1981). In addition, the teaching of parenting may reduce the incidence of child abuse, since parents who abuse their children seem to lack this skill, dealing with problems in an insensitive manner and using physical coercion to get their way (Reid, Taplin and Lorber, 1981; Browne, in press).

These non-biological treatments seem to offer more hope of controlling individual aggression than the use of drugs, hormones and psychosurgery which have only limited uses and which often suppress symptoms rather than dealing with the cause of the aggression.

12.8 OVERVIEW

Our intention in writing this chapter was to see whether biologists have anything to contribute to our understanding of aggression in the human species, and if so, how useful this contribution might be. Can the concepts used to describe and theories used to explain agonistic behaviour in animals be applied to people?

12.8.1 Another disclaimer

It is most important to stress that biological observations have no moral implications. Biology tells us, for example, that (notwithstanding homicide rates and football hooligans) humans are not unique among animals in sometimes killing each other (page 44) and that in some species it makes adaptive sense for males of low status to copulate forcibly (page 306). However, these observations are all totally irrelevant when it comes to deciding, as concerned people and not as biologists, whether murder or rape are acceptable forms of behaviour. Nor does an explanation in functional terms mean that the behaviour concerned is inaccessible to modification. Many environmental factors influence when and how often fighting occurs in non-human animals (page 208). Humans are uniquely susceptible to such influences (especially those stemming from our cultural background) and also have the ability to alter their environment, and thus to modify their aggressive responses.

12.8.2 How comparable are the behavioural phenomena?

The kinds of situation in which humans and animals come into conflict are comparable (page 328); thus we see infanticide, sibling rivalry, fighting over mates and conflict between the sexes over whether mating should occur. So is the form that agonistic interactions take; people sometimes defend territories or establish dominance hierarchies. In humans, as in animals, there is a tendency for males to be more aggressive than females, but females do fight when it is in their interest to do so (page 352). However, similar behavioural phenomena may have quite different underlying mechanisms and functions, so there is no reason to assume that because, for example, defence of an area in people looks like territoriality in other animals it is necessarily comparable in terms of its causes and consequences.

In addition, some manifestations of conflict in humans have no obvious parallels in animals. These mostly stem from human language and the complex cultural, moral and religious phenomena that this allows. Neither verbal aggression nor justification of hurting others have a simple counterpart in animals. Perhaps aggression during hand-to-hand fighting in humans reflects behavioural predispositions inherited from our primate ancestors, and perhaps small-scale inter-tribal wars are only different in degree from group defence of territories in primates. However, warfare between nation states is largely a political phenomenon; feelings of patriotism may have their roots in group identity as seen in primates, but wars are started by political and economic events and not by biological ones.

12.8.3 Biological explanations of the causes of human aggression

People are motivated to fight by internal factors as well as by external events, but these do not have the properties of an inevitably accumulating drive (page 336). One proximate cause of aggression in both human and non-human animals is exposure to aversive or frustrating situations, although in neither case does this inevitably cause fighting. The situations that trigger fights in higher primates are complex, but the sophisticated, cognitive influences on agonistic decisions (page 335) that our language allows probably have no counterparts among animals.

Such data as we have on the behavioural effects of testosterone (page 339) suggest that this hormone may enhance certain aspects of aggression and that testosterone secretion is subject to behaviourally mediated feed-back effects in human males as in many vertebrates. However, these effects are weak and all the social and environmental influences that complicate hormone–behaviour relationships in other animals are there with a vengeance in humans. There are very few satisfactory data relating differences in human aggression to neuro-

biological variation, although abnormally violent behaviour is sometimes associated with high levels of general cerebral arousal, perhaps
mediated by low serotonin levels.

12.8.4 Biology and the development of human aggression

A small proportion of naturally occurring variability in personality
traits which influence conflict behaviour in people depends on genetic
differences (page 345). Very young babies show aggressive responses
to aversive stimulation, and almost as soon as they can walk, children
show effective and organized aggression which is often structured
along dominance lines (page 347). However, from the very beginning
of their lives, young children experience complex patterns of positive
and negative reinforcement for aggressive behaviour in their relations
with their parents, with their peers and with unrelated adults. In
humans as in other animals, this experience promotes learning and
thus influences how individuals respond to conflict, but in humans
such effects are uniquely important in determining adult patterns of
aggressive behaviour. In addition, certain aspects of learning (for
example, through hearing verbal comments on fights, real or filmed)
are effectively special to people, as is the range of non-aggressive
alternatives available.

12.8.5 Sociobiology, behavioural ecology and human aggression

Sociobiology has a bad reputation as little more than oversimple
extrapolation to humans of facts discovered about animals. However,
good biological data, including many collected by scientists who
would describe themselves as sociobiologists, have a real contribution
to make in helping us to understand our own behaviour. For example,
the cost/benefit approach to behaviour, together with the idea of
inclusive fitness maximization (page 314) has revolutionized the way
biologists study the adaptive significance of animal behaviour. Some
of the more interesting contributions of biology to the human sciences
have been attempts to see whether this general approach (but *not*
specific facts about why animals fight) can explain any of the enormous variability that exists in human social behaviour (page 350).
Cost/benefit studies have helped us to understand the occurrence of
territoriality in human cultures and, possibly, the occurrence of
infanticide and rape (pages 352, 355 and 357).

The value of this approach to human behaviour is limited by the fact
that the selective forces impinging on people in the past may not be the

same as those acting today; indeed, as a result of complex modern culture, they may not be acting at all. In addition, and at the risk of being repetitive, we emphasize the fact that these studies only concern possible functions of human behaviour, not its causes and certainly not whether it is justified. Regardless of their adaptive consequences, defence of space and rape have complex proximate causes and developmental histories, involving cultural, sociological and political factors; their moral acceptability is also a completely separate issue.

12.8.6 Biology and the control of human aggression

Biologists have shown that aggression is not necessarily an aberrant or abnormal form of behaviour, but that it is part of the adaptive behavioural equipment with which an animal copes with its environment. So humans are not uniquely aggressive among animals and there is no need to search for the evolutionary origins of bloodthirstiness, for example, by evoking the switch from omnivorous scrounger to hunter. Biological studies have also shown that there is no necessary motivational link between overt aggressiveness and other, possibly desirable traits such as competitiveness and dominance. These provide background information which may help when moral, religious and political decisions are being made about the degree of control of aggression which is desirable.

It is clear that aggression does not inevitably accumulate and that vicarious violence and aggressive sport have the opposite of a cathartic effect; this precludes certain, frequently proposed, ways of reducing aggression. Some biological factors (level of physiological arousal, of testosterone or particular patterns of brain activity) contribute to individual variation in human aggressiveness and thus potentially provide a route for prediction and control of this behaviour. However, the effect of these organic factors is small compared to the effects of social and economic circumstances, for example. This is especially true when group aggression rather than individual aggression is concerned. The greatest change in behaviour will come from modification of the causal factor with the strongest effect, so if we seriously want to reduce levels of aggression in society, we should look to sociological, economic and political solutions and not to biology.

References

Abel, G.G., Becker, J.V., Blanchard, E. and Flanagan, B. (1981) The behavioural assessment of rapists. pp. 211–30. In *Violence and the Violent Individual* (eds J.R. Hays, T.K. Roberts and K.S. Solway) S.P. Medical and Scientific Books, New York.

Adams, D.B. (1979) Brain mechanisms for offense, defense and submission. *Behav. Brain Sci*, **2**, 201–41.

Adams, D.B. (1981) Motivational systems of social behaviour in male rats and monkeys: are they homologous? *Aggr. Behav.*, **7**, 5–18.

Adams, D.B. (1983) Hormone-brain interactions and their influence on agonistic behaviour. In *Hormones and Aggressive Behaviour* (ed B. Svare), Plenum, New York, pp. 223–45.

Adkins-Regan, E. (1983) Sex steroids and the differentiation and activation of avian reproductive behaviour. In *Hormones and Behaviour in Higher Vertebrates* (eds J. Balthazart, E. Pröve and R. Gilles) Springer-Verlag, Berlin, pp. 218–28.

Aggleton, J.P. and Passingham, R.E. (1981) Syndrome produced by lesions of the amygdala in monkeys (*Macacca mulatta*). *J. Comp. Physiol. Psychol.*, **95**, 961–72.

Ahmed, S.M. (1982) Factors affecting frustrating and aggression relationships. *J. Soc. Psychol.*, **116**, 173–77.

Aiello, J.R., Nicosia, G. and Thompson, D.E. (1979) Physiological, social and behavioural consequences of crowding on children and adolescents. *Child. Dev.*, **50**, 195–202.

Akerman, B. (1966) Behavioural effects of electrical stimulation of the forebrain in pigeons. *Behaviour*, **26**, 323–50.

Alatalo, R.V., Lundberg, A. and Stählbrandt, K. (1982) Why do pied flycatcher females mate with already-mated males? *Anim. Behav.*, **30**, 585–93.

Albert, D.J. and Chew, G.I. (1982) The septal forebrain and inhibitory modulation of attack and defense in the rat, a review. *Behav. Neur. Biol.*, **30,**, 357–88.

Albert, D.J. and Walsh, M.L. (1984) Neural systems and the inhibitory modulation of agonistic behaviour; a comparison of mammalian species. *Neurosci. Biobehav. Rev.*, **8**, 5–24.

Albone, E.S. and Shirley, S.G. (1984) *Mammalian Semiochemistry*. John Wiley and Sons, Chichester.

Alcock, J. (1979) The evolution of intraspecific diversity in male reproductive strategies in some bees and wasps. In *Sexual Selection and Reproductive Competition in Insects* (eds M.S. Blum and N.A. Blum), Academic Press, New York, pp. 381–402.

Alcock, J. (1983) Post copulatory male guarding by males of the damselfly (*Hetaerina vulnerata*) *Anim. Behav.*, **30**, 99–107.

Alcock, J. and Schaefer, J.E. (1983) Hilltop territoriality in a sonoran desert botfly. *Anim. Behav.*, **31**, 518–25.

Alexander, B.K. and Roth, E.M. (1970) The effects of acute crowding on aggressive behaviour of Japanese macaques. *Behaviour*, **39**, 73–81.

Alexander, R.D. (1961) Aggressiveness, territoriality and sexual behaviour in field crickets. *Behaviour*, **17**, 130–223.

Andersson, M. (1976) Social behaviour and communication in the great skua. *Behaviour*, **58**, 40–77.

Andersson, M. (1978) Optimal foraging area: size and allocation of search effort. *Theoret. Pop. Biol.*, **13**, 397–409.

Andersson, M., Wicklund, G. and Rundgren, H. (1980) Parental defence of offspring: a model and an example. *Anim. Behav.*, **28**, 536–42.

Andrew, R. (1972) The information potentially available in mammalian displays. In *Non-verbal Communication* (ed. R.A. Hinde) Cambridge University Press, pp.179–203.

Appleby, D.C. (1980) Social rank and food access in red deer stags. *Behaviour*, **74**, 294–309.

Apsey, W.P. (1975) Ontogeny of display in immature *Schizocosa crassipes*. *Psyche*, **82**, 174–80.

Apsey, W.P. (1977) Wolf-spider sociobiology: I Agonistic display and dominance–subordinance relations in adult male *Schistocosa crassipes*. *Behaviour*, **62**, 103–41.

Archer, J. (1976) The organisation of aggression and fear in vertebrates. In *Perspectives in Ethology, Vol. 2* (eds P.P.G. Bateson and P. Klopfer) Plenum, New York, pp. 231–98.

Archer, J. and Westeman, K. (1981) Sex differences in aggressive behaviour of school children. *Br. J. Soc. Psychol.*, **20**, 31–36.

Archer, M.E. (1985) Population dynamics of the social wasps *Vespula vulgaris* and *Vespula germanica* in England. *J. Anim. Ecol.*, **54**, 473–86.

Arens, W. (1979) The man-eating myth. Anthropology and Anthropophagy. Oxford University Press, New York.

Arnold, S.J. (1976) Sexual behaviour, sexual interference and sexual defence in salamanders, *Ambystoma maculatum*, *Ambystoma tigrinum* and *Plethodon jordani*. *Zeit. Tierpsychol.*, **42**, 247–300.

Arnold, S.J. (1977) The evolution of courtship behaviour in New World salamanders with some comments on Old World salamanders. In *The Reproductive Biology of Amphibians* (eds D.H. Taylor and S.I. Guttman) Plenum, New York, pp. 141–83.

Arthur, W. (1982) The evolutionary consequences of interspecific competition. *Adv. Ecol. Res.*, **12**, 127–87.

Austad, S.N. (1983) A game theoretical interpretation of male combat in the bowl and doily spider *Frontinella pyramitela*. *Anim. Behav.*, **31**, 59–73.

Averill, J.R. (1982) *Anger and Aggression: an Essay in Emotion*, Springer-Verlag, New York.

Avis, H.H. and Peeke, H.V. (1979) The effects of pargyline, scopolamine and imipramine on territorial aggression in the convict cichlid (*Cichlasoma nigrofasciatum*). *Psychopharmacology*, **66**, 1–2.

Ayer, M.L. and Whitsett, J.M. (1980) Aggressive behaviour of female prairie deermice in laboratory populations. *Anim. Behav.*, **28**, 763–71.

Ayre, D.J. (1982) Inter-genotype aggression in solitary sea anemones, *Actinia tenebrosa*. *Marine Biol.*, **68**, 199–206.

Baerends, G.P. (1984) The organisation of the pre-spawning behaviour of the cichild fish *Adequidens portalegrensis*. *Neth. Jnl. Zool.*, **34**, 233–366.

Bailey, W.J. and Stoddart, J.A. (1982) A method for the construction of dominance hierarchies in the field cricket *Teleogryllus oceanicus* (Le Guillou). *Anim. Behav.*, **30**, 216–20.

Baker, A.E.M. (1981) Gene flow in house mice: behaviour in a population cage. *Behav. Ecol. Sociobiol.*, **8**, 83–90.

Baker, M.C., Belcher, C.S. Deutsch, L.C. *et al.* (1981) Foraging success in junco flocks and the effects of social hierarchy. *Anim. Behav.*, **29**, 137–42.

Bakker, T.C.M. (1986) Aggressiveness in sticklebacks (*Gasterosteus aculeatus*): a behaviour genetic study. *Behaviour*, **98**, 1–144.

Baldwin, J.D. and Baldwin, J.I. (1977) The role of learning phenomena in the ontogeny of exploration and play. In *Primate Biosocial Development* (eds S. Chevalier-Skolmikoff and F.E. Poirier) Garland Publications, New York, pp. 343–408.

Balph, M.H., Balph, D.F. and Romesberg, H.C. (1979) Social status signalling in winter flocking birds: an examination of a current hypothesis. *Auk*, **96**, 78–93.

Bandler, R. (1982a) Identification of neuronal cell bodies mediating components of quiet biting attack in the cat: induction of jaw opening following microinjections of glutamate into the hypothalamus. *Brain Res.*, **245**, 192–97.

Bandler, R. (1982b) Neural control of aggressive behaviour. *Trends Neurosci.*, **5**, 390–94.

Bandura, A. (1983) Psychological mechanisms of aggression. In *Aggression. Theoretical and Empirical Reviews Vol. 1* (eds R.G. Green and E.I. Donnerstein, Academic Press, New York. pp. 1–40.

Banks, E.M. and Allee, W.C. (1957) Some relations betweens flock size and agonistic behaviour in domestic hens. *Physiol. Zool.*, **30**, 255–68.

Barette, C. (1977) Fighting behaviour of muntjac and the evolution of antlers. *Evolution*, **31**, 169–76.

Barker, D.G., Murphy, J.B. and Smith, K.W. (1979) Social behaviour in a captive group of Indian pythons, *Python molurus* with formation of a linear social hierarchy. *Copeia*, 466–71.

Barkley, M.S. and Goldman, B.D. (1977) The effects of castration and silastic implants of testosterone on intermale aggression in the mouse. *Horm. Behav.*, **9**, 32–48.

Barkley, M.S. and Goldman, B.D. (1978) Studies of opponent status and steroid mediation of aggression in female mice. *Behav. Biol.*, **23**, 118–23.

Barlow, G.W. (1974). Derivation of threat display in the gray reef shark. *Mar. Behav. Physiol*, **3**, 71–81.

Barlow, G.W. and Rowell, T. (1984) The contribution of game theory to animal behaviour. *Brain Behav. Sci.*, **7**, 101–03.

Barlow, G.W., Rogers, W. and Bond, A.B. (1984) Dummy-elicited aggressive behaviour in the polychromatic midas cichlid. *Biol. Behav.*, **9**, 115–30.

Barnard, C.J. (1984) *Producers and Scroungers* Croom-Helm, London.

Barnard, C.J. and Brown, C.A.J. (1984) A pay-off asymmetry in resident-resident disputes between shrews. *Anim. Behav.*, **24**, 302–04.

Barnard, C.J. and Sibly, R.M. (1981) Producers and scroungers: a general model and its applications to captive flocks of house sparrows. *Anim. Behav.*, **29**, 543–50.

Barner-Barry, C. (1980) The structure of young children's authority relationships. In *Dominance Relationships* (eds D.R. Omark, F.F. Strayer and D.G. Freedman) Garland STPM Press, New York, pp. 177–89.

Barnett, S.A., Hicking, W.E., Munroe, K.M.H. and Walker, K.Z. (1975) Socially induced renal pathology of captive wild rats (*Rattus villosissimus*). *Aggr. Behav.*, **1**, 123–34.

Barr, G.A. (1981) Effects of different housing conditions on intraspecies fighting between male Long-Evans hooded rats. *Physiol. Behav.*, **27**, 1041–44.

Barr, G.A., Gibbons, J.L. and Moyer, K.E. (1976) Male-female differences and the influence of neonatal and adult testosterone on intraspecific aggression in rats. *J. Comp. Physiol. Psychol.*, **90**, 1169–83.

Barrett, B.E. (1979) A naturalistic study of sex differences in children's aggression. Merril-Palmer Quarterly, 25, 193–203.

Barrett, P. and Bateson, P.P.G. (1978) The development of behaviour in cats. *Behaviour*, **66**, 106–20.

Bartos, L. (1980) The date of antler casting, age and social hierarchy relationships in the red deer stag. *Behav. Proc.*, **5**, 293–99.

Bartos, L. (1983) Some observations on the relationship between preorbital gland opening and social interactions in red deer. *Aggr. Behav.*, **9**, 59–67.

Basch, P.F. (1970) Competitive exclusion in parasites. *Exp. Parasitol.*, **27**, 193–216.

Bateman, A.J. (1948) Intra-sexual selection in Drosophila. *Heredity*, **2**, 349–68.

Bateson, P.P.G. (1981) Ontogeny of behaviour. *Brit. Med. Bull.*, **37,** 159–64.

Bayliss, J. (1981) The evolution of parental care in fishes, with reference to Darwin's rule of sexual selection. *Env. Biol. Fish.*, **6,** 223–51.

Beach, F., Buehler, M.C. and Dunbar, I.F. (1982) Competitive behaviour in male, female and pseudo-hermaphrodite female dogs. *J. Comp. Physiol. Psychol.*, **96,** 855–75.

Bean, N.J. (1982) Modulation of agonistic behaviour by the dual olfactory system in male mice. *Physiol. Behav.*, **29,** 433–37.

Bean, N.J. and Connor, R. (1978) Central hormonal replacement and home cage dominance in castrated rats. *Horm. Behav.*, **11,** 100–09.

Beden, S.N. and Brain, P.F. (1982) Studies on the effect of social stress on measures of disease resistance in laboratory mice. *Aggr. Behav.*, **8,** 126–29.

Bediz, G.M. and Whitsett, J.M. (1979) Social inhibition of sexual maturation in male prairie deer mice. *J. Comp. Physiol. Psychol.*, **93,** 493–500.

Beecher, M.D. and Beecher, I.M. (1979) Sociobiology of bank swallows: reproductive strategy of the male. *Science*, **205,** 1283–85.

Belcher, D.L. (1983) The evolution of agonistic behaviour in Amblyopsid fishes. *Behav. Ecol. Sociobiol.*, **12,** 35–42.

Bell, W.J. (1978) Directional cues in tactile stimuli involved in agonistic encounters in cockroaches. *Physiol. Entomol.*, **3,** 1–6.

Belson, W.A. (1978) *Television Violence and the Adolescent Boy.* Saxon House Westmead.

Belyaev, D.K. and Trut, L.N. (1975) Some genetic and endocrine effects of selection for domestication in silver foxes. In *Wild Canids* (ed. M.W. Fox) Van Nostrand, New York, pp. 416–26.

Bentley, D.R. and Hoy, R.R. (1972) Genetic control of the neural network generating cricket (*Teleogryllus gryllus*) calls. *Anim. Behav.*, **20,** 478–92.

Benton, D. (1981) The measurement of aggression in the laboratory. In *The Biology of Aggression* (eds P.F. Brain and D. Benton) Sijthoff and Noordhoff, Alphen aan den Rijn, Netherlands, pp. 487–502.

Berg, R.A., Shanin, R.D. and Hull, E.M. (1975) Early isolation in the gerbil (*Meriones unguiculatus*): behavioural and physiological effects. *Physiol. Psychol.*, **3,** 35–38.

Berger, J. (1980) The ecology, structure and functions of play in bighorn sheep (*Ovis canadensis*). *J. Zool.*, **192,** 531–42.

Berkowitz, L. (1981) On the difference between internal and external reactions to legitimate and illigitimate frustrations: a demonstration. *Aggr. Behav.*, **7,** 83–96.

Berkowitz, L. (1982) Aversive conditions as stimuli to aggression. *Adv. Exp. Soc. Psychol.*, **15,** 249–88.

Berkowitz, L. (1983) Aversively stimulated aggression. Some parallels and differences in research with animals and humans. *Amer. Psychol.*, **38,** 1135–44.

Berkowitz, L. and Powers, P.C. (1979) Effects of timing and justification of witnessed aggression on the observer's punitiveness. *J. Res. Pers.*, **13**, 71–80.

Berkowitz, L., Cochran, S.T. and Embree, M.C. (1981) Physical pain and the goal of aversively stimulated aggression. J. Pers. Soc. Psychol., **40**, 687–700.

Berman, C.M. (1980) Early agonistic experience and rank acquistion among free ranging rhesus monkeys. *Int. J. Primatol.*, **1**, 153–70.

Bermond, B. (1978) Neuro-hormonal regulation of aggressive behaviour in the rat. Functions of testosterone and the area preoptica hypothalamus anterior. Ph. D. thesis, University of Amsterdam.

Bernard, B.K., Findelstein, E.R. and Everett, G.M. (1975) Alterations in mouse aggressive behaviour and brain monoamine dynamics as a function of age. *Physiol. Behav.*, **15**, 731–36.

Bernstein, I.S. (1985) The adaptive value of maladaptive behaviour. *Ethol. Sociobiol.*, **85**, 297–303.

Bernstein, I.S. and Ehardt, C.L. (1985) Agonistic aiding: kinship, rank, age and sex influences. *Amer. J. Primatol.*, **8**, 37–52

Bernstein, I.S. and Ehardt, C.L. (1986) The influence of kinship and socialization on aggressive behaviour in rhesus monkeys *Macaca mulatta*. *Anim. Behav.*, **34**, 739–47.

Berril, M. (1976) Aggressive behaviour of post-puerulus larvae of the western rock lobster, *Palinurus longipes*. *Austr. J. Marine Fresh. Res.*, **27**, 83–88.

Berry, J.F. and Shine, R. (1980) Sexual size dimorphism and sexual selection in turtles. *Oecologia*, **44**, 185–91

Berthold, A.A. (1849) Transplantation der Hoden. *Arch. Anat. Physiol.*, **16**, 42–46.

Bertilson, H.S. (1983) Methodology in the study of aggression. In *Aggression. Theoretical and Empirical Reviews. Vol. 1* (ed R.G. Green and E.J. Donnerstein) Academic Press, New York, pp. 213–45.

Bertram, B.C.R. (1975) Social factors influencing reproduction in wild lions. *J. Zool.*, **177**, 462–82.

Betts, B.J. (1976) Behaviour in a population of columbian ground squirrels, *Spermophilus columbianus columbianus*. *Anim. Behav.*, **24**, 652–80.

Bhatia, K. and Golin, S. (1978) Role of locus of control in frustration-produced aggression. *J. Comp. Physiol. Psychol.*, **46**, 364–65.

Biben, M. (1983) Comparative ontogeny of social behaviour in three South American canids, the maned wolf, the crab eating fox and bush dog: implications for sociality. *Anim. Behav.*, **31**, 814–26.

Billy, A.J. and Liley, N.R. (1985) The effects of early and late androgen treatments on the behaviour of *Sarotherodon mossambicus*. *Horm. Behav.*, **19**, 311–30.

Birkhead, T.T. (1982) Timing and duration of mate guarding in magpies, *Pica pica*. *Anim. Behav.*, **30**, 277–83.

Bisazza, A. (1982) Hereditary differences in social behaviour of male mice (*Mus musculus*). *Boll. Zool.*, **49**, 207–11.

Bjerke, T. (1981) Aggression in young organisms: why does it occur despite aversive stimulation? In *Behavioural Development* (eds K. Immelmann, G.W. Barlow, L. Petrinovich and M.Main) Cambridge University Press, pp. 593–602.

Black-Cleworth, P. (1970) The role of electric discharges in the non-reproductive social behaviour of *Gymnotus carapo*. *Anim. Behav. Monog.*, **3**, 1–77.

Blanchard, D.C., Blanchard, R.J., Lee, E.N.C. and Nakamura, S. (1979) Defensive behaviour in rats following septal and septal-amygdala lesion. *J. Comp. Physiol. Psychol.*, **93**, 378–90.

Blanchard, R.J. and Blanchard, D.C. (1981) The organisation and modelling of animal aggression. In *The Biology of Aggression* (eds P.F. Brain and D. Benton) Sijthoff and Noordhoff, Alphen aan den Rhijn, Netherlands, pp. 529–61.

Blanchard, R.J., Takahashi, L.K., Fukunaga, K.K. and Blanchard, D.C. (1977) Functions of the vibrissae in the defensive and aggressive behaviour of the rat. *Aggr. Behav.*, **3**, 231–40.

Blum, N.S. and Blum, M.A. (1979) *Reproductive Competition and Sexual Selection in Insects*. Academic Press, New York.

Blurton-Jones, N.B. (1972) Categories of child-child interaction. In *Ethological Studies of Child Behaviour* (ed. N. Blurton-Jones) Cambridge University Press, pp. 97–127.

Blurton-Jones, N.B. and Konner, M. (1973) Sex differences in behaviour of London and Bushman children. In *Comparative Ecology and Behaviour of Primates* (eds. R.P. Michael and J.H. Crook) Academic Press, London, pp. 690–749.

Boag, D.A. and Alway, J.H. (1980) Effects of social environment within the brood and dominance rank in gallinaceous birds. *Can. J. Zool.*, **58**, 44–49.

Boag, D.A. and Alway, J.H. (1981) Heritability of dominance status among Japanese quail: a preliminary report. *Can. J. Zool.*, **59**, 441–44.

Bols, R.J. (1977) Display reinforcement in the Siamese fighting fish *Betta splendens*: aggressive motivation or curiosity. *J. Comp. Physiol. Psychol.*, **91**, 233–44.

Boonstra, R. (1978) Effect of adult Townsend voles (*Microtus townsendii*) on survival of young. *Ecology*, **59**, 242–48.

Boonstra, R. (1984) Aggressive behaviour of adult meadow voles (*Microtus pennsylvanicus*) towards young. *Oceologia*, **62**, 126–31.

Borowsky, R. (1978) Social inhibition of maturation in natural populations of *Xiphophorus variegatus*. *Science*, **201**, 933.

Bouissou, M.F. (1983) Hormonal influences in aggressive behaviour in ungulates. In *Hormones and Aggressive Behaviour* (ed. B. Svare) Plenum, New York, pp. 507–33.

Bouissou, M.F. and Gaudioso, V. (1982) Effect of early androgen treatment on subsequent social behaviour in heifers. *Horm. Behav.*, **16** 132–46.

Bourguigon, E. (1979) *Psychological Anthropology. An Introduction to Human Nature and Cultural Differences*. Holt, Rinehart and Winston, New York.

Bowman, L.A., Dilley, S.R. and Keverne, E.B. (1978) Suppression of oestrogen-induced LH surges by social subordination in talapoin monkeys. *Nature*, **275**, 56–57.

Brace, R.C. (1981) Intraspecific aggression in the colour morphs of the anemone *Phymactis clematis* from Chile. Marine Biol., **64**, 85–93.

Brace, R.C. and Pavey, J. (1978) Size-dependent dominance and hierarchy in the anemone *Actinia equina*. *Nature*, **273**, 752–53.

Brace, R.C., Pavey, J. and Quicke, D.J. (1979) Intraspecific aggression in the colour morphs of the anemone *Actinia equina*: the 'convention' governing dominance ranking. *Anim. Behav.*, **27**, 553–61.

Bradley, A.J., McDonald, I.R. and Lee, A,K. (1976) Corticosteroid binding globulin and mortality in a dasyurid marsupial. *J. Endocrinol.*, **70**, 323–24.

Brain, P.F. (1980) Adaptive aspects of humoral correlates of attack and defeat in laboratory mice: a study in ethobiology. *Rec. Progr. Br. Res.*, **53**, 391–413.

Brain, P.F. (1981) Hormones and aggression in infra-human vertebrates. In *The Biology of Aggression* (eds P.F. Brain and D. Benton) Sijthoff and Nordhoff, Alphen aan den Rijn, Netherlands, pp. 181–214.

Brain, P.F. (1983) Pituitary-gonadal influences on social aggression. In *Hormones and Aggression* (ed. B.B. Svare) Plenum, New York, pp. 3–25.

Brain, P.F. (1984) Biological explanations of human aggression and resulting therapies offered by such approaches: a critical evaulation. In *Advances in the Study of Aggression. Vol. 1* (eds. R.J. Blanchard and C.D. Blanchard) Academic Press, Orlando, Florida, pp. 63–102.

Brain, P.F. (1986) *Alcohol and Aggression*. Croom Helm, London.

Brain, P.F., Haug, M. and Kamis, A (1983) Hormones and different tests for aggression with particular reference to the effects of testosterone metabolites. In *Hormones and Behaviour in Higher Vertebrates* (eds J. Balthazart, E. Pröve and R. Gilles) Springer-Verlag, Berlin, pp. 290–304.

Branch, G.M. (1975) Mechanisms reducing intraspecific competition in *Patella* spp.: migration, differentiation and territorial behaviour. *J. Anim. Ecol.*, **44**, 575–600.

Breed, M.D. (1980) Behavioural strategies during inter-male agonistic interactions in a cockroach. *Anim. Behav.*, **28**, 1063–69.

Breed, M.D. and Bell, W.J. (1983) Hormonal influences on invertebrate aggressive behaviour. In *Hormones and Aggressive Behaviour* (ed. B.B. Svare) Plenum, New York, pp. 577–90.

Breed, M.D., Smith, S.K. and Gall, B.G. (1980) Systems of mate selection in a cockroach species with male dominance hierarchies. *Anim. Behav.*, **28**, 130–34.

Brett, R.A. (1986) The ecology and behavior of the naked mole-rat (*Heterocephalus glaber* Ruppell) (Rodentia: Bathyergidae). Ph D. thesis, University of London.

Brian, M.V. (1979) Caste differentiation and division of labour. In *Social Insects, Vol 1* (ed. H.R. Harman) Academic Press, New York, pp. 121–222.

Brinck, C. and Hoffmeyer, I. (1984) Marking urine and preputial gland secretion of male bank voles *Clethrionomys glareolus J. Chem. Ecol.*, **10**, 1295–1307.

Brockman, J. (1984) The evolution of social behaviour in insects. In *Behavioural Ecology, 2nd ed* (Eds J.R. Krebs and N.B. Davies) Blackwell, Oxford, pp. 340–60.

Brodzinsky, D.M., Messa, S.B. and Tew, J.D. (1979) Sex differences in children's expression and control of fantasy and overt aggression. *Child Dev.*, **50**, 372–79.

Broida, J., Michael, S.D. and Svare, B. (1981) Plasma prolactin levels are not related to the initiation, maintenance and decline of postpartum aggression in mice. *Behav. Neur. Biol.*, **32**, 121–28.

Bronson, F.H. (1973) Establishment of social rank among grouped mice: relative effects on circulating FSH, LH and corticosterone. *Physiol. Behav.*, **10**, 947–51.

Bronson, F.H. (1976) Urine marking in mice: causes and effects. In *Mammalian Olfaction, Reproductive Processes, and Behaviour* (ed R.L. Doty) Academic Press, New York, pp. 119–41.

Bronson, F.H., Stetson, M.H. and Stiff, M.E. (1973) Serum FSH and LH in male mice following aggressive and non-aggressive interactions. *Physiol. Behav.*, **10**, 369–92.

Bronstein, P.M. (1981) Social reinforcement in *Betta splendens*: a reconsideration. *J. Comp. Physiol. Psychol.*, **81**, 943–50.

Bronstein, P.M. (1985a) Predictors of dominance in male *Betta splendens*. *J. Comp. Psychol.*, **99**, 47–55.

Bronstein, P.M. (1985b) Prior-residence effect in *Betta splendens*. *J. Comp. Psychol.*, **99**, 56–59.

Brooke, M. de L. (1981) Size as a factor influencing ownership of copulation burrows by the Ghost crab (*Ocypode ceratophthalamus*). *Zeit. Tierpsychol.*, **55**, 63–78.

Broom, D.M. and Leaver, J.D. (1978) Effects of group-rearing or partial isolation on later social behaviour of calves. *Anim. Behav.*, **26**, 1255–63.

Brown, J.L. (1963) Aggressiveness, dominance and social organisation in the Stellar jay. *Condor*, **65**, 460–84.

Brown, J.L. (1964) The evolution of diversity in avian territorial systems. *Wils. Bull.*, **76**, 160–69.

Brown, J. and Hunsperger, R.W. (1963) Neuroethology and the motivation of agonistic behaviour. *Anim. Behav.*, **11**, 439–48.

Brown, R.D., Chao, C.C. and Faulkner, L.W. (1983) The endocrine control of the initiation and growth of antlers in the white-tailed deer, *Odocoileus virginiatus. Acta End.*, **103**, 138–44.

Browne, K.D. (In press) Sequential analysis of social behaviour patterns shown by mother and infant in abusing and non-abusing families. In *Proc. 10th. Int. Congr. Assoc. Child and Adol. Psychiatry, 1982*, (ed. J. Anthony).

Bryant, D.M. (1979) Reproductive costs in the house martin. *J. Anim. Ecol.*, **48**, 655–75.

Brzoska, J. (1982) Vocal response of male European water frogs (*Rana esculenta* complex) to mating and territorial calls. *Behav. Proc.*, **7**, 37–47.

Bucher, T.L., Ryan, M.J. and Bartholomew, C.A. (1981) Oxygen consumption during resting, calling and nest building in the frog *Physalaemus pustulosus. Physiol. Zool.*, **55**, 10–22.

Buckle, G.R. and Greenberg, L. (1981) Nestmate recognition in sweat bees (*Lasioglossum zephyrum*): does an individual recognise its own odour or only odours of its nestmates? *Anim. Behav.*, **29**, 802–09.

Buhot-Averseng, M-C. and Goyens, J. (1981) Effects of cohabitation with another female on aggressive behaviour in pregnant mice. *Aggr. Behav.*, **7**, 111–22.

Bullock, D. and Merrill, L. (1980) The impact of personal preference on consistency through time: the case of childhood aggression. *Child Dev.*, **51**, 808–14.

Bunnell, B.N., Sodetz, F.J. and Shalloway, D.I. (1970) Amygdaloid lesions and social behaviour in the golden hamster. *Physiol. Behav.*, **5**, 153–61.

Burger, A.E. and Millar, R.P. (1980) Seasonal changes in sexual and territorial behaviour and plasma testosterone levels in male lesser sheathbills (*Chionis minor*). *Zeit. Tierpsychol.*, **52**, 397–406.

Burger, J. (1980) Territory size differences in relation to reproductive stage and type of intruder in Herring gulls (*Larus argentatus*). *Auk*, **97**, 733–741.

Burger, J. (1981) Super territories – a comment. *Amer. Nat.*, **118**, 578–80.

Burgess, J.W. and Uetz, C.W. (1982) Social spacing strategies in spiders. In *Spider Communication. Mechanisms and Ecological Significance* (ed. P.N. Witt and J.S. Rovner) Princeton University Press, pp. 317–51.

Burghardt, G. (1978) Behavioural ontogeny in reptiles. In *The Development of Behaviour* (eds G. Burghart and M. Bekoff) Garland STMP, New York, pp. 149–74.

Burghardt, G. and Rand, A.S (1982) *Iguanas of the World*. Noyes Publishers, Park Ridge, New Jersey.

Buschinger, A., Winter, U. and Faber, W. (1983) The biology of *Myrmoxenus gardiagini. Psyche*, **90**, 335–42.

Buskirk, R.E. (1975) Aggressive display and orb defence in a communal spider, *Metabus gravidus. Anim. Behav.*, **23**, 560–67.

Buskirk, R.E., Frolich, C. and Ross, K.G. (1984) The natural selection of sexual cannibalism. *Amer. Nat.*, **123**, 612–25.

Byers, J.A. (1980) Play partner preference in Siberian ibex, *Capra ibex sibirica. Zeit. Tierpsychol.*, **53**, 23–40.

Byers, J.A. (1983) Social interactions of juvenile collared peccaries, *Tayassu tajucu*, *J. Zool.*, **201**, 83–96.

Byers, J.A. (1984) Play in ungulates. In *Play in Animals* (ed. P.K. Smith) Blackwells, Oxford, pp. 43–65.

Cade, W. (1979) The evolution of alternative male reproductive strategies in field crickets. In *Sexual Selection and Reproductive Competition in Insects* (eds M.S. Blum and N.S. Blum) Academic Press, New York, pp. 343–79.

Cade, W. (1981) Alternative mating strategies: genetic differences in crickets. *Science*, **212**, 563–64.

Cairns, R.B., Hood, K.E. and Midlarn, J. (1985) On fighting in mice: is there a sensitive period for isolation? *Anim. Behav.*, **33**, 166–80.

Cairns, R.B., MacCombie, D.J. and Hood, K.E. (1983) A developmental-genetic analysis of aggressive behaviour in mice. 1. Behavioural outcomes. *J. Comp. Psychol.*, **97**, 69–89.

Caldwell, R.C. (1979) Cavity occupation and defensive behaviour in the stomatopod (*Gonodactylus festai*): evidence for chemically mediated individual recognition. *Anim. Behav.*, **27**, 197–201.

Calhoun, J.B. (1962) Population density and social pathology. *Sci. Amer.*, **206**, 138–48.

Camhi, J.M. (1984) *Neuroethology*. Sinauer Assoc, Sunderland, Mass.

Cammaerts, M.C., Inwood, M.R., Morgan, E.D., *et al.* (1978) Comparative studies of the pheromones emitted by workers of the ants *Myrmica rubra* and *M. sabrinodis*. *J. Insect Physiol.*, **24**, 207–14.

Campbell, D.J. and Shipp, E. (1979) Regulation of spatial pattern in populations of the field cricket *Teleogryllus commodus*. *Zeit. Tierpsychol.*, **51**, 260–68.

Campbell, M., Cohen, I.L. and Small, A.M. (1982) Drugs in aggressive behaviour. *J. Amer. A. Chil.*, **21**, 107–17.

Capanna, E., Corti, M., Mainardi, D., *et al.* (1984) Karyotype and intermale aggression in wild house mice: ecology and speciation. *Behav. Genet.*, **14**, 195–208.

Carli, G., Diprisco, C.L., Martelli, G. and Viti, A. (1982) Hormonal changes in soccer players during an agonistic season. *J. Sport Med.*, **22**, 489–94.

Caro, T. (1981) Predatory behaviour and social play in cats. *Behaviour*, **76**, 1–24.

Caro, T. and Alawi, R.M. (1985) Comparative aspects of behavioural development in two species of free-living hyrax. *Behaviour*, **95**, 87–109.

Caro, T.M. and Bateson P. (1986) Organization and ontogeny of alternative tactics. *Anim. Behav.*, **34**, 1483–99.

Caro, T. and Collins, D.A. (1986) Male cheetahs of the Serengeti. *National Geographic Research*, **2**, 75–86.

Carpenter, C.C. (1977), A survey of stereotyped reptilian behaviour patterns. In *Biology of the Reptilia, Vol 7.* (eds C. Gans and D. Tinkle). London, Academic Press, pp. 335–404.

Carpenter, C.C. (1982) The aggressive displays of iguanine lizards. In

Iguanas of the World (eds G.M. Burghard and A.S. Rand) Noyes Publications, Park Ridge, New Jersey, pp. 215–31.

Carpenter, C.C., Gillingham, J.C. and Murphy, J.B. (1976) The combat ritual of the rock rattlesnake *Crotalus cepidus*. *Copeia*, 764–80.

Carpenter, F.L. and Macmillen, R.E. (1976) Energetic cost of feeding territories in an Hawaiian honeycreeper. *Oecologia*, **26**, 213–23.

Carpenter, F.L., Paton D.C. and Hixon, M.A. (1983) Weight gain and adjustment of feeding territory size in migrant birds. *Proc. Natl. Acad. Sci.*, **80**, 7259–63.

Carroll, D., and O'Callaghan, M.A.J. (1981) Psychosurgery and the control of aggression. In *The Biology of Aggression* (eds P.F. Brain and D. Benton), Noordhoof and Sijthoff, Alphen aan den Rijn, pp. 457–472.

Carroll, J.C. (1977) The intergenerational transmission of family violence: the long-term effects of aggressive behaviour. *Aggr. Behav.*, **3**, 289–99.

Caryl, P. (1982) Animal signals: a reply to Hinde. *Anim. Behav.*, **30**, 240–44.

Cavallin, B.A. and Houston, B.K. (1980) Aggressiveness, maladjustment, body experience and the protective function of personal space. *J. Clin. Psychol.*, **36**, 170–76.

Chagnon, N.A. and Burgos, P.E. (1979) Kin selection and conflict; an analysis of Yanomano ax fights. In *Evolutionary Biology and Human Social Behaviour* (eds N.A. Chagnon and W. Irons). North Scituate, Mass. Duxbury Press, pp. 213–320.

Chagnon, N.A. and Irons, W. (1979) *Evolutionary Biology and Human Social Behaviour*. Duxbury Press, North Scituate, Mass.

Chagnon, N.A., Finn, M.V. and Melancon, T.F. (1979) Sex ratio variation among the Yanomano indians. In *Evolutionary Biology and Human Social Behaviour*. (Eds. N.A. Chagnon and W. Irons). North Scituate, Mass, Duxbury Press, pp. 290–320.

Chalmers, N. (1984) Social play in monkeys: theories and data. In *Play in Animals* (ed. P.K. Smith) Blackwell, Oxford, pp. 141–81.

Chamove, A.S. (1980) Non-genetic induction of acquired levels of aggression. *J. Abn. Psychol.*, **89**, 469–88.

Chao, L. and Levin, B.R. (1981) Structured habitats and the evolution of anti-competitor toxins in bacteria. *Proc. Natl. Acad. Sci.*, **78**, 6324–28.

Chapais, B. (1983) Reproductive activity in relation to male dominance and the likelihood of ovulation in Rhesus monkeys. *Behav. Ecol. Sociobiol.*, **12**, 215–28.

Chauvin-Muckensturm, B. (1978) Les réactions aux leurres des Epinoches vivant en groupe. *Biol. Behav.*, **3**, 233–42.

Chauvin-Muckensturm, B. (1979) Le problèm de la fixation génotypique du comportement aggressif chez les épinoches. *C. R. Acad. Sci. Paris D*, **289**, 935–38.

Cheng, K.M. Burns, J.T., McKinney, F. (1982) Forced copulation in captive mallards *Anas platyrhynchos*: II. Temporal factors. *Anim. Behav.*, **30**, 695–99.

Chevalier-Skolnikoff, S. (1973) Facial expression of emotion in nonhuman primates. In *Darwin and Facial Expression* (ed. P. Ekman) Academic Press, New York, pp. 11–89.

Cheyne, J.A. (1978) Communication, affect and social behaviour. In *Aggression Dominance, and Individual Spacing.* (eds L. Krames, P. Pliner and T. Alloway) Plenum, New York, pp. 129–53.

Chizar, D., Drake, R.W. and Windell, J.T. (1975) Aggressive behaviour of rainbow trout (*Salmo gairdneri*) of two ages. *Behav. Biol*, **13**, 425–431.

Chizar, D., Ashe, V., Seixas, S. and Henderson, D. (1976) Social–aggressive behaviour after various intervals of social isolation in bluegill sunfish (*Lepomis macrochirus*) in different states of reproductive readiness. *Behav. Biol.*, **16**, 475–87.

Christensen, M.T. (1977) Aggressive behaviour between wild and genetically sterile male Norway rats (*Rattus norvegicus*). *Aggr. Behav.*, **3**, 17–31.

Christian, J.J. (1971) Population density and reproductive efficiency, *Biol. Reprod.*, **4**, 248–294.

Christian, J.J. (1980) Endocrine factors in population regulation. In *Biosocial Mechanisms of Population Regulation* (eds M.N. Cohen, R.S. Malpass and H.G. Klein), Yale University Press, New Haven, pp. 55–116.

Christiansen, K.O. (1977a) A review of studies of criminality among twins. In *Biosocial Bases of Criminal Behaviour.* (eds S.A. Mednick and K.O. Christiansen) Gardner Press, New York, pp. 45–88.

Christiansen, K.O. (1977b) A preliminary study of criminality among twins. In *Biosocial Bases of Criminal Behaviour* (eds S.A. Mednick and K.O. Christiansen) Gardner Press, New York, pp. 89–108.

Christie, M.H. (1979) Effects of aromatisable androgens on aggressive behaviour among rats. *Horm. Behav.*, **13**, 85–92.

Christie, M.H. and Barfield, R.J. (1979) Effects of castration and home cage residency on aggressive behaviour in rats. *Horm. Behav.*, **13**, 85–91.

Ciaranello, R.D. (1979) Genetic regulation of the catecholamine synthesising enzymes. In *Genetic Variation in Hormone Systems*, Volume II (ed. R. Shire) CRC. Press, Boca Raton, Florida.

Ciaranello, R.D., Lipsky, A. and Axelrod, J. (1974) Association between fighting behaviour and catecholamine biosynthetic enzyme activity in two inbred mouse sublines. *Proc. Natl. Acad. Sci.*, **71**, 3006–08.

Clancy, A.N., Coquelin, A., Macrides, F., Gorski, R.A. and Noble, E.P. (1984) Sexual behaviour and aggression in male mice: involvement of the vomeronasal system. *J. Neurosci*, **4**, 2222–29.

Clark, C.R. and Nowell, N.W. (1979) The effect of anti-oestrogen C1-628 on androgen-induced aggressive behaviour in castrated male mice. *Horm. Behav.*, **12**, 205–10.

Clarke, J.R. (1953) The effect of fighting on the adrenals, thymus and spleen of the vole *Microtus agrestis*. *J. Endocrinol.*, **9**, 114–26.

Cloudsley-Thompson, T.L. (1958) *Spiders, Scorpions, Centipedes and Mites.* Pergamon, New York.

Clutton-Brock, T.H. and Albon, S.D. (1979) The roaring of red deer and the evolution of honest advertisement. *Behaviour.*, **69**, 145–70.

Clutton-Brock, T.H. and Albon, S.D. (1985) Competition and population regulation of social mammals. In *Behavioural Ecology* (ed. R. Sibly and R.H. Smith) Blackwell Scientific Publications, Oxford, pp. 557–75.

Clutton-Brock, T.H., Albon, S.D., and Guinness, F.E. (1986) Great expectations: dominance, breeding success and offspring sex ratios in red deer. *Anim. Behav.*, **34**, 460–71.

Clutton-Brock, T.H., Guinness, F.E. and Albon, S.D. (1982) *Red Deer. Behaviour and Ecology of Two Sexes*, Edinburgh University Press.

Coates, D. (1980) Prey-size intake in humbug damselfish, *Dascyllus arunus* (Pisces, Pomacentridae) living within social groups. *J. Anim. Ecol.*, **49**,. 335–40.

Cochran, C.A. and Perachio, A.A. (1977) Dihydrotestosterone propionate effects on dominance and sexual behaviours in gonadectomised male and female rhesus monkeys. *Horm. Behav.*, **8**, 175–87.

Coe, C.L., Mendoza, S.P. and Levine, S. (1979) Social status constrains the stress response in the squirrel monkey. *Physiol. Behav.*, **23**, 633–38.

Cole, K.S. and Noakes, D.L.G. (1980) Development of early social behaviour of rainbow trout, *Salmo gairdneri. Behav. Proc.*, **5**, 97–112.

Colgan, P.W. and Gross, M.R. (1977) Dynamics of aggression in male pumpkinseed sunfish (*Lepomis gibbosus*) over the reproductive phase. *Zeit. Tierpsychol.*, **43**, 139–51.

Cook, W.T. and Siegel, P.B. (1974) Social variables and divergent selection for mating behaviour of male chickens. *Anim. Behav.*, **22**, 390–96.

Cooper, W.E. (1977) Information analysis of agonistic behavioural sequences in male iguanid lizards, *Anolis carolinensis. Copeia*, 721–35.

Copenhaver, C. and Ewald, P.W. (1980) Cost of territory establishment in hummingbirds. *Oecologia*, **46**, 155–60.

Cox, C.R. (1981) Agonistic encounters among male elephant seals: frequency, context and the role of female preference. *Amer. Zool.*, **21**, 197–209.

Craig, J.L. and Douglas, M.E. (1986) Resource distribution, aggressive asymmetries and variable access to resources in the nectar feeding bellbird. *Behav. Ecol. Sociobiol.*, **18**, 231–40.

Cram, A. and Evans, S.M. (1980) Stability and lability in the evolution of behaviour in nereid polychaetes. *Anim. Behav.*, **28**, 483–90.

Crane, J. (1949) Comparative biology of salticid spiders III. Systematics and behaviour in representative species. *Zoologica*, **34**, 31–52.

Crane, J. (1957) Basic patterns of display in fiddler crabs. *Zoologica*, **42**, 69–82.

Crane, J. (1966) Combat, display and ritualization in fiddler crabs. *Phil. Trans. Roy. Soc. B*, **251**, 459–72.

Crapon de Caprona, M.D. (1982) The influence of early experience on preferences for optical and chemical cues produced by both sexes in the cichlid fish *Haplochromis burtoni. Zeit. Tierpsychol.*, **58,** 329–61.

Crassini, B., Law, H.G. and Wilson, E. (1979) Sex differences in assertive behaviour. *Austr. J. Psychol.*, **31,** 15–20.

Crews, D. (1983) Regulation of reptilian reproductive behaviour. In *Advances in Vertebrate Neuroethology* (eds J.P. Ewer, R.R. Caprianica and D. Ingle), Plenum, New York, pp. 997–1032.

Crews, D. and Greenberg, N. (1981) Function and causation of social signals in lizards. *Amer. Zool.*, **21,** 273–94.

Crews, D. and Moore, M.C. (1986) Evolution of mechanisms controlling mating behaviour. *Science*, **231,** 121–25.

Crews, D., Gustafson, S. and Tokartz, R.R. (1983) Psychobiology of parthenogenesis. In *Lizard Ecology* (eds R.B. Huey, E.R. Pianka and T.W. Schrew) Harvard University Press, Cambridge, Mass, pp. 205–31.

Croft, D.B. (1981) Behaviour of red kangaroos *Macropus rufus* in Northwestern New South Wales, Australia. *Austr. Mamm.*, **4,** 5–58.

Cronin, C.L. *et al.* (1980) Dominance relations and females. In *Dominance Relations* (eds D.R. Omark, F.F. Strayer and D.G. Freedman) Garland STPM Press, New York, pp. 299–318.

Crook, J.R. and Shields, W.M. (1985) Sexually selected infanticide by adult male barn swallows. *Anim. Behav.*, **33,** 754–62.

Crump, M.L (1983) Opportunistic cannibalism by amphibian larvae in temporary aquatic environments. *Amer. Nat.*, **121,** 281–89.

Dabelsteen, T. (1982) Variation in response of breeding blackbirds *Turdus merula* to playback of song. *Zeit. Tierpsychol.*, **58,** 311–28.

Daev, E.V. (1981) Meiotic abnormalities in young males of the house mouse *Mus musculus* under the effect of exogenous metabolites of sexually mature animals. *Zoologich. Zh.*, **60,** 1268–71.

Daly, J.C. (1981) Effects of social organisation and environmental diversity on determining the genetic structure of a population of the wild rabbit, *Oryctolagus cuniculus. Evolution*, **35,** 689–706.

Daly, M. and Wilson, M. (1981) Abuse and neglect of children in evolutionary perspective. In *Natural Selection and Social Behaviour.* (eds R.D. Alexander and D.W. Tinkle) Chiron Press, New York, pp. 405–16.

Daly, M. and Wilson, M. (1982) Homicide and kinship. *Amer. Anthrop.*, **84,** 372–78.

Daly, M. and Wilson, M. (1984) A sociobiological analysis of human infanticide. In *Infanticide: Comparative and Evolutionary Perspecives* (eds G. Hausfater and S.B. Hrdy) Aldine, New York, pp. 487–502.

Datta, S. (1983) Relative power and the acquisition of rank. In *Primate Social Relationships* (ed R.A. Hinde), Blackwell Scientific Publications, Oxford, pp. 93–103.

David, J.H.M. (1978) Observations on social organisation of springbok, *Antidorcas marsupialis*, in the Bontebok National Park, Swellendam. *Zool. Afric.*, **13,** 115–22.

Davies, N.B. (1978) Territorial defence in the speckled wood butterfly. (*Pararge aegea*): the resident always wins. *Anim. Behav.*, **26**, 138–47.

Davies, N.B. (1982) Behaviour and competition for scarce resources. In *Current Problems in Sociobiology* (eds King's College Sociobiology Group) Cambridge University Press, pp. 363–80.

Davies, N.B. (1983) Polyandry, cloacal pecking and sperm competition in dunnocks. *Nature* **362**, 334–36.

Davies, N.B. (1985) Cooperation and conflict among dunnocks *Prunella modularis* in a variable mating system. *Anim. Behav.*, **33**, 628–48.

Davies, N.B. (1986) Reproductive success of dunnocks, *Prunella modularis* in a variable mating system. I. Factors influencing provisioning rate, nestling weight and fledgling success. *J. Anim. Ecol.*, **55**, 123–38.

Davies, N.B. and Halliday, T.R. (1978) Deep croaks and fighting assessment in toads *Bufo bufo*. *Nature*, **274**, 683–85.

Davies, N.B. and Halliday, T.R. (1979) Competitive mate searching in male common toads, *Bufo bufo*. *Anim. Behav.*, **27**, 1253–67.

Davies, N.B. and Houston, A.I. (1981) Owners and satellites: the economics of territorial defence in the pied wagtail, *Motacilla alba*. *J. Anim. Ecol.*, **50**, 157–80.

Davies, N.B. and Houston, A.I. (1983) Time allocation between territories and flocks: owner-satellite conflict in foraging pied wagtails, *Motacilla alba*. *J. Anim. Ecol.*, **52**, 621–34.

Davies, N.B. and Houston, A.I. (1984) Territory economics. In *Behavioural Ecology* (eds J.R. Krebs and N.B. Davies) 2nd edn, Blackwell Scientific Publications Oxford pp. 148–69.

Davies, N.B. and Houston, A.I. (1986) Reproductive success of dunnocks *Prunella modularis* in a variable mating system. II. Conflicts of interest among breeding birds. *J. Anim. Ecol.*, **55**, 139–54.

Davis, R.E. and Kassel, J. (1975) The ontogeny of agonistic behaviour and the onset of sexual maturation in the paradise fish *Macropodus opercularis*. *Behav. Biol.*, **14**, 31–39.

Davis, S.F. (1978) Shock-elicited attack and biting as a function of chronic versus acute insulin injection. *Bull. Psychon. Soc.*, **12**, 149–51.

Davis, W.J. (1982) Territory size in *Megaceryle alcyon* along a stream habitat. *Auk*, **99**, 353–62.

Dawkins, R. (1976) *The Selfish Gene*, Oxford University Press, Oxford.

Dawkins, R. and Carlisle, T.R. (1976) Parental investment, mate desertion and a fallacy. *Nature*, **262**, 131–33.

Day, H.D., Seay, B.M., Hale, P. and Hendricks, D. (1982) Social deprivation and the ontogeny of unrestricted social behaviour in the laboratory rat. *Devel Psychobiol.*, **15**, 47–59.

DeBlois, C.S. and Stewart, M.A. (1980) Aggressiveness and antisocial behaviour in children: their relationships to other dimensions of behaviour. *Comm. Pyschol. Psychiatr. Behav.*, **5**, 303–12.

De Boer, B.A. (1980) A causal analysis of the territorial and courtship behaviour of *Chromis cyanea*. *Behaviour*, **73**, 1–50.

De Bold, J.F. and Miczek, K.A. (1981) Sexual dimorphism in the hormonal control of aggressive behaviour of rats. *Pharmacol. Biochem. Behav.*, **14**, Suppl 1, 89–93.

De Bruin, J.P.C. (1983) Neural correlates of motivated behaviour in fish. In *Advances in Vertebrate Neuroethology* (eds J.P. Ewer, R.R. Caprianica and D.J. Ingle) Plenum, New York, pp. 969–95.

Deiker, T.E. and Hoffeld, D.R. (1973) Interference with ritualized threat behaviour in *Cichlasoma nigrofasciatum*. *Anim. Behav.*, **21**, 607–12.

De Laet, J.V. (1985) Dominance and aggression in juvenile great tits, *Parus major major*, in relation to dispersal. In *Behavioural Ecology*, (eds R.M. Sibly and R.H. Smith) Blackwell Scientific Publications, Oxford, pp. 375–79.

Deleporte, P. (1978) Ontogenèse des relations interindividuelles chez *Periplanetsa americana* i. étude longitudinale par confrontations de male aux differente stades de dévelopement. *Biol. Behav.*, **3**, 259–72.

Deleporte, P. (1982) Relations sociales et utilisation du milieu par des larves males du premier stade chez *Periplaneta americana*. *Ins. Soc.*, **29**, 485–510.

Delgado, J.M.R. (1979) Neurophysiological mechanisms of aggressive behaviour. In *Aggression and Behaviour Change* (eds S. Feshbach and A. Fraczek) Praeger, New York, pp. 54–65.

Deligne, J., Quennedey, A. and Blum, M.S (1981) The enemies and defense mechanisms of termites. In *Social Insects*, Vol. II (ed H.R Herman) Academic Press, New York, pp. 1–76.

Deluty, R.H. (1981) Adaptiveness of aggression, assertive and submissive behaviour for children. *J. Clin. Child Psychol.* **10**, 155–58.

Demski, L.S. (1973) Feeding and aggressive behaviour evoked by hypothalamic stimulation in a cichlid fish. *Comp. Biochem. Physiol.*, **44**, 685–92.

Demski, L.S. and Gerald, J.W. (1974) Sound production and other effects of midbrain stimulation in free swimming toadfish. *Brain Behav. Evol.*, **9**, 41–59.

Demski, L.S. and Knigge, K.M. (1971) The telencephalon and hypothalamus of the bluegill (*Lepomis macrochirus*): evoked feeding, aggressive and reproductive behaviour with representative frontal sections. *J. Comp. Neurol.*, **143**, 1–16.

De Montellano, B.R.O. (1978) Aztec cannibalism: an ecological necessity? *Science*, **200**, 611–17.

Dessi-Fulgheri, F., Lucarini, N. and Lupo di Prisco, C. (1976) Relationship between testosterone metabolism in the brain and other endocrine variables and intermale aggression in mice. *Aggr. Behav.*, **2**, 223–31.

Deviche, P. and Balthazart, J. (1976) Inhibition of adult social behaviour in ducks induced by juvenile administration of gonadal hormone. *Behav. Proc.*, **1**, 41–56.

Devinsky, O. and Bear, D. (1984) Varieties of aggressive behaviour in temporal lobe epilepsy. *Amer. J. Psychiatr.*, **141**, 651–55.

De Weid, D. (1980) Behavioural actions of neurohypophysial peptides. *Proc. Roy. Soc.B*, **210**, 183–85.

Dewsbury, D.A. (1981) Social dominance, copulatory behaviour and differential reproduction in deermice (*Peromyscus maniculatus*). *J. Comp. Physiol. Psychol.*, **95**, 880–95.

Dewsbury, D.A. (1982) Dominance rank and differential reproduction. *Q.Rev.Biol.*, **57**, 135–59.

Dhondt, A.A. and Hublé, J. (1968) Fledgling date and sex in relation to dispersal in young great tits. *Bird Study*, **15**, 127–34.

Dhondt, A.A. and Schillemans, J. (1983) Reproductive success of the great tit in relation to its territorial status. *Anim. Behav.*, **31**, 902–12.

Diakow, C. and Raimondi, D. (1981) Physiology of *Rana pipiens* reproductive behaviour: a proposed mechanism for inhibition of the release call. *Amer. Zool.*, **21**, 295–304.

Diamond, J.M. (1982) Mimicry of friarbirds by orioles. *Auk*, **99**, 187–96.

Dickemann, M. (1979) Female infanticide, reproductive strategies and social stratification: a preliminary model. In *Evolutionary Biology and Human Social Behaviour* (eds N.A. Chagnon and W. Irons) Duxbury Press, North Scituate, Mass, pp. 321–68.

Dill, L.M. (1977) Development of behaviour in alevins of Atlantic salmon. *Anim. Behav.*, **25**, 116–21.

Dill, L.M., Ydenberg, R.C. and Fraser, A.H.G. (1981) Food abundance and territory size in juvenile coho salmon (*Oncorhynchus kisutch*). *Can.J. Zool.*, **59**, 1801–09.

Dimond, S.J. (1980) *Neuropsychology*, Butterworths, London.

Dingle, H. (1972) Aggressive behaviour in stomatopods and the use of information theory in the analysis of animal communication. In *Behaviour of Marine Animals* (eds H.E. Winn and B. Olla) Plenum Press, New York pp. 126–56.

Dingle, H. (1983) Strategies of agonistic behaviour in crustacea. In *Studies in Adaptation. The Behaviour of Higher Crustacea* (eds S. Rebach and D.W. Dunham) John Wiley, New York, pp. 85–111.

Dingle, H. and Caldwell, R.L. (1978) Ecology and morphology of feeding and agonistic behaviour in mudflat stomatopods. *Biol. Bull.*, **155**, 134–49.

Dini, F. and Luponni, P. (1982) The inheritance of the mate-killer trait in *Euplotes crassus*. *Protistology*, **18**, 179–84.

Dinnock, R.V. (1983) In defense of the harem-intraspecific aggression by male water mites. *Ann. Ent. Soc. Amer.*, **76**, 463–65.

Distel, H. (1978) Behaviour and electrical brain stimulation in the green iguana *Iguana iguana* II Stimulation effects. *Exp. Brain Res.*, **31**, 353–67.

Distel, H. and Veazey, J. (1982) The behavioural inventory of the green iguana (*Iguana iguana*). In *Igunanas of the World* (eds G.M. Burghardt and A.S. Rand) Noyes Publications, Park Ridge, New Jersey, pp. 252–69.

Dittus, W.P.J. (1977) The social regulation of population density and age-sex distribution in the toque monkey. *Behaviour*, **63**, 281–322.

Dittus, W.P.J. (1979) The evolution of behaviours regulating density and age-specific sex ratio in a primate population. *Behaviour*, **69**, 265–302.

Dixson, A.F. (1980) Androgens and aggressive behaviour in primates: a review. *Aggr. Behav.*, **6**, 37–67.

Dixson, A.F. and Herbert, J. (1977) Testosterone, aggressive behaviour and dominance rank in captive adult male Talapoin monkeys, *Physiol. Behav.*, **18**, 539–43.

Dodge, K.A. (1980) Social cognition and children's aggressive behaviour. *Child Dev.*, **51**, 162–170.

Doherty, P.J. (1983) Tropical territorial damselfish: is density limited by aggression or recruitment? *Ecology*, **64**, 176–90.

Done, B.S. and Heatwole, H. (1977) Effects of hormones on aggressive behaviour and social organization of the scincid lizard *Sphenomorphus kosciuskoi*. *Zeit. Tierpsychol.*, **44**, 1–12.

Dow, M., Ewing, A.W. and Sutherland, I. (1976) Studies on the behaviour of cyrpinodont fish III The temporal patterning of aggression in *Aphyosemion striatum*. *Behaviour*, **59**, 252–68.

Drent, P.J. (1978) Territory-occupancy and reproduction. *Inst. Roy. Neth. Acad. Arts. Sci.*, **71**, 274–79.

Drews, D.R., Forand, K.J., Todd, G.G., Chellel, L.D. and Gay, R.L. (1982) A descriptive study of social development in family groups of rats. *Bull. Psychonom. Soc.*, **19**, 177–180.

Drickamer, L.C. (1974) A ten year summary of reproductive data for free ranging *Macaca mulatta*. *Folia Prim.*, **21**, 61–80.

D'Udine, B., Robinson, D.J. and Oliverio, A. (1982) An analysis of single gene effects on audible and ultrasonic vocalisation in the mouse. *Behav. Neur. Biol.*, **36**, 197–203.

Duerden, J.E. (1905) On the habits and reactions of crabs bearing actinians in their claws. *Proc. Zool. Soc. Lond.*, **2**, 494–511.

Dugan, B. and Wiewandt, T.V. (1982) Socio-ecological determinants of mating strategies in iguanid lizards. In *Iguanas of the World* (eds G. Burghardt and A.S. Rand) Noyes Publications, Park Ridge, New Jersey, pp. 303–19.

Dumpert, K. (1978) *The Social Biology of Ants* (translated by C. Johnson, 1981). Pitman's Advanced Publications, Boston.

Dunbar, R.M. (1982) Intra-specific variations in mating strategy. In *Perspectives in Ethology* (eds P.P.G. Bateson and P. Klopfer) Plenum Press, New York, pp. 385–431.

Dunbar, R.M. and Dunbar, E. (1977) Dominance and reproductive success among female gelada baboons. *Nature*, **266**, 351–52.

Dunford, C. (1977) Kin selection for ground squirrel alarm calls. *Amer. Nat.*, **111**, 782–85.

Dunning, E., Murphy, P. and Williams, J. (In press) 'Casuals', 'Terrace Crews' and 'Fighting Firms': towards a sociological explanation of football hooligan behaviour. In *The Anthropology of Aggression*, Blackwell Scientific Publications, Oxford.

Dunning, E.G., Maguire, J.A. Murphy, P.J. and Williams, J.M. (1982) The social roots of football hooligan violence. *Leisure Studies*, **1**, 139–56.

Durham, W.H. (1976) Resource competition and human aggression. *Quart. Rev. Biol.*, **51**, 385–415.

Duvall, D. Herskowitz, R. and Trupiano-Duvall, J. (1980) Responses of five-lined skinks *Eumeces fasciatus* and ground skinks *Scincella lateralis* to conspecific and interspecific chemical cues. *J. Herp.* **14**, 121–27.

Dyson-Hudson, R. and Smith, E.A. (1980); Human territoriality: an ecological reassessment. In *Selected Readings in Sociobiology*. (ed J.H. Hunt) McGraw-Hill, New York, pp. 367–393.

Eaton, G.G., Modahl, K.B. and Johnson, D.F. (1981) Aggressive behaviour in a confined troop of Japanese macaques: effects of density, season and gender. *Aggr. Behav.*, **7**, 145–64.

Eaton, R.L. (1974) The biology and social behaviour and reproduction in the lion. In *The World's Cats* (ed R.L. Eaton) Vol. 2, Woodland Park Zoo, Seattle, pp. 3–55.

Eberhard, W. (1979) The function of horns in *Podischnus agenor* and other beetles. In *Reproductive Competition and Sexual Selection in Insects* (eds N.S. Blum and M.A. Blum) Academic Press, New York, pp. 231–58.

Eberhart, J.A., Keverne, E.B. and Meller, R.E. (1980) Social influences on plasma testosterone levels in male talapoin monkeys. *Horm. Behav.*, **14**, 246–66.

Ebersole, J.P. (1980) Food density and territory size in fish. *Amer. Nat.*, **115**, 492–509.

Ebert, P.D. and Hyde, J.S. (1976) Selection for agonistic behaviour in wild female *Mus musculus*. *Behav. Genet.*, **6**, 291–304.

Ebling, F.J. (1977). Hormonal control of mammalian skin glands. In *Chemical Signals in Vertebrates*. (eds D. Müller-Schwartze and M.M. Mozell). Plenum Press, New York, pp. 17–33.

Edmunds, G. and Kendrick, D.C. (1980) *The Measurement of Human Aggression*, Ellis Horwood, Chichester.

Edwards, D.A. (1969) Early androgen stimulation and aggressive behaviour in male and female mice. *Physiol. Behav.*, **4**, 333–38.

Edwards, P.J. (1982) Plumage variation, territoriality and breeding displays of the golden plover, *Pluvialis apricaria* in southwest Scotland. *Ibis*, **124**, 88–96.

Ehrenkrantz, J., Bliss, E. and Sheard, M.H. (1974) Plasma testosterone: correlation with aggressive behaviour and social dominance in man. *Psychosom. Med.*, **36**, 469–75.

Eibl-Eibesfeldt, I. (1961) The fighting behaviour of animals. *Sci. Amer.*, **205**, 112–21.

Eibl-Eibesfeldt, I. (1980) Human ethology – concepts and implications for the science of man. *Brain Behav. Sci.*, **2**, 1–57.

Eichelman, B. (1979) Role of biogenic amines in aggressive behaviour. In *Psychopharmacology of Aggression* (ed. M. Sandler) Raven Press, New York, pp. 61–93.

Eichelman, B., Elliot, G.R. and Barchas, J.D. (1981) Biochemical, pharmacological and genetic aspects of aggression. In *Biobehavioural Aspects of Aggression*, (eds D.A. Hamburg and M.B. Trudeau,) Alan Liss, New York, pp. 51–84.

Eickwort, K.R. (1973) Cannibalism and kin selection in *Labidomera clivicollis* Coleoptera: Chrysomelidae. *Amer. Nat.*, **107**, 452–453.

Eisenberg, J.F. (1981) *The Mammalian Radiations*, Athlone, London.

Eiserer, L.A., Emerling, M.R. and Scardina, S.J. (1976) Stimulus factors in the isolation-induced aggression of Bobwhite quail and Kharki-Campbell ducklings. *Aggr. Behav.*, **2**, 285–93.

Ekman, J. (1984) Density dependent mortality and population fluctuations of the temperate zone willow tit (*Parus montanus*). *J.Anim. Ecol.*, **53**, 119–34.

Eleftheriou, B.E., Bailey, D.W. and Denenberg, V. (1974) Genetic analysis of fighting in mice. *Physiol. Behav.*, **13**, 773–77.

Elias, M. (1981) Serum cortisol, testosterone, and testoesterone-binding globulin responses to competitive fighting in human males. *Aggr. Behav.*, **7**, 215–24.

Elliot, F.A., (1983) Biological roots of violence. *Proc. Amer. Phil. Soc.*, **127**, 84–94.

Elliott, J.P. and Cowan, I. McT. (1978) Territoriality, density and prey of the lion in the Ngorongoro Crater, Tanzania. *Can. J. Zool.*, **56**, 1726–34.

Ellis, D.H. (1979) Development of behaviour in the golden eagle. *Wildlife Monogr.*, **70**, 1–94.

Ellis, M.E. and Kesner, R.P. (1983) The noradrenergic system of the amygdala and aversive information processing. *Behav. Neurosci.*, **97**, 399–415.

Elwood, R. (1980) The development, inhibition and disinhibition of pup-cannibalism in the Mongolian gerbil. *Anim. Behav.*, **28**, 1188–94.

Ely, D.L. and Henry, J.P. (1978) Neuroendocrine response patterns in dominant and subordinate mice. *Horm. Behav.*, **10**, 156–69.

Emlen, S.T. (1971) The role of song in individual recognition in the indigo bunting (*Passerina cyanea*). *Zeit. Tierpsychol.*, **28**, 241–46.

Emlen, S.T. (1984) Cooperative breeding in birds and mammals. In *Behavioural Ecology* (eds J.R. Krebs and N.B. Davies) 2nd edn, Blackwell Scientific Publications, Oxford, pp. 305–339.

Engbretson, G.A. and Livzey, R.L. (1972) The effects of aggressive display on body temperature in the fence lizard, *Sceloporus occidentalis occidentalis*. *Physiol. Zool.*, **45**, 247–54.

Enoksson, B. and Nilsson, S.G. (1983) Territory size and population density in relation to food supply in the nuthatch, *Sitta europaea*. *J. Anim. Ecol.*, **52**, 927–35.

Enquist, M. (1985) Communication during aggressive interactions with special reference to variation in choice of behaviour. *Anim. Behav.*, **33**, 1152–61.

Enquist, M. Plane, E. and Röed, J. (1985) Aggressive communication in fulmars (*Fulmarus glacialis*) competing for food. *Anim. Behav.*, **33**, 1007–20.

Ens, B.J. and Goss-Custard, J.D. (1984) Interference among oystercatchers, *Haematopus ostralegus*, feeding on mussels, *Mytilus edulis*, on the Exe estuary. *J.Anim. Ecol.*, **53**, 217–32.

Epple, G. (1978) Lack of effects of castration on scent marking, displays and aggression in a South American primate (*Saguinus fuscicollis*). *Horm. Behav.*, **11**, 139–50.

Epple, G. (1981) Effects of prepubertal castration on the development of scent glands, scent marking and aggression in the saddle-backed tamarin (*Saguinus fuscicollis*). *Horm. Behav.* **15**, 54–67.

Erlinge, S., Sandell, M. and Brinke, C. (1982) Scent marking and its territorial significance in stoats, *Mustela erminea. Anim. Behav.*, **30**, 811–18.

Eron, L.D. (1982) Parent-child interaction, television violence and aggression in children. *Amer. Psychol.*, **37**, 197–211.

Essock-Vitale, S.M. and McGuire, M.T. (1980) Predictions derived from the theories of kin selection and reciprocation assessed by anthropological data. *Ethol. Sociobiol.*, **1**, 233–43.

Essman, E.J. and Valzelli, L. (1984) Regional brain serotonin receptor changes in differentially housed mice: effects of amphetamine, *Pharmacol. Res. Comm.*, **16**, 401–08.

Evans, C.S. and Goy, R.W. (1968) Social behaviour and reproductive cycles in captive ring tailed lemurs (*Lemur catta*) *J. Zool.*, **156**, 181–97.

Evans, H.E. (1966) *The Comparative Ethology and Evolution of the Sandwasps*, Harvard University Press.

Evans L.T. (1952) Endocrine relationships in turtles. III Some effects of male hormone in turtles. *Herpetologica*, **8**, 11–14.

Evans, S.M. (1973) A study of fighting reactions in some nereid polychaetes. *Anim. Behav.*, **21**, 138–46.

Ewald, P.W. (1985) Influence of asymmetries in resource quality and age on aggression and dominance in black-chinned hummingbirds. *Anim. Behav.*, **33**, 705–19.

Ewald, P.W., Hunt, G.L. and Warner, M. (1980) Territory size in western gulls: importance of intrusion pressure, defence investments and vegetation structure. *Ecology*, **61**, 80–87.

Ewing, A. (1975) Studies of the behaviour of cyprinodont fish II. The evolution of aggressive behaviour in Old World rivulins. *Behaviour*, **52**, 172–95.

Ewing, L.S. (1967) Fighting and death from stress in a cockroach. *Science*, **155**, 1035–36.

Fabre, J.H. (1907) *Souvenirs entomologiques*, Series 9, Paris.

Fagen, R.M. (1980) When doves conspire: evolution of non-damaging fighting tactics in a non-random encounter minimal-conflict model. *Amer. Nat.*, **115**, 858–69.

Fagen, R.M. (1981) *Animal Play Behaviour*, Oxford University Press, New York.

Fairbanks, L.A. and McGuire, M.T. (1985) Relationships of vervet mothers with sons and daughters from one through three years of age. *Anim. Behav.*, **33**, 40–50.

Falconer, D.S. (1981) *Introduction to Quantitative Genetics*, 2nd edn, Longmans, London.

Falls, J.B. (1969) Functions of territorial song in the white-throated sparrow. In *Bird Vocalisation* (ed R.A. Hinde) Cambridge University Press, pp. 207–32.

Fält, B. (1978) Differences in aggressiveness between brooded and non-brooded domestic chicks. *App. Anim. Ethol.*, **4**, 211–21.

Farr, J.A. (1983) The inheritance of quantitative fitness traits in guppies, *Poecilia reticulata*. *Evolution*, **36**, 1193–1209.

Farragher, K. and Crews, D. (1979) The role of the basal hypothalamus in the regulation of reproductive behaviour in the lizard, *Anolis carolinensis*: lesion studies. *Horm. Behav.*, **13**, 185–206.

Farrington, D. (1979) The family background of aggressive youths. In *Aggression and Anti-social Behavior in Childhood and Adolesence*, (eds L.A. Hershov and M. Berger) Pergamon, Oxford, pp. 73–93.

Fass, B. and Stevens, D.A. (1977) Pheromonal influences on rodent agonistic behaviour. In *Chemical Signals in Vertebrates*. (eds D. Müller-Schwartze and M.M. Mozell) Plenum Press, New York, pp. 185–206.

Feinsinger, P. and Chaplin, S.B. (1975) On the relationship between wing disc loading and foraging strategy in hummingbirds. *Amer. Nat.*, **109**, 217–24.

Felson, R.B. (1983) Aggression and violence between siblings. *Soc. Psychol. Quart.*. **46**, 271–84.

Fenchel, T. (1975) Character displacement and coexistence in mud snails (Hydrobiidae). *Oecologia*, **20**, 19–32.

Ferguson, M.M. and Noakes, D.L.G. (1982) Genetics and social behaviour in char (*Salvelinus* spp.) *Anim. Behav.*, **30**, 128–34.

Ferguson, M.M. and Noakes, D.L.G. (1983) Behaviour-genetics of lake char (*Salvelinus namaycush*) and brook charr (*S. fontinalis*): observations of backcross and F2 generations. *Zeit. Tierpsychol.*, **62**, 72–86.

Ferguson, T.J., Rule, B.G. and Lindsay, R.C. (1982) The effects of caffeine and provocation on aggression. *J.Res.Pers.*, **16**, 60–71.

Fernald, R.D. (1976) The effect of testosterone on the behaviour and colouration of adult male cichlid fish (*Haplochromis burtoni*). *Horm. Res.*, **7**, 172–78.

Fernald, R.D. (1977) Quantitative behavioural observations of *Haplochromis burtoni* under semi-natural conditions. *Anim.Behav.*, **25**, 643–53.

Fernald, R.D. (1980) Responses of male cichlid fish (*Haplochromis burtoni*) reared in isolation to models of conspecifics. *Zeit. Tierpsychol.*, **54**, 85–93.

Fernald, R.D. and Hirata, N.R. (1979) The ontogeny of social behaviour and body coloration in the African cichlid fish (*Haplochromis burtoni*). *Zeit. Tierpsychol.*, **50**, 180–87.

Ferno, A. (1978) The effect of social isolation on the aggressive and sexual behaviour in a cichlid fish, (*Haplochromis burtoni*). *Behaviour*, **65**, 43–61.

Feshbach, S. (1978) Cognitive processes in the development and regulation of aggression. In *Determinants and Origins of Aggression* (eds W.W. Hartup and J. de Wit) Mouton, The Hague, pp. 163–87.

Ficken, M. (1977) Avian Play. *Auk*, **94**, 573–82.

Ficken, R.W., Ficken, M.S. and Hailman, J.P. (1978) Differential aggression in genetically different morphs of the white-throated sparrow (*Zonotrichia albicollis*). *Zeit. Tierpsychol.*, **46**, 43–57.

Figler, M.H. and Einhorn, D.M. (1983) The territorial prior-residence effect in convict cichlids: temporal effects of establishment and proximate mechanisms. *Behaviour*, **85**, 157–83.

Finley, L.M. and Haley, L.E. (1983) The genetics of aggression in the juvenile american lobster, *Homarus Americanus. Aquaculture.* **33**, 135–39.

Fisher, R.C. (1970) Aspects of the physiology of endoparasitic Hymenoptera. *Biol. Rev.*, **46**, 243–78.

Flood, J.F., Jarvik, M.E., Bennett, E.L. and Orme, A.E. (1976) Effects of ACTH peptide fragments on memory formation. *Neuropept. Pharmacol. Biochem. Behav.*, **5**, Suppl. 1, 41–51.

Floody, O.R. and Pfaff, D.W. (1977) Aggressive behaviour in female hamsters: the hormonal basis for fluctuations in female aggressiveness correlated with oestrus state. *J. Comp. Physiol. Psychol.*, **91**, 443–64.

Flynn, J.P., Smith, D., Coleman, K. and Opsahl, C.A. (1979) Anatomical pathways for attack behaviour in the cat. In *Human Ethology. Claims and Limits of a New Discipline.* (eds M. Von Cranach, K. Foppa, W. LePenies and D. Ploog). Cambridge University Press, London, pp. 301–15.

Forster, L. (1982) Visual communication in jumping spiders (Salticidae) In *Spider Communication* (eds. P.N. Witt and J.S. Rovner) Princeton University Press, Princeton New Jersey, pp. 163–212.

Foster, M.S. (1981) Cooperative behavior and social organization of the swallow-tailed manakin (*Chiroxiphia caudata*) *Behav. Ecol. Sociobiol.*, **9**, 167–77.

Fox, M.F. (1971) *Integrative Development of Brain and Behaviour in the Dog*, University of Chicago Press.

Fox, M.F. (1978) *The Wild Canids*. Van Nostrand, New York.

Francis, L. (1973) Intraspecific aggression and its effect on the distribution of *Anthopleura elegantissima* and some related sea anemones. *Biol. Bull.*, **144**, 73–92.

Francis, L. (1976) Social organisation within clones of the sea anemone, *Anthopleura elegantissima*. *Biol. Bull.*, **150**, 361–76.

Francis, R.C. (1984) The effects of bidirectional selection for social dominance on agonistic behaviour and sex ratios in the paradise fish (*Macropodus opercularis*). *Behaviour*, **90**, 25–45.

Franck, D., Marek, J. and Winter, W. (1978) Untersuchungen zum Aggressionsverhalten van *Poecilia spenops*. *Zeit. Tierpsychol.*, **46**, 170–83.

Franks, N.R. and Scovell, E. (1983) Dominance and reproductive success among slave-making ants. *Nature*, **304**, 724–25.

Franzblau, M.A. and Collins, T.P. (1980) Test of a hypothesis of territory regulation in an insectivorous bird by experimentally increasing prey abundance. *Oecologia*, **46**, 164–70.

Frayer, D.W. and Wolpoff, M.H. (1985) Sexual dimorphism *Ann. Rev. Anthrop.*, **14**, 429–73.

Fredericksen, L.W. and Peterson, G.L. (1977) Schedule-induced aggression in humans and animals: a comparative parametric review. *Aggr. Behav.*, **3**, 57–76.

Fredericksen, L.W. and Rainwater, N. (1981) Explosive behaviour: a skill development approach to treatment. In *Violent Behaviour* (ed R.B. Stuart) Brunner-Mazel, New York.

Freedman, J.L. (1978) Crowding and behaviour: the effects of high density on human behaviour and emotions. In *Aggression, Dominance and Individual Spacing*. (eds L. Krames, P. Pliner and T. Alloway) Plenum Press, New York, pp. 91–106.

Frey, D.F. and Miller, R.J. (1972) The establishment of dominance relationships in the blue gourami. *Behaviour*, **42**, 8–62.

Fricke, H.W. (1979) Mating system, resource defence and sex change in the anemone fish *Amphiprion akallopisos*. *Zeit. Tierpsychol.*, **50**, 313–27.

Frodi, A. (1978) Experiential and physiological responses associated with anger and aggression in women and men. *J. Res. Pers.*, **12**, 335–49.

Fugle, G.N., Rothstein, S.T., Osenberg, C.W. and McGinley, M.D. (1984) Signals of status in wintering white-crowned sparrows, *Zontrichia leucophrys gambelii*. *Anim. Behav.*, **32**, 86–93.

Gaines, M.S. and McClenaghan, L.R.W. (1980) Dispersal in small mammals. *Ann. Rev. Ecol. Syst.*, **11**, 163–96.

Galusha, J.G. and Stout, J.F. (1977) Aggressive communication by *Larus glaucescens*. IV Experiments on visual communication. *Behaviour*, **62**, 222–35.

Gandelman, R. (1983) Hormones and infanticide. In *Hormones and Aggressive Behaviour* (ed. B.B. Svare) Plenum Press, New York, pp. 105–18.

Gandelman, R., Rosenthal, C. and Howard, S.M. (1980) Exposure of female mouse foetuses of various ages to testosterone and the later activation of intraspecies fighting. *Physiol. Behav.*, **25**, 333–35.

Gandelman, R. Vom Saal, F.S. and Reinisch, J. (1977) Contiguity to male foetuses affects morphology and behaviour in female mice. *Nature*, **266**, 722–24.

Gans, C. and Tinkle, D.W. (1977) *The Biology of the Reptilia*, Vol. 7, Academic Press, London.

Ganzer, J. Schmidtmayer, J. and Laudien, H. (1984) Das Kampfverhalten der Männchen von *Betta splendens* Regan-eine etho-physiologische Studie. *Zeit Tierpsychol. Suppl*, **26**, 1–103.

Gargett, V. (1978) Sibling aggression in the black eagle in the Matapos, Rhodesia. *Ostrich*, **49**, 57–63.

Garnett, M.C. (1981) Body size, its heritability and influence on juvenile survival among great tits, *Parus major. Ibis*, **123**, 31–41.

Garrett, T.W. and Campbell, C.S. (1980) Changes in social behaviour of male golden hamsters accompanying photoperiodic changes in reproduction. *Horm. Behav.*, **14**, 303–18.

Gärtner, K., Wnakel, B. and Gaudszuhn, D. (1981) The hierarchy of copulation competition and its correlation with paternity in grouped male laboratory rats. *Zeit. Tierpscyhol.*, **56**, 243–54.

Gass, C.L., Angehr, G. and Centa, J. (1976) Regulation of food supply by feeding territoriality in the rufous hummingbird. *Can. J. Zool.*, **54**, 2046–54.

Gaston, A.J. (1978) The evolution of group territorial behaviour and co-operative breeding. *Amer. Nat.*, **112**, 1091–1100.

Geen, R.G. (1978) Effects of attack and uncontrollable noise on aggression. *J. Res. Pers.*, **12**, 15–29.

Geen, R.G. and Quantry, N.B. (1977) The catharsis of aggression: an evaluation of an hypothesis. *Adv. Exp. Soc. Psychol.*, **10**, 1–37.

Geen, R.G., Beattly, W.W. and Arkin, R.M. (1984) *Human motivation. Physiological, Behavioural and Social Approaches*, Allyn and Bacon, Boston.

Geist, V. (1966) The evolution of horn-like organs. *Behaviour*, **27**, 175–214.

Geist, V. (1971) *The Mountain Sheep*, University of Chicago Press.

Geist, V. (1978) On weapons, combat and ecology. In *Aggression, Dominance and Individual Spacing*, (eds L. Krames, P. Pliner and T. Alloway) Plenum Press, New York.

Geist, V. (1986) New evidence for high frequency of antler wounding in cervids. *Can. J. Zool.*, **64**, 380–384.

Gelles, R.J. (1980) Violence in the family – a review of research in the '70s. *J. Marriage*, **42**, 873–85.

George, C. and Main, M. (1979) Social interactions of young abused children: approach, avoidance and aggression. *Child Dev.*, **50**, 306–18.

Gerard, J.J. Driessen, A. Th. van Raalte and Gerrit. J. de Bruyn (In press) Cannibalism in the red wood ant, *Formica polyctena* (Hymenoptera: Formicidae). Publ. Meijendel-comité, new series, No. 69.

Getty, T. (1979) On the benefits of aggression: the adaptiveness of inhibition and superterritories. *Amer. Nat.*, 605–09.

Ghosh, P.R., Sahn, A. and Maiti, B.R. (1983) Leucocyte responses to fighting in the adult male bandicoot rat. *Acta Anat.*, **115**, 263–65.

Gill, F.B. (1978) Proximate costs of competition for nectar. *Amer. Zool.*, **18**, 639–49.

Gill, F.B. and Wolf, L.L. (1975) Economics of feeding territoriality in the golden-winged sunbird. *Ecology*, **56**, 33–45.

Gill, F.B. and Wolf, L.L. (1979) Nectar loss by golden-winged sunbirds to competitors. *Auk*, **96**, 448–61.

Gill, F.B., Mack, A.L. and Ray, R.T. (1982) Competition between hermit hummingbirds *Phaethorninae* and insects for nectar in a Costa Rican rainforest. *Ibis*, **124**, 44–49.

Gipps, J.H.W. (1984) The behaviour of mature and immature male bank voles, (*Clethrionomys glareolus*). *Anim Behav.*, **32**, 836–39.

Glantz, R.M. (1974) Defense reflex and motor detector responsiveness to approaching targets: the motion detector trigger to the defensive reflex pathways. *J. Comp. Physiol.*, **95**, 297–314.

Glanzman, D.L. and Krasne, F.B. (1983) Serotonin and octopamine have opposite effects on the crayfish's lateral giant escape reaction. *J. Neurosci*, **3**, 2263–69.

Gleason, P.E., Michael, S.D. and Christian, J.J. (1979) Effects of gonadal steroids on agonistic behaviour of female *Peromyscus leucopus*. *Horm. Behaviour*, **12**, 30–39.

Gleason, P.E., Michael, S.D. and Christian, J.J. (1980) Aggressive behaviour during the reproductive cycle of female *Peromyscus leucopus*: effects of encounter site. *Behav. Neur. Biol.*, **29**, 506–11.

Glover, D.G., Smith, M.H., Ames, L., Joule, J. and Dubach, J.M. (1977) Genetic variation in pika populations. *Can. J. Zool.*, **55**, 1841–45.

Goldfoot, D.A. (1979) Sex-specific, behaviour-specific actions of dihydrotestosterone: activation of aggression but not mounting in ovariectomized guinea pigs. *Horm. Behav.*, **13**, 241–55.

Golub, M.S., Sassenrath, E.N. and Goo, G.P. (1979) Plasma cortisol levels and dominance in peer groups of rhesus monkey weanlings. *Horm. Behav.*, **12**, 50–59.

Goodall, J. Bandora, J.A., Bergman, E., *et al.* (1979) Inter-community interactions in the chimpanzee population of the Gombe National Park. In *The Great Apes* (eds. D. Hamburg, and E. McGowan) Vol. 5, Benjamin and Cummings, Menlo Park, pp. 13–53.

Goodwin, F.K. and Post, R.M. (1983) 5-Hydroxytryptamine and depression:

a model for the interaction of normal variance with pathology, *Brit. J. Clin. Pharmacol.*, **15**, 3938–4058.

Gordon, T.P., Rose, R.M., Grady, C.L. and Bernstein, I. (1979) Effects of increased testosterone secretion on the behaviour of adult male rhesus monkeys living in a social group. *Folia Primatol.*, **32**, 149–60.

Gorton, R.E. and Gerhardt, K.M. (1978) The comparative ontogeny of social interactions in two species of cockroach. *Behaviour*, **70**, 172–84.

Gorton, R.E., Gerhardt, K.M., Colliander, K. and Bell, W.J. (1983) Social behaviour as a function of context in a cockroach. *Anim. Behav.*, **31**, 152–59.

Goss-Custard, J.D. (1980) Competition for food and interference among waders. *Ardea*, **68**, 31–52.

Goss-Custard, J.D., Durell, S.E.A. Le V dit, and Ens, B.J. (1982) Individual differences in aggressiveness and food stealing among wintering oyster-catchers, *Haematopus ostralagus*. *Anim. Behav.*, **30**, 917–28.

Goss-Custard, J.D., Clarke, R.T. and Durell, S.E.A. le V. dit (1984) Rates of food intake and aggression of oystercatchers *Haematopus ostralegus* on the most and least preferred mussel *Mytilus edulis* beds of the Exe estuary. *J. Anim. Ecol.*, **53**, 233–45.

Goss-Custard, J.D., Durell, S.E.A. Le V. dit., McGrorty, S., *et al.* (1981) Factors affecting the occupation of mussel *Mytilus edulis* beds by oyster-catchers *Haematopus ostralegus* on the Exe estuary, S. Devon. In *Feeding and Survival Strategies of Esturine Organisms* (eds N.V. Jones and W.J. Wolff), London, Plenum, pp. 217–29.

Gouzoules, H. Gouzoules, S. and Fedigan, L. (1982) Behavioural dominance and reproductive success in female Japanese monkeys (*Macacca fuscata*). *Anim. Behav.*, **30**, 1138–50.

Gouzoules, S., Gouzoules, H. and Marler, P. (1984) Rhesus monkey (*Macacca mulatta*) screams: representational signalling in the recruitment of agonistic aid. *Anim. Behav.*, **32**, 182–93.

Govind, C.K. (1985) Neural control of bilateral asymmetry in crustacean cheliped muscles. In *Coordination of Motor Behaviour* (ed. B.M.H. Bush and F. Clarec) Cambridge University Press, London, pp.11–32.

Goyens, J. and Sevenster, P. (1976) Influence du facteur hereditaire et de la densité de population sur l'ontogenèse de l'agressivité chez l'épinoche (*Gasterosteus aculeatus*). *Neth. J. Zool.*, **26**, 427–31.

Grafen, A. (1979) The hawk-dove game played between relatives. *Anim. Behav.*, **27**, pp. 905–907.

Grafen, A. and Ridley, M. (1983) A model of mate guarding. *J. Theoret. Biol.*, **102**, 549–67.

Grant, E.C., Mackintosh, J.H. and Lerwill, C.J. (1970) The effect of a visual stimulus on the agonistic behaviour of the golden hamster. *Zeit. Tierpsychol.*, **27**, 73–77.

Green, R.E. (1983) Spring dispersal and agonistic behaviour of the red-legged partridge, *Alectoris rufa. J. Zool.*, **201**, 541–56.

Greenberg, N. (1983) Central and autonomic aspects of aggression and dominance in reptiles. In *Advances in Vertebrate Neuroethology*. (eds. J.P. Ewert, R.R. Caprianica and D.J. Ingle) Plenum Press, New York, pp. 1134–43.

Greenberg, N. and Crews, D. (1983) Physiological ethology and aggression in amphibians and reptiles. In *Hormones and Aggressive Behaviour* (ed B.B. Svare) Plenum Press, New York, pp. 469–506.

Greig-Smith, P.W. (1980) Parental investment in nest defence by stonechats (*Saxicola torquata*). *Anim. Behav.*, **28**, 604–19.

Grünbaum, H., Bergman, G., Abbot, D.P. and Odgen, J.C. (1978) Intra-specific agonistic behaviour in the rock-boring sea urchin, *Echinometra lucunter*). *Bull. Mar. Sci.*, **28**, 181–88.

Gustafsson, T., Andersson, B. and Meurling, P. (1980) Effects of social rank on the growth of the preputial glands in male bank voles, *Clethrionomys glareolus. Physiol. Behav.*, **24**, 689–92.

Hadfield, M.G. (1983) Dopamine: mesocortical versus nigrostriatal uptake in isolated fighting mice and controls. *Behav. Brain Res.*, **7**, 269–81.

Hahn, M.E. and Haber, S.B. (1982) The inheritance of agonistic behaviour in male mice: a diallel analysis. *Aggr. Behav.*, **8**, 19–38.

Hails, C.J. and Bryant, D.M. (1979) Reproductive energetics of a free-living bird. *J. Anim. Ecol.*, **48**, 471–82.

Hamilton, W.D. (1964) The genetical theory of social behaviour. *J. Theoret. Biol.*, **7**, 1–52.

Hamilton, W.D. (1979) Wingless and fighting males in fig wasps and other insects. In *Sexual Selection and Reproductive Competition in Insects* (eds M.S. Blum and N.A. Blum) Academic Press, London, pp.167–220.

Hamilton, W.J. and Busse, C. (1982) Social dominance and predatory behaviour of chacma baboons. *J. Human. Evol.*, **11**, 567–73.

Hannes, R.P. and Franck, D. (1983) The effects of social isolation on androgen and corticosteriod levels in a cichlid fish (*Haplochromis burtoni*) and in swordtails (*Xiphophorus helleri*). *Horm. Behav.*, **17**, 292–301.

Hannes, R.P., Franck, D. and Liemann, F. (1984) Effects of rank order fights on whole-body and blood concentrations of androgens and corticosteroids in the male swordtail (*Xiphophorus helleri*). *Zeit. Tierpsychol.*, **65**, 53–64.

Harding, C.F. (1983) Hormonal influences on avian aggressive behaviour. In *Hormones and Aggressive Behaviour* (ed B.B. Svare) Plenum Press, New York, pp. 435–67.

Harding, C.F. and Follett, B.K. (1979) Hormone changes triggered by aggression in a natural population of blackbirds. *Science*, **203**, 918–20.

Hardy, K.G. (1975) Colicinogeny and related phenomena. *Bacteriol Rev.*, **39**, 464–515.

Harper, D.G.C. (1982) Competitive foraging in mallards: ideal free ducks. *Anim. Behav.*, **30**, 575–85.

Harrel, A. (1980) Aggression by high school basketball players: an observational study of the effects of opponents' aggression and frustration-inducing factors. *Int. J. Sport*, **11**, 290–98.

Harrison, M.J.S. (1983) Territorial behaviour in the green monkey *Cercopithecus sabaeus*: seasonal defense of local food supplies. *Behav. Ecol. Sociobiol.*, **12**, 85–94.

Hart, B.L. and Barrett, R.E. (1973) Effects of castration on fighting, roaming and urine spraying in adult male cats. *J. Amer. Vet. Med. Assoc.*, **163**, 290–92.

Hassell, M.P. (1978) *Dynamics of Arthopod Predator–prey Systems*. Princeton University Press, Princeton, New Jersey.

Haug, M. and Brain, P.F. (1978) Attack directed by groups of castrated male mice towards lactating or non-lactating intruders: a urine-dependent phenomenon? *Physiol. Behav.*, **21**, 549–52.

Haug, M. and Brain, P.F. (1979) Effect of treatment with testosterone and oestradiol on the attack directed by groups of gonadectomized male and female mice towards lactating intruders. *Physiol. Behav.*, **23**, 397–400.

Hausfater, G. (1975) Dominance and reproduction in baboons, *Papio cynocephalus*. A quantitative analysis. *Contr. Primatol.*, **7**. 1–150.

Hausfater, G. and Hrdy, S.B. (1984) *Infanticide: Comparative and Evolutionary Perspectives*. Aldine, Hawthorne, New York.

Hay, D. (1985) *Essentials of Behaviour Genetics*. Blackwell Scientific Publications, Oxford.

Hazlett, B.A. (1970) Tactile stimuli in the social behaviour of *Pagurus bernhardus*. *Behaviour*, **36**, 20–48.

Hazlett, B.A. (1972) Stereotypy of agonistic movements in the spider crab, *Microphrys bicornutus*. *Behaviour*, **42**, 270–78.

Hazlett, B.A. (1979) The meral spot of *Gonodactylus oerstedii* as a visual stimulus. *Crustaceana*, **36**, 196–98.

Hazlett, B.A. and Provenzano, A.J. (1965) Development of behaviour in laboratory reared hermit crabs. *Bull. Marine Sci.*, **15**, 617–33.

Hazlett, B.A. and Winn, H.E. (1962) Sound production and associated behaviour of Bermudan crustaceans. *Crustaceana*, **4**, 25–38.

Hegner, J.P. (1979) A gene-imposed nervous system difference influencing behavioural covariance. *Behav. Genet.*, **9**, 165–75.

Hegner, R.E. (1985) Dominance and anti-predator behaviour in blue tits (*Parus caeruleus*). *Anim. Behav.*, **33**, 762–68.

Heiligenberg, W. (1976) A probabilistic approach to the study of motivation. In *Simpler Networks and Behaviour* (ed. J. Fentress) Sinauer, Sunderland, Mass, pp.301–13.

Heinzeller, T. and Raab, A. (1982) Reactions of the pituitary to acute and chronic intermale aggression in the tree shrew. *Aggr. Behav.*, **8**, 178–81.

Henry, J.D. and Herrero, S.M. (1974) Social play in the American black bear: its similarity to canid social play and an examination of its identifying characteristics. *Amer. Zool.*, **14**, 371–89.

Highsmith, R.C. (1983) Sex reversal and fighting behaviour: coevolved phenomena in a tanaid crustacean. *Ecology*, **64**, 719–26.

Hildemann, W.H., Raison, R.L., Cheung, G., Hull, C.J., Akaka, L. and Okamoto, J. (1977) Immunological specificity and memory in a scleratinian coral *Nature*, **270**, 219–23.

Hinde, R.A. (1981) Animal signals: ethological and games-theory approaches are not incompatible. *Anim. Behav.*, **29**, 535–42.

Hinton, J.W. (1981a) Adrenal and medullary hormones and their psychophysiological correlates in violent and psychopathic offenders. In *The Biology of Aggression* (eds P.F. Brain and D. Benton) Sijthoff and Noordhoff, Alphen aan den Rijn, The Netherlands, pp. 291–300.

Hinton, J.W. (1981b) Biological approaches to criminality. In *Multidisciplinary Approaches to Aggression Research* (eds P.F. Brain and D. Benton) Elsevier/North Holland, Amsterdam, pp. 447–62.

Hixon, M. (1980) Food production and competitor density as determinants of feeding territory size. *Amer. Nat.*, **115**, 510–30.

Hodapp, A. and Frey, D. (1982) Optimal foraging by firemouth cichlids, *Cichlasoma meeki*, in a social context. *Anim. Behav.*, **30**, 983–89.

Hogstad, O. (1978) Sexual dimorphism in relation to winter foraging and territorial behaviour of the three-toed woodpecker *Picoides tridactylus* and three *Dendrocopus* species. *Ibis*, **120**, 198–203.

Hohn, E.O. (1970) Gonadal hormone concentrations in northern phalaropes in relation to nuptial plumage. *Can.J.Zool.*, **48**, 400–01.

Hole, G.J. and Einon, D. (1984) Play in rodents. In *Play in Animals* (ed P.K. Smith) Blackwell Scientific Publications, Oxford, pp. 95–117.

Hölldobler, B. (1979) Territories of the African weaver ant (*Oecophylla longinoda*): a field study. *Zeit. Tierpsychol.*, **51**, 201–13.

Hölldobler, B. (1984) Evolution of insect communication. In *Insect Communication* (ed T. Lewis) Academic Press, London, pp. 349–77.

Hood, K. (1981) Aggression among female rats over the oestrus cycle. In *The Biology of Aggression* (eds P.F. Brain and D. Benton) Sijthoff and Noordhoff, Alphen aan den Rijn, The Netherlands, pp. 349–77.

Hoogland, J.L. (1985) Infanticide in Prairiedogs: lactating females kill offspring of close kin. *Science*, **231**, 1037–38.

Hopkins, S.E., Schubert, T.A. and Hart, B.L. (1976) The effects of castration on urine marking, aggression and roaming in dogs. *J.Amer. Vet. Med. Assn*, **168**, 1108–10.

Horel, A., Roland, C. and Leborgne, R. (1979) Mise en evidence d'une tendence au groupement chez les jeunes de l'araignée solitaire *Coelotes terrestris*. *Rev. Arachnol.*, **2**, 157–64.

Hörmann-Heck, S.V. (1957) Untersuchungen über der Erbang einiger Verhaltensweisen bei Gryllenbastarden (*Gryllus campestris* × *G. bimaculatus*). *Zeit. Tierpsychol.*, **14**, 137–83.

Horrocks, J. and Hunte, W. (1984) Maternal rank and offspring rank in vervet monkeys: an appraisal of the mechanisms of rank acquisition. *Anim. Behav.*, **31**, 712–82.

Houston, A.I. and Davies, N.B. (1985) The evolution of cooperation and life histories in the dunnock, *Prunella modularis*. In *Behavioural Ecology*. (eds R.M. Sibly and R.H. Smith) Blackwell Scientific Publications, Oxford, pp. 471–88.

Houston, A.I., McCleery, R.H. and Davies, N.B. (1985) Territory size, prey renewal and feeding rate: interpretation of observations on the pied wagtail (*M. alba*) by simulation. *J. Anim. Ecol.*, **54**, 227–40.

Howard, R.D. (1978) The influence of male-defended oviposition sites on embryo mortality in bullfrogs. *Ecology*, **59**, 789–98.

Hoyle, G. (1985) Generation of behaviour: the orchestration hypothesis. In *Feedback and Control in Invertebrates and Vertebrates* (eds W.J.P. Barnes and M.N. Gladden) Croom Helm, London, pp. 57–75.

Hrudecky, P. (1985) Possible pheromonal regulation of reproduction in wild carnivores. *J. Chem. Ecol.*, **11**, 241–50.

Hrdy, S.B. (1979) Infanticide among animals: a review, classification and examination of the implications for the reproductive strategies of females. *Ethol. Sociobiol.*, **1**, 13–40.

Huck, U.W., Carter, C.S. and Banks, E.M. (1979) Oestrogen and progesterone interactions influencing sexual and social behaviour in the brown lemming, *Lemmus trimucronatus*. *Horm. Behav.*, **12**, 40–49.

Hunt, G.L. and Hunt, M.W. (1976) Gull chick survival: the significance of growth rates, timing of breeding and territory size. *Ecology*, **57**, 62–75.

Hunter, A.J. and Dixson, A.F. (1983) Anosmia and aggression in male owl monkeys (*Aotus trivirgatus*). *Physiol. Behav.*, **30**, 875–79.

Huntingford, F.A. (1984) *The Study of Animal Behaviour*, Chapman and Hall, London.

Hutchison, J.B. and Steimer, T. (1983) Hormone-mediated behavioural transitions: a role for brain aromatase. In *Hormones and Behaviour in Higher Vertebrates* (eds J. Balthazart, E. Prove and R. Gilles) Springer-Verlag, Berlin, pp. 261–74.

Hutchison, J.B., Wingfield, J.C. and Hutchison, R.E. (1984) Sex differences in plasma concentrations of steroids during the sensitive period for brain differentiation in the zebra finch, *J. Endocr.*, **103**, 363–69.

Hutchinson, P.R. (1983) The pain aggression relationship and its expression in naturalistic settings. *Aggr. Behav.*, **9**, 229–42.

Hyatt, G.W. (1977) Quantitative analysis of size-dependent variation in the fiddler crab wave display (*Uca pugilator*). *Marine Behav. Physiol.*, **5**, 19–36.

Hyatt, G.W. and Salmon, M. (1978) Combat in the fiddler crabs *Uca pugilator* and *U. pugnax*: a quantitative analysis. *Behaviour*, **65**, 182–211.

Hyde, J.S. and Sawyer, T.F. (1979) Correlated characters in selection for aggressiveness in female mice. II Maternal aggressiveness. *Behav. Genet.*, **9**, 571–77.

Hyde, J.S. and Sawyer, T.F. (1980) Selection for agonistic behaviour in wild female mice. *Behav. Genet.*, **10**, 349–59.

Illius, A.W., Haynes, N.B. and Lamming, G.E. (1978) Effects of ewe proximity on peripheral plasma testosterone levels and behaviour in the ram. *J. Rep. Fert.*, **48**, 25–32.

Ingram, D.K. and Corfman, T.P. (1980) An overview of neurobiological comparisons in mouse strains. *Neurosci. Biobehav. Rev.*, **4**, 421–35.

Isaacson, R.L. (1982) *The Limbic System*, 2nd. edn, Plenum Press, New York.

Izard, C.E. (1977) *Human Emotions*, Plenum Press, New York.

Jacoby, C.A. (1983) Ontogeny of behaviour in the larval instars of the Dungeness crab, *Cancer magister, Zeit. Tierpsychol.*, **63**, 1–16.

Jachowski, R.L. (1974) Agonistic behaviour of the blue crab *Callinectes sapidus. Behaviour*, **50**, 232–53.

Jacques, A.M. and Dill, L.M. (1980) Zebra spiders may use uncorrelated asymmetries to settle contests. *Amer. Nat.*, **116**, 899–901.

Jaeger, R.G. (1984) Agonistic behavior of the red-backed salamander. *Copeia*, 309–14.

Jaeger, R.G. and Gergits, W.F. (1979) Intra- and inter-specific communication in salamanders through chemical signals on the substrate. *Anim. Behav.*, **27**, 150–56.

Jaeger, R.G., Kalvarsky, D. and Shimizu, N. (1982) Territorial behaviour of the red-backed salamander: expulsion of intruders. *Anim. Behav.*, **30**, 490–96.

Jaeger, R.G., Nishikawa, K.C.B. and Barnard, D.E. (1983) Foraging tactics of a terrestrial salamander: costs of territorial defence. *Anim. Behav.*, **31**, 191–98.

Jaisson, P. (1985) Social Behaviour. In *Comprehensive Insect Physiology, Biochemistry and Pharmacology* (eds G.A. Kerkat and L.I. Gilbert) Vol. 9, Pergamon Press, Oxford, pp. 673–94.

Jakobsson, S., Radesater, T. and Jarvi, T. (1979) On the fighting behaviour of *Nannacara anomala* (Pisces, Cichlidae). *Zeit. Tierpsychol.*, **49**, 210–20.

Janis, C. (1982) Evolution of horns in ungulates: ecology and paleoecology. *Biol. Rev.*, **57**, 261–318.

Jannett, F.J., Jr. (1981) Scent mediation of intraspecific, interspecific, and intergeneric agonistic behavior among sympatric species of voles (Microtinae). *Behav. Ecol. Sociobiol.*, **8**, 293–96.

Jansson, A. and Vuoristo, T. (1979) Significance of stridulation in larval Hydropsychidae. *Behaviour*, **71**, 167–86.

Jarman, M.V. (1979) Impala social behaviour: territory, hierarchy, mating, and the use of space. *Adv. Ethol.*, **21**, 1–93.

Jarvis, J.U.M. (1981) Eusociality in a mammal: cooperative breeding in naked mole-rat colonies. *Science*, **212**, 571–73.

Jarvis, J.U.M. (In press) Reproduction. In *The Biology of the Naked Mole-rat* (eds J.U.S. Jarvis, P.W. Sherman and R.M. Alexander).

Jenssen, T.A. (1977) Evolution of anoline lizard display behaviour. *Amer. Zool.*, **17**, 203–15.

Johns, L.S. and Liley, N.R. (1970) The effects of gonadectomy and testosterone treatment on the reproductive behaviour of the male blue gourami. *Can. J. Zool.*, **48**, 977–87.

Johnson, C. (1964) The evolution of territoriality of the Odonata. *Evolution*, **18**, 89–92.

Johnson, D.E., Modahl, K.B. and Epton, G.G. (1982) Dominance status of adult male Japanese macaques: relationship to female dominance status, male status, male mating behaviour, seasonal changes and developmental changes. *Anim. Behav.*, **30**, 382–92.

Johnson, L.K. (1982) Sexual selection in a breutid weevil. *Evolution*, **36**, 251–62.

Johnson, R.H. and Nelson, D.R. (1973) Agonistic display in the gray reef shark, *Carcharhinus menisorrah*, and its relationship to attacks on man. *Copeia*, 76–84.

Johnston, R.E. (1981) Testosterone dependence of scent marking by male hamsters (*Mesocricetus auratus*). *Behav. Neur. Biol.*, **31**, 96–99.

Johnston, R.F. and Fleischer, R.C. (1981) Overwinter mortality and sexual size dimorphism in the house sparrow. *Auk*, **98**, 503–11.

Jones, A.R. (1980) Chela injuries in the fiddler crab, *Uca burgersi*. *Marine. Behav. Physiol.*, **7**, 47–56.

Jones, G.P. and Norman, M.D. (1986) Feeding selectivity in relation to territory size in a herbivorous reef fish *Oecologia*, **68**, 549–56.

Jones, R.B. and Nowell, N.W. (1973) Familiar visual and olfactory stimulii alter aggression in mice. *Physiol. Behav.*, **10**, 221–23.

Jouventin, P. (1982) Visual and vocal signals in Penguins, their evolution and adaptive characters. *Zeit. Tierpsychol.*, **24**, 1–149.

Jurgens, U. (1983) Control of vocal aggression in squirrel monkeys. In *Advances in Vertebrate Neuroethology*. (eds J.P. Ewert, R.R. Caprianica and D. Ingle) Plenum Press, New York, pp. 1086–1102.

Kahl, M.P. (1964) Food ecology of the white stork *Mycteria americana* in Florida. *Ecol. Mon.*, **34**, 97–117.

Kahn, M.W. and the Behavioural Health Technicians Staff (1980) Wife beating and cultural context: prevalence in an Aboriginal and islander community in northern Australia. *Amer. J. Commun. Psychol.*, **8**, 727–31.

Kamel, F., Mock, E.J., Wright, W.W. and Frankel, A.I. (1975) Alterations in plasma concentrations of testosterone, LH and prolactin associated with mating in the male rat. *Horm. Behav.*, **6**, 277–88.

Kantak, K.M., Hegstrand, L.R. and Eichelman, B. (1984) Regional changes in monoamines and metabolites following defensive aggression in the rat. *Brain Res. Bull.*, **12**, 227–32.

Karli, P. (1981) Conceptual and methodological problems associated with the study of brain mechanisms underlying aggressive behaviour. In *The Biology of Aggression* (ed P.F. Brain and D. Benton), Sijthoff and Noordhoff, Alphan aan den Rijn, The Netherlands. pp. 323–61.

Kasparyan, D.R. (1981) The functional aspect of evolution of the sting in the Hymenoptera. *Entomol Rev.*, **59**, 49–54.

Kasuya, E., Hibino, Y., and Ito, Y. (1980) On intercolonial cannibalism in Japanese paper wasps, *Polistes chinensis antennalis peres*, and *P. jagdwigae* Torre (Hymenoptera: Vespidae). *Res. Pop. Ecol.*, (Kyoto), **22**, 255–62.

Kaufman, T.H. (1962) Ecology and social behaviour of the coati, *Nasua nasica* on Barro Colorado Island, Panama. *Univ. Calif. Publ. Zool.*, **60**, 95–222.

Kawai, A. (1978) Sibling cannibalism in the first instar larvae of *Harmonia axyrides* (Coleoptera, Coccinellidae). *Kontyu*, **46**, 14–19.

Keating, E.G., Korman, L.A. and Horel, J.A. (1970) The behavioural effects of stimulating and ablating the reptilian amygdala (*Caiman sklerops*). *Physiol Behav.*, **5**, 55–59.

Keegan-Rogers, V. and Schultz, R.J. (1984) Differences in courtship aggression in six clones of unisexual fish. *Anim. Behav.*, **32**, 1040–44.

Keenleyside, M.H.A. (1971) Aggressive behaviour of male longear sunfish. (*Lepomis megalotis*). *Zeit. Tierpsychol.*, **28**, 227–40.

Kelley, D.B. (1980) Auditory and vocal nuclei of frog brain concentrate sex hormones. *Science*, **207**, 553–55.

Kendrick, C. and Dunn, J. (1983) Sibling quarrels and maternal responses. *Devel. Psychol.*, **19**, 62–70.

Kessler, S. Elliot, G.R., Orenberg, E.K. and Barchas, J.D. (1977) A genetic analysis of aggressive behavior in two strains of mice. *Behav. Gen.*, **7**, 313–21.

Keverne, E.B. (1979) Sexual and aggressive behavior in social groups of talapoin monkeys. *CIBA Found. Symp.*, **62**, 271–86.

Keverne, E.B., Meller, R.E. and Martinez-Arias, A.M. (1978) Dominance, aggression and sexual behaviour in social groups of talapoin monkeys. In *Recent Advances in Primatology*, (eds D.J. Chivers and I. Herbert) Vol. 1, Academic Press, New York, pp. 533–42.

King, G.E. (1976) Society and territory in human evolution. *J. Human Evol.*, **3**, 323–32.

King, J.A. (1958) Maternal behavior and behavior development in two subspecies of *Peromyscus maniculatus*. *J. Mammal.*, **39**, 171–90.

Kingston, D.I. (1980) Eye colour changes during aggressive displays in the godeid fish. *Copeia*, 169–71.

Kipling, C. (1983) Changes in the growth of pike (*Esox lucius*) in Windermere. *J. Anim. Ecol.*, **52**, 647–58.

Kleiman, D.G. (1979) Parent–offspring conflict and sibling competition in a monogamous primate. *Amer. Nat.*, **114**, 753–60.

Kling, A. (1972) Effects of amygdalectomy on social affiliative behaviour in non-human primates. In *The Neurobiology of the Amygdala*. (ed. B.E. Eleftheriou) Plenum Press, New York, pp. 511–36.

Kluge, A.G. (1981) The life history, social organisation and parental behaviour of *Hyla rosenbergi* a nest-building gladiator frog. *Misc. Publ. Mus.
Zool. Univ. Mich.*, **160,** 1–170.

Knapton, R.W. (1979) Optimal size of territory in the clay-coloured sparrow,
Spizella pallida. Can. J. Zool., **57,** 1358–70.

Knowlton, N. (1980) Sexual selection and dimorphism in two demes of a
symbiotic pair-bonding snapping shrimp. *Evolution*, **34,** 161–73.

Knowlton, N. (1982) Parental care and sex role reversal. In *Current Problems in
Sociobiology* (ed. Kings College Sociobiology Group, Cambridge), Cambridge University Press, pp. 203–22.

Kodric-Brown, H. and Brown, J.H. (1978) Influence of economics, interspecific competition and sexual dimorphism on territoriality of migrant
rufous hummingbirds. *Ecology*, **59,** 285–96.

Kohda, M. (1981) Interspecific territoriality and agonistic behaviour of a
temperate pomacentrid fish, *Eupomacentrus altus. Zeit. Tierpsychol.*, **56,**
205–16.

Koolhaas, J.M. (1978) Hypothalamically induced intraspecific aggressive
behaviour in the rat. *Exp. Brain Res.*, **32,** 365–75.

Kortmulder, K. (1982) Etho-ecology of 17 *Barbus* spp. (Pisces, cyprinidae).
Neth. J. Zool., **32,** 144–68.

Kostowski, W. and Tarchalska, B. (1977) The effects of some drugs affecting
brain 5HT on the aggressive behavior and spontaneous electrical activity of
the CNS of the ant, *Formica rufa. Brain Res.*, **38,** 143–49.

Kostowski, W., Tarchalska, B. and Markowska, L. (1975) Aggressive
behavior and brain serotonin and catecholamines in ants (*Formica rufa*).
Pharmacol. Biochem. Behav., **3,** 717–19.

Kraftt, B. (1983) The significance and complexity of communication in
spiders. In *Spider Communication* (eds P.N. Witt and J.S. Rovner) Princeton
University Press, pp.15–66.

Kramer, H. and Lemon, R.E. (1983) Dynamics of territorial singing between
neighbouring song sparrows (*Melospiza melodia*). *Behaviour*, **85,** 198–223.

Kravitz, E.A., Beltz, B.S., Glusman, S., *et al.* (1983) Neurohormones and
lobsters: biochemistry to behavior. *Trends Biochem. Sci.*, **6,** 345–49.

Krebs, C.J. (1978) A review of the Chitty hypothesis of population regulation.
Can. J. Zool., **56,** 2463–80.

Krebs, C.J. (1985) Do changes in spacing behaviour drive population cycles in
small mammals? In *Behavioural Ecology* (ed R.M. Sibly and R.H. Smith)
Blackwell Scientific Publications, Oxford, pp. 295–312.

Krebs, J.R. (1971) Territory and breeding density in the great tit *Parus major.
Ecology*, **52,** 2–22.

Krebs, J.R. (1977) Song and territory in the Great Tit. In *Evolutionary Ecology*
(eds B. Stonehouse and C.M. Perrins), Macmillan, London, pp. 47–62.

Krebs, J.R. (1982) Territorial defense in great tits: do residents always win?
Behav. Ecol. Sociobiol., **4,** 185–94.

Kruijt, J.P. (1964) Ontogeny of social behaviour in Burmese jungle fowl (*Gallus gallus spacticeus*). *Behaviour*, **12**, 1–201.

Kruk, M.R., van der Poel, A.M., Meelis, W., *et al.* (1983) Discriminant analysis of the localisation of aggression-inducing electrode placements in the hypothalamus of male rats. *Brain Res.*, **260**, 61–79.

Kugler, C. (1979) Evolution of the sting apparatus in the myrmicine ants. *Evolution*, **33**, 117–30.

Kummer, H. Gotz, W., and Angst, W. (1974) Triadic differentiation: an inhibitory process protecting pair-bonds in baboons. *Behaviour*, **49**, 62–87.

Lacey, E.A. and Sherman, P.W. (in press) Social organisation of naked mole-rat (*Heterocephalus glaber*) colonies: Evidence for division of labour. In *The Biology of the Naked Mole-Rat*, (eds. J.U.M. Jarvis, P.W. Sherman and R.D. Alexander) Publisher under negotiation.

Lagerspetz, K.Y.H. (1979) Modification of aggressiveness in mice. In *Aggression and Behaviour Change* (eds S. Feshbach and A. Fraczek), Praeger, New York, pp. 66–82.

Lagerspetz, K.Y.H. and Heino, T. (1970). Changes in social reactions resulting from early experience with other species. *Psychol. Rep.*, **27**, 255–59.

Lagerspetz, K.Y.H., Tirri, R. and Lagerspetz, K.M.J. (1968) Neurochemical and endocrinological studies of mice selectivily bred for aggressiveness. *Scand. J. Psychol.*, **9**, 157–60.

Lambert, W.W. (1978) Cross-cultural exploration of children's aggressive strategies. In *Origins of Aggression* (eds W.W. Hartup and J. De Wit), Mouton, The Hague, pp. 189–212.

Lambert, W.W. (1980) Towards an integrative theory of children's aggression. *Colloq. Instit. Psychol.*, 153–64.

Landis, W.G. (1981) The ecology, role of the killer trait, and interactions by 5 species of the *Paramecium aurelia* complex inhabiting the littoral zone. *Can. J. Zool.*, **59**, 1734–43.

Lando, H.A. and Donnerstein, E.I. (1978) The effects of a success or failure on subsequent aggressive behaviour. *J. Res. Pers.*, **12**, 225–34.

Lang, F. Govind, C.K., Costello, W.J. and Green, S.I. (1977) Developmental changes in escape behaviour during growth of the lobster. *Science*, **197**, 682–85.

Lau, P. and Miczek, K.A. (1977) Differential effects of septal lesions on attack and defensive–submissive reactions during intraspecies aggression in rats. *Physiol. Behav.*, **18**, 479–85.

Lazarus, J. and Inglis, I. (1986) Shared and unshared parental investment, parent-offspring conflict and brood size. *Anim. Behav.*, **34**, 1791–1804.

Leader-Williams, N. (1979) Age-related changes in the testicular and antler cycles of reindeer, *Rangifer tarandus*. *J. Rep. Fert.*, **57**, 117–26.

Le Boeuf, B.J. (1977) Male–male competition and reproductive success in elephant seals. *Amer. Zool.*, **14**, 163–76.

Lee, C.T. and Ingersoll, D.W. (1983) Pheromonal influences on aggressive behaviour. In *Hormones and Aggressive Behaviour* (ed B.B. Svare) Plenum Press, New York, pp. 373–91.

Lee, P.C. (1983) Play as a means of developing relationships. In *Primate Social Relationships.* (ed. R.A. Hinde) Blackwell Scientific Publications, Oxford, pp. 82–89.

Lehman, M.N. and Adams, D.B. (1977) A statistical and motivational analysis of the social behaviors of the male laboratory rat. *Behaviour*, **61**, 229–75.

Leimar, O. and Enquist, M. (1984) Effects of asymmetries in owner–intruder conflicts. *J. Theoret. Biol.*, **111**, 475–91.

Lenington, S. (1981) Child abuse: the limits of sociobiology. *Ethol. Sociobiol.*, **2**, 17–29.

Leppelsack, H.J. (1983) Analysis of song in the auditory pathway of song birds. In *Advances in Vertebrate Neuroethology*, (eds J.P. Ewert, R.R. Caprianica and D. Ingle) Plenum Press, New York, pp. 783–99.

Leshner, A.I. (1978) *Introduction to Behavioural Endocrinology*. Oxford University Press, New York.

Leshner, A.I. (1983) Pituitary adrenocortical effects on inter male agonistic behaviour. In *Hormones and Aggressive Behaviour* (ed B.B. Svare), Plenum Press, New York, pp. 27–83.

Leshner, A.I. and Potitch, J.A. (1979) Hormonal control of submissiveness in mice: irrelevance of the androgens and relevance of the pituitary–adrenal hormones. *Physiol. Behav.*, **22**, 531–34.

Leventhal, H. (1980) Toward a comprehensive theory of emotion. *Ann. Rev. Psychol.*, **13**, 140–207.

Levine, S. and Broadhurst, P.L. (1963). Genetic and ontogenetic determinants of adult behaviour in the rat. *J.Comp.Physiol.Psychol.*, **56**, 423–28.

Levinson, D.M., Reeves, D.L. and Buchanan, D.R. (1980) Reductions in aggression and dominance status in guinea pigs following bilateral lesion in the basolateral amygdala or lateral septum. *Physiol. Behav.*, **25**, 963–71.

Lewis, T. (ed.) (1983) *Insect Communication*, Academic Press, London.

Liberman, R.P., Marshall, B.D., jr, and K.L. Burke. (1981) Drug, and environmental interventions for aggressive psychiatric patients. In *Violent Behaviour* (ed. R.B. Stuart) Brunner/Muzet, New York, pp. 227–64.

Lightcap, J.L., Kurland, J.A. and Burgess, R.L. (1982) Child abuse: a test of some predictions from evolutionary theory. *Ethol. Sociobiol.*, **3**, 61–67.

Liley, N.R. (1969) Hormones and reproductive behaviour in fishes. *Fish Physiol.*, **3**, 73–116.

Liley, N.R. (1982) Chemical communication in fish. *Can. J. Fish. Aqu. Sci.*, **39**, 22–35.

Lima, S.L. (1984) Territoriality in variable environments: a simple model. *Amer. Nat.*, **124**, 641–55.

Lincoln, G.A. (1971) The seasonal reproductive changes in the red deer stag (*Cervus elephas.*) *J. Zool.*, **163**, 105–23.

Lincoln, G.A. and Davidson, W. (1971) The relationship between sexual and aggressive behavior and pituitary and testicular activity during the seasonal cycle of rams and the influence of photoperiod. *J. Rep. Fert.*, **49**, 267–78.

Lincoln, G.A., Guinness, F. and Short, R. (1972) The way in which testosterone controls social and sexual behaviour of the red deer stag (*Cervus elaphus*). *Horm. Behav.*, **3**, 375–96.

Lindberg, W.J. and Frydenberg, R.B. (1980) Resource centred agonism of *Pilumnus sayi*, an associate of the bryozoan *Schizoporella pungeus*. *Behaviour*, **75**, 235–50.

Lipcius, R.N. and Herrnkind, W.F. (1982) Moult cycle alterations in behavior, feeding and diel rhythms of a decapod crustacean, the spiny lobster *Panulirus argus*. *Marine Biol.*, **68**, 241–52.

Lipp, M.P. and Hunsperger, R.W. (1978) Threat, attack and flight elicited by electrical stimulation of the ventromedial hypothalamus of the marmoset monkey, *Callithrix jacchus*. *Brain, Behav. Evol.*, **15**, 260–93.

Little, M. (1979) *Mischocyttarus flavitarsis* in Arizona: social and nesting biology of a polistine wasp. *Zeit. Tierpsychol.*, **50**, 282–312.

Looney, T.A. and Cohen, P.S. (1982) Aggression induced by intermittent positive reinforcement. *Neurosci. Biobehav. Rev.*, **6**, 15–37.

Lore, R.K. and Stipo-Flaherty, S. (1984) Postweaning social experience and adult aggression in rats. *Physiol. Behav.*, **33**, 571–74.

Lore, R.K. Flannelly, K. and Farina, P. (1976) Ultrasounds produced by rats accompany decreases in intraspecific fighting. *Aggr. Behav.*, **2**, 175–81.

Lorenz, K. (1966) *On Aggression*, Methuen, London.

Losey, G.S. (1978) Information theory and communication: In *Quantitative Ethology* (ed P.W. Colgan), Wiley, New York, pp. 43–78.

Lott, D.F. (1979) Dominance relations and breeding rate in mature male American bison. *Zeit. Tierpsychol.* **49**, 418–32.

Lyon, M., Goldman, L., Hoage, R. (1985) Parent–offspring conflict following a birth in the primate, *Callimico goeldi. Anim. Behav.*, **33**, 1364–65.

Maccoby, E.E. and Jacklin, C.N. (1974) *The Psychology of Sex Differences* Stanford University Press, Stanford, California.

Maccoby, E.E. and Jacklin, C.N. (1980) Sex differences in aggression: a rejoinder and reprise. Child Dev., **51**, 964–80.

MacGintie, N. (1939) The natural history of the blind goby, *Typhlogobius californiensis. Amer. Midl, Nat.*, **21**, 489–501.

Machida, T., Yonezawa, Y. and Noumara, T. (1981) Age associated changes in plasma testosterone levels in male mice and their relation to social dominance or subordinance. *Horm. Behav.*, **15**, 238–45.

Mackintosh, J.H. (1970) Territory formation in laboratory mice. *Anim. Behav.*, **18**, 177–83.

Mackintosh, J.H. (1981) Behaviour of the house mouse. *Symp.Zool. Soc. Lond.*, **47**, 337–65.

Maclean, S.F. and Seastedt, T.R. (1979) Avian territoriality: sufficient resources or interference competition. *Amer. Nat.*, **114**, 308–12.

Maeda, H., Sato, T. and Maki, S. (1985) Effects of dopamine agonists on hypothalamic defensive attack in rats. *Physiol. Behav.*, **35**, 89–92.

Maier, D. and Roe, P. (1983) Preliminary investigations of burrow defense and intraspecific aggression in the sea urchin, *Strongylocentrotus purpuratus*. *Pacific Sci.*, **37**, 145–50.

Malamuth, N.M. and Donnerstein, E. (1982) The effects of aggressive–pornographic mass media stimuli. *Adv. Exp. Psychol.*, **15**, 103–35.

Maley, M.J. (1969) Electrical stimulation of agonistic behaviour in the mallard. *Behaviour*, **34**, 138–60.

Mallory, F.F. and Brooks, R.J. (1978) Infanticide and other reproductive strategies in the collared lemming, *Dicrostonyx groenlandicus*. *Nature*, **273**, 144–46.

Mallow, G.C. (1981) The relationship between aggressive behaviour and menstrual cycle stage in female Rhesus monkeys (*Macacca mulatta*). *Horm. Behav.*, **15**, 239–69.

Malmberg, T. (1980) *Human Territoriality: Survey of Behavioural Territories in Man with Preliminary Analysis of Meaning*. Mouton, The Hague.

Malmquist, B. (1983) Breeding behaviour of brook lampreys *Lampetra planeri*: experiments on mate choice. *Oikos*, **41**, 143–48.

Malon, D. and Stephens, P.J. (1979) The motor organisation of claw closer muscles in the snapping shrimp. *J. Comp. Physiol.*, **132A**, 109–15.

Mandelzys, N., Lane, E.B. and Marceau, R. (1981) The relationship of violence to alpha levels in a biofeedback training paradigm. *J. Clin. Psychol.*, **37**, 202–09.

Mangan, R.L. (1979) Reproductive behaviour of the cactus fly, *Odontoloxosus longicornis*, male territoriality and female guarding as adaptive strategies. *Behav. Ecol. Sociobiol.*, **4**, 265–78.

Mann, L., Newton, J.W. and Innes, J.M. (1982) A test between deindividuation and emergent norm theories of crowd aggression. *J. Pers. Soc. Psychol.*, **42**, 260–72.

Mann, M.A. and Svare, B. (1983) Prenatal testosterone exposure elevates maternal aggression in mice. *Physiol. Behav.*, **30**, 583–87.

Manning, A. (1979) *An Introduction to Animal Behaviour*, Edward Arnold, London.

Marcellini, D.L. (1978) The acoustic behaviour of lizards. In *Behaviour and Neurology of Lizards* (eds N. Greenberg and P.D. Maclean) National Institute of Mental Health, Rockville, Maryland.

Marsteller, F.A., Siegel, P.B. and Gross, W.B. (1980) Agonistic behaviour, the development of the social hierarchy and stress in genetically diverse flocks of chickens. *Behav. Proc.*, **5**, 339–54.

Martin, F.D. and Hengstebeck, M.F. (1981) Eye colour and aggression in juvenile guppies *Poecilia reticulata*. *Anim. Behav.*, **29**, 325–31.

Massey, A. (1977) Agonistic aids and kinship in a group of pigtail macaques. *Behav. Ecol. Sociobiol.*, **2**, 31–40.

Matte, A.C. and Turnow, H. (1979) Dissociated effects of inhibition of catecholamine synthesis on motor activity, 'emotionality' and aggression in mice. *Research*, **4**, 239–45.

Matthews R. and Matthews, J. (1980) *Insect Behaviour*, John Wiley, New York.

Maxson, S.C. (1981) The genetics of aggression in vertebrates. In *The Biology of Aggression* (eds P.F. Brain and D. Benton) Sijthoff and Nordhoff, Alphen aan den Rijn, The Netherlands, pp. 69–103.

Maxson, S.C., Schrenker, P. and Vigue L. (1983) Genetics, hormones and aggression. In *Hormones and Aggressive Behaviour* (ed. B.B. Svare), Plenum Press, New York, pp. 179–96.

Maxson, S.C., Platt, T., Schrenker, P. and Trattner, A (1982) The influence of the Y-chromosome of Rb/1Bg mice on agonistic behaviours. *Aggr. Behav.*, **8**, 285–91.

Maynard Smith, J. (1974) The theory of games and the evolution of animal conflicts. *J. Theoret. Biol.*, **47**, 209–21.

Maynard, Smith, J. (1977) Parental investment; a prospective analysis. *Anim. Behav.*, **25**, 1–9.

Maynard Smith, J. (1981) Will a sexual population evolve to an ESS? *Amer. Nat.*, **117**, 1015–18.

Maynard Smith, J. (1982) *Evolution and the Theory of Games.* Cambridge University Press.

Maynard Smith, J. (1984) Game theory and the evolution of behaviour. *Brain Behav. Sci.*, **7**, 95–125.

Maynard Smith, J. and Parker, G.A. (1976) The logic of asymmetrical contests. *Anim. Behav.*, **24**, 159–75.

Maynard Smith, J. and Riechert, S.E. (1984) A conflicting-tendency model of spider agonistic behaviour: hybrid-pure population line comparisons. *Anim. Behav.*, **32**, 564–78.

Mazur, A. (1983) Hormones, aggression and dominance in humans. In *Hormones and Aggressive Behaviour* (ed B.B. Svare) Plenum Press, New York, pp. 563–76.

Mazur, A. and Lamb, T.A. (1980) Testosterone, status and mood in human males. *Horm. Behav.*, **14**, 236–46.

McCann, T.S. (1980) Territoriality and breeding behaviour of adult male Antarctic Fur Seal, *Arctocephalus gazella*. *J. Zool.*, **192**, 295–310.

McCarthy, R. and Southwick, C.H. (1979) Parental environment: effects on survival, growth and aggressive behaviours of two rodent species. *Devel. Psychobiol.*, **12**, 269–77.

McCauley, D.E. (1982) The behavioural components of sexual selection in the milkweed beetle *Tetraopus tetraophthalmus*. *Anim. Behav.*, **30**, 23–8.

McCleery, R.H. and Perrins, C.M. (1985) Territory size, reproductive success and population dynamics in the great tit, *Parus major*. In *Behavioural Ecology* (eds R.M. Sibly and R.H. Smith) Blackwell Scientific Publications, Oxford, pp. 353–73.

McGrew, W.C. (1972) *An Ethological Study of Children's Behaviour*. Academic Press, New York.

McGuire, M.T., Raleigh, M.J. and Brammer, G.L. (1984) Adaptation, selection and benefit–cost balances; implications of behavioural–physiological studies of social dominance in male vervet monkeys. *Ethol. Sociobiol.*, **5**, 269–77.

McIlwain, H. and Bachelard, H.S. (1985) *Biochemistry and the Central Nervous System*, 5th edn, Churchill–Livingstone, Edinburgh.

McIntyre, D.C. and Chew, G. (1983) Relation between social rank, submissive behaviour and brain catecholamine levels in ring-necked pheasants. (*Phasianus colchicus*). *Behav. Neurosci.*, **97**, 595.

McIntyre, D.C., Healy, L.M. and Saari, M. (1979) Intraspecies aggression and monoamine levels in rainbow trout (*Salmo gairdneri*) fingerlings. *Behav. Neural Biol.*, **25**, 90–8.

McKillup, S.C. (1983) A behavioural polymorphism in the marine snail *Nassarius pauperatus*: geographic variation correlated with food availability, and differences in competitive ability between morphs. *Oecologia*, **56**, 58–60.

McKinney, F. (1961) An analysis of the displays of the European Eider (*Somateria mollissima mollissima*) and the Pacific Eider (*S.m.nigra.*) *Behaviour*, **7**, 1–124.

McKinney, F. and Stolen, P. (1982) Extra-pair bond courtship and forced copulation among captive green-winged teal (*Anas crecca carolinensis*). *Anim. Behav.*, **30**, 461–74.

McNicol, R.E. and Noakes, D.L.G. (1984) Environmental influences on territoriality of juvenile brook charr, *Salvelinus funtinalis* in a stream environment. *Environ. Biol. Fishes.*, **10**, 29–42.

Meaney, M.J. and Stewart, J. (1981). A descriptive study of social development in the rat. *Anim. Behav.*, **29**, 34–45.

Mebs, T. (1964) Zur biologie und populations dynamik des maussebussards (*Buteo buteo*). *J. Ornithol.*, **105**, 247–306.

Mednik, S.A., Gabrielli, W.F. Jr and Hutchings, B. (1984) Genetic influences in criminal convictions: evidence from an adoption cohort. *Science*, **224**, 891–94.

Mefferd, R.B. Jr, Lennon., J.M. and Dawson, N.E. (1981) Violence – an ultimate non-coping behavior. In *Violence and the Violent Individual* (eds J.R. Hays, T.K. Roberts and K.S. Solway). S.P. Medical and Scientific Books, New York, pp. 273–296.

Meggitt, M. (1977) *Blood is their Argument*. Macmillan Press, Toronto.

Metcalf, R.A. and Whitt, G.S. (1977) Relative inclusive fitness in the social wasp *Polistes metricus*. *Behav. Ecol. Sociobiol.*, **2**, 353–60.

Meyer, H.J.H. (1983) Steroid influences upon the discharge frequency of a weakly electric fish. *J. Comp. Physiol.*, **153**, 29–37.

Meyer-Bahlberg, H.F.L. (1981a) Androgens and human aggression. In *The Biology of Aggression* (eds P.F. Brain and D. Benton) Sijthoff and Nordhoff, Alphen aan den Rijn, The Netherlands, pp. 263–89.

Meyer-Bahlberg, H.F.L. (1981b) Sex chromosomes and aggression in humans. In *The Biology of Aggression* (eds P.F. Brain and D. Benton) Sijthoff and Nordhoof, Alphen aan den Rijn, The Netherlands, pp. 109–24.

Michard, C. and Carlier, M. (1985) Les conduites d'aggression intraspecifique chez la souri domestique. Differences individuelles et analyses génétique. *Biol. Behav.*, **10**, 123–46.

Michener, G.R. (1981) Ontogeny of spatial relationships and social behaviour in juvenile Richardson's ground squirrels. *Can. J. Zool.*, **59**, 1666–76.

Miczek, K.A. (1983) Etho-pharmacology of aggression, defense and defeat. In *Aggressive Behavior: Genetic and Neural Approaches* (eds E.C. Simmel, M.E. Hahn, and J.K. Walters) Lawrence Erlbaum Association, Hillsdale, New Jersey, pp. 147–66.

Miczek, K.A. and Krsiak, M. (1981) Pharmacological analysis of attack and flight. In *Multidisciplinary Approaches to Aggression Research*. (eds P.F. Brain and D. Benton) North Holland Biomedical Press, Amsterdam, pp. 341–54.

Milavsky, J.R., Kessler, M.C., Stripp, H.M. and Rubens, W.S. (1982) *Television and Violence – A Panel Study*. Academic Press, New York.

Miller, G.R., Watson, A. and Jenkins, D. (1970) Responses of red grouse populations to experimental improvement of their food. In *Animal Populations in Relation to their Food Resources* (ed. A. Watson) Symposia of the British Ecological Society, 10, Blackwell Scientific Publications, Oxford, pp. 323–34.

Miller, L.C. (1980) Aspects of the ecology and sociobiology of the parasite *Moniliformes dubius* (Acanthocephala). *Diss. Abstr.*, **41**, (06), 2030B.

Miller, R.J. (1978) Agonistic behaviour in fishes and terrestrial vertebrates. In *Contrasts in Behaviour* (eds E. Reese and F. Lighter) Wiley, New York, pp. 281–311.

Miller, R.J. and Robinson, H.W. (1974) Reproductive behaviour and phylogeny in the genus *Trichogaster*. *Zeit. Tierpsychol.*, **34**, 484–99.

Missakian, E.A. (1980) Gender differences in agonistic behaviour and dominance relations of Synanon communally reared children. In *Dominance Relations* (eds D.R. Omark, F.F. Strayer and D.G. Freedman) Garland STPM Press, New York, pp. 397–413.

Moglich, M.H.J. and Alpert, G.D. (1979) Stone dropping by *Conomyrma bicolor* (Hymenoptera: Formicidae). A new technique of interference competition *Behav. Ecol. Sociobiol.*, **6**, 105–14.

Molenock, J. (1976) Agonistic interactions of the crab *Petrolisthes*. *Zeit. Tierpsychol.*, **41**, 277–94.

Møller, A.P. (1985) Mixed reproductive strategy and mate guarding in a semi-colonial passerine, the swallow *Hirundo rustica*. *Behav. Ecol. Sociobiol.*, **17**, 401–08.

Monaghan, P. (1980) Dominance and dispersal between feeding sites in the herring gull (*Larus argentatus*). *Anim. Behav.*, **28**, 521–27.

Monroe, R.R. (1981) Episodic discontrol. In *Biological Psychiatry* (eds C. Perrins, G. Struwe and B. Jansson) Elsevier, Amsterdam, pp.524–27.

Montgomery, E.L. and Caldwell, R.C. (1984) Aggressive brood defense by females in the stomatopod *Gonodactylus bredini*. *Behav. Ecol. Sociobiol.*, **14**, 247–52.

Moore, F.L. and Zoeller, R.I. (1985) Stress-induced inhibition of reproduction; evidence of supressed secretion of LH–RH in an amphibian. *Gen. Comp. End.*, **60**, 252–58.

Moore, M.C. (1984) Changes in territorial defence produced by changes in circulatory testosterone: a possible hormonal basis for mate-guarding in white-crowned sparrows. *Behaviour*, **88**, 215–26.

Moors, P.J. (1980) Sexual dimorphism in the body size of Mustelids. (Carnivora). *Oikos*, **34**, 147–58.

Morand, C., Young, S.N. and Ervin, F.R. (1983) Clinical response of aggressive schizophrenics to oral tryptophan. *Biol. Psychiatry*, **18**, 575–78.

Mori, A. (1979) Analysis of population changes by measurement of body weight in the Koshima troop of Japanese monkeys. *Primates*, **30**, 371–98.

Moritz, R.F.A. and Hillesteim, E. (1985) Inheritance of dominance in honeybees. *Behav. Ecol. Sociobiol.*, **17**, 87–89.

Morrison, D.W. and Morrison, S.H. (1981) Economics of harem maintenance by a neotropical bat. *Evolution*, **62**, 864–66.

Morse, D.H. (1967) Foraging relationships of brown-headed nuthatches and pine warblers. *Ecology*, **48**, 94–103.

Moss, R. and Watson, A. (1985) Adaptive value of spacing behaviour in population cycles of red grouse and other animals. In *Behavioural Ecology* (eds R.M. Sibly and R.H. Smith) Blackwell Scientific Publications, Oxford, pp. 275–93.

Moss, R., Watson, A., Rothery, P. and Glennie, W. (1982) Inheritance of dominance and aggressiveness in captive red grouse *Lagopus lagopus scoticus*. *Anim. Behav.*, **8**, 1–18.

Mossing, T. and Damber, J-E. Rutting behavior and androgen variation in reindeer (*Rangifer tarandus* L.) *J. Chem. Ecol.*, **7**, 377–89.

Moyer, J.T. and Yogo, Y. (1980) The lek-like mating system of *Halichoeres melanochir* (Pisces: Labridae) at Miyate jima, Japan. *Zeit. Tierpsychol.*, **60**, 209–18.

Moynihan, M. (1961) Hostile and sexual behaviour patterns of South American and Pacific Laridae. *Behav. Suppl.*, **8**, 1–365.

Moynihan, M. and Rodaniche, A.F. (1982) The behaviour and natural history of the Caribbean Reef Squid *Sepioteuthus sepioidea*. *Zeit. Tierpsychol. Suppl.*, **25**, 1–156.

Mugford, R.A. and Nowell, N.W. (1970) Pheromones and their effect on aggression in mice. *Nature*, **226**, 967–68.

Mugford, R.A. and Nowell, N.W. (1971) The preputial gland as a souce of aggression-promoting odours in mice. *Physiol. Behav.*, **6**, 247–49.

Müller-Schwartze, D., Stagge, B. and Müller-Schwartze, C., (1982) Play behaviour persistence, decrease and energetic compensation during food shortage in deer fawns. *Science*, **215**, 85–87.

Munro, A.D. and Pitcher, T.J. (1983) Hormones and agonistic behaviour in teleosts. In *Control Processes in Fish Physiology* (eds J.C. Rankin, T.J. Pitcher and P.T. Duggan) Croom Helm, London, pp. 155–75.

Murdoch, W.W. and Sih, A. (1978) Age-dependent interference in a predatory insect. *J. Anim. Ecol.*, **47**, 581–92.

Murton, R.K., Isaacson, A.J. and Westwood, N.J. (1971) The significance of gregarious feeding behaviour and adrenal stress in a population of wood pigeons (*Columba palumbus*). *J. Zool.*, **165**, 53–84.

Myers, J.P., Connors, P.G. and Pitelka, F.A. (1979) Territoriality in non-breeding shorebirds. *Stud. Avian Biol.*, **2**, 231–46.

Myers, J.P., Connors, P.G. and Pitelka, F.A. (1981) Optimal territory size and the sanderling: compromise in a variable environment. In *Foraging Behaviour: Ecological, Ethological and Psychological Approaches* (eds A.C. Kamil and T.D. Sargent) Garland STPM Press, New York, pp. 135–58.

Myhre, G., Ursin, M. and Hanssen, I. (1981) Corticosterone and body temperature during acquisition of social hierarchy in the captive willow ptarmigan (*Lagopus lagopus*). *Zeit. Tierpsychol.*, **57**, 123–30.

Nagamine, C.M. and Knight, A.W. (1980) Development, motivation and function of some sexually dimorphic structures of the Malaysian prawn *Macrobrachium rosenbergi*. *Crustaceana*, **39**, 141–52.

Namikas, J. and Wehmer, F. (1978) Gender composition of the litter affects behaviour of male mice. *Behav. Biol.*, **23**, 219–24.

Neapolitan, J. (1981) Parental influences on aggressive behavior: a social learning approach. *Adolescence*, **16**, 832–40.

Ncil, S.J. (1984) Field studies of the behavioural ecology and agonistic behavior of *Cichlasoma meeki* (Pisces: Cichlidae). *Environ. Biol. Fishes*, **10**, 59–68.

Nelson, D.R. (1981) Aggression in sharks: is the gray reef shark different? *Oceanus*, **24**, 45–55.

Nelson, J.B. (1978) *The Sulidae*, Oxford University Press, Oxford.

Nettleship, M.A. and Esser, A.H. (1976) Introduction, Definitions In *Discussions on War and Human Aggression* (eds R.D. Gwens and H.A. Nettleship) Mouton, The Hague, pp. 1–31.

Nevo, E., Naftall, G. and Guttman, R. (1975) Aggression patterns and speciation. *Proc. Nat. Acad. Sci. USA*, **72**, 3250–54.

Newman, E.I. (1983) Interactions between plants. In *Physiological Plant Ecology* (eds O.L. Lamge, P.S. Nobel, C.B. Osmond, and H. Ziegler) Vol 3, Springer-Verlag, Berlin.

Newton, I. (1978) *Population Ecology of Raptors*. T&D Poyser, Birkhamsted.

Nievergeldt, B. (1974). A comparison of rutting behaviour and grouping in the Ethiopian and Alpine Ibex (*Capra wallie*). In *The Behaviour of Ungulates in Relation to Management* (eds V. Geist and F. Walther) IUCN Publication, Morges, Switzerland, pp. 324–40.

Niewenhuys, R., Voodg, J. and Van Huijzen, C. (1981) *The Human Central Nervous System*, 2nd edn Springer-Verlag, Berlin.

Nikoletseas, M. and Lore, R. (1981) Aggression in domesticated rats in a burrow digging environment. *Aggr. Behav.*, **7**, 245–52.

Noakes, D.L.G. and McNicol R.E. (1982) Geometry of the eccentric territory. *Can. J. Zool.*, **60**, 1776–79.

Nottebohm, F. (1980) Brain pathways for vocal learning in birds: a review of the first 10 years. *Progr. Psychobiol. Physiol. Psychol*, **9**, 86–128.

Nowell, N.W., Thody, A.J. and Woodley, R. (1980) A melanocyte-stimulating hormone and aggressive behaviour in the male mouse. *Physiol. Behav.*, **24**, 5–9.

Nowicki, S. and Armitage, K.B. (1979) Behaviour of juvenile yellow-bellied marmots: play and social integration. *Zeit. Tierpsychol.*, **51**, 85–105.

Oakeshott, O.G. (1974) Social dominance and mating success in the house mouse. *Oecologia*, **15**, 143–58.

O'Connell, M.E., Reboulleau, C., Feder, H.H. and Silver, R. (1981) Social interactions and androgen levels in birds. 1. Female characteristics associated with increased plasma androgen levels in the male ring dove (*Streptopelia risoria*). *Gen. Comp. Endocrinol.*, **44**, 454–63.

O'Connor, R.J. (1978) Brood reduction in birds: selection for fratricide, infanticide and suicide. *Anim. Behav.*, **26**, 79–96.

O'Donnell, V., Blanchard, R.J. and Blanchard, D.C. (1981) Mouse aggression increases after 24 hours of isolation or housing with female. *Behav. Neur. Biol.*, **32**, 89–103.

Ogawa, S. and Makino, J. (1984) Aggressive behaviour in inbred strains of mice during pregnancy. *Behav. Neur. Biol.*, **40**, 195–204.

Ohno, S., Geller, L.N. and Lai, D.J. (1974) Tfm mutation and masculinisation versus feminisation of the mouse CNS. *Cell*, **3**, 235–42.

Olivier, B., Olivier-Aardema, R. and Wiepkema, P.R. (1983) Effects of anterior hypothalamic and mammillary area lesions on territorial aggressive behaviour in male rats. *Behav. Brain Res.*, **9**, 59–81.

Olivierio, A. (1983) Genes and behaviour: an evolutionary perspective. *Adv. St. Behav.*, **13**, 191–217.

Olsen, K.L. (1979) Induction of male mating behaviour in androgen-insensitive (Tfm) and normal (King–Woltzman) male rats: effects of testosterone propionate, estradiol benzoate and dihydroxytestosterol. *Horm. Behav.*, **13**, 66–72.

Olweus, D. (1978) Personality factors and aggression; with special reference to violence within peer group. In *Origin of Aggression* (eds W.W. Hartup and J. deWit), Mouton, Paris, pp. 272–77.

Olweus, D. (1984) Development of stable aggressive reaction patterns in males. In *Advances in the Study of Aggression*, (eds R.J. Blanchard and D.C. Blanchard) Vol. 1 Academic Press, Orlando, Florida, pp. 103–37.

Orenberg, E.K., Renson, J., Elliott, G.R., *et al.* (1975) Genetic determination of aggressive behavior and brain cyclic AMP. *Psychopharmacol. Commun.*, **1**, 99–107.

Oring, L.W. (1982) Avian mating systems. In *Avian Biology* (eds D.S. Farner and J.R. King) Vol. 6, Academic Press, New York, pp. 1–92.

Otte, D. and Stayman, K. (1979) Beetle horns: some patterns in functional morphology. In *Sexual Selection and Reproductive Competition in Insects* (eds M.S. Blum and N.A. Blum), Academic Press, New York, pp. 259–92.

Ottinger, M. (1983) Sexual behaviour and endocrine changes during reproductive maturation and aging in the avian male. In *Hormones and Behaviour in Higher Vertebrates* (eds J. Balthazart, E. Prove and R. Gilles), Springer-Verlag, Berlin, pp. 350–67.

Owen-Smith, N. (1975) The social ecology of the white rhinoceros (*Ceratotherium simum*). *Zeit. Tierpsychol*, **38**, 337–84.

Packard, A. and Sanders, G.D. (1971) Body patterns of *Octopus vulgaris* and maturation of the response to disturbance. *Anim. Behav.*, **19**, 780–90.

Packer, C. (1977) Reciprocal altruism in *Papio anubis*. *Nature*, **265**, 411–43.

Packer, C. (1979) Male dominance and reproductive activity in *Papio anubis*. *Anim. Behav.*, **27**, 37–45.

Packer, C. (1983) Sexual dimorphism: the horns of African antelopes. *Science*, **221**, 1191–93.

Palmour, R.M. (1983) Genetic models for the study of aggressive behavior, *Prog. Neuro-Psychopharmacol. Biol. Psychiat.*, **7**, 513–17.

Parens, M. (1979) *The Development of Aggression in Early Childhood*, Jason Aronson, New York.

Parke, R.P., Berkowitz, L., Layens, J.P., *et al* (1977) Some effects of violent and nonviolent movies on the behaviour of juvenile delinquents. *Adv. Exp. Soc. Psychol.*, **10**, 135–72.

Parker, G.A. (1979) Sexual selection and sexual conflict. In *Sexual Selection and Reproductive Competition in Insects* (eds M.S. Blum and N.A. Blum), Academic Press, New York, pp. 123–66.

Parker, G.A. (1984) Evolutionarily stable strategies. In *Behavioural Ecology* (eds J.R. Krebs and N.B. Davies), Blackwell Scientific Publications, Oxford, pp. 30–61.

Parker, G.A. (1985) Models of parent–offspring conflict. V. Effects of the behaviour of the two parents. *Anim. Behav.*, **33**, 519–33.

Parker, G.A. and Rubenstein, D.I. (1981) Role assessment, reserve strategy, and acquisition of information in asymmetric animal conflicts. *Anim. Behav.*, **29**, 221–40.

Parmigiani, S. and Brain, P.F. (1983) Effects of residence, aggressive experience and intruder familiarity on attack shown by male mice. *Behav. Proc.*, **8**, 45–57.

Parsons, J. (1971) Cannibalism in Herring Gulls. *Brit. Birds*, **64**, 528–37.

Parzefal, J. (1969) Zur vergleichenden Ethologie verschiedener Mollienesia-arten einschliesslich einer hühlenform von *M. sphenops. Behaviour*, **33**, 1–37.

Parzefal, J. (1979) Zur Genetik und biologischen Bedeutung des Aggressions-verhaltens von *Poecilia sphenops. Zeit. Tierpsychol.*, **50**, 399–422.

Patterson, A.T., Rickerby, J., Simpson, J. and Vickers, C. (1980) Possible interaction of melanocyte-stimulating hormone (MSH) and the pineal in the control of territorial aggression in mice. *Physiol. Behav.*, **24**, 843–48.

Patterson, I.J. (1980) Territorial behaviour and the limitation of population density. *Ardea*, **68**, 53–62.

Patterson, I.J. (1985) Limitation of breeding density through territorial behaviour: experiments with convict cichlids, *Cichlasoma nigrofasciatum*. In *Behavioural Ecology* (eds R.M. Sibly and R.H. Smith), Blackwell Scientific Publications, Oxford, pp. 393–407.

Patton, J.L. and Feder, J.H. (1981) Microspatial genetic heterogeneity in pocket gophers; non-random breeding and genetic drift. *Evolution*, **35**, 912–20.

Payne, A.P. (1977). Changes in aggressive and sexual responses of male golden hamsters after neonatal androgen administration. *J. Endocrinol.*, **73**, 331–37.

Payne, R.B. and Payne, K. (1977) Social organisation and mating success in local song populations of village indigobirds, *Vidua chalybeata. Zeit Tierpsychol.*, **45**, 113–73.

Peeke, F.W. (1972) An experimental study of the territorial function of vocal and visual display in the red-winged blackbird (*Agelaius phoeniceus*). *Anim. Behav.*, **20**, 112–18.

Peeke, M.V.S. (1982) Stimulus and motivation-specific habituations of aggression in the three-spined stickleback. *J. Comp. Physiol. Psychol.*, **96**, 816–22.

Perachio, A.A. (1978) Hypothalamic regulation of behavioural and hormonal aspects of aggression and sexual performance. In *Recent Advances in Primatology* (eds D.J. Chivers and J. Herbert), Academic Press, London, pp. 549–65.

Perrill, S.A., Gerhardt, H.C. and Daniel, R. (1978) Sexual parasitism in the green tree frog (*Hyla cinerea*). *Science*, **200**, 1179–80.

Perron, R., Desjeux, D., Mathon, T. and Mises, R. (1980) Les interactions agressives chez l'enfant. Etude par observation directe dans differents groupes do vie. *Enfance*, 157–78.

Peterson, N. (1975) Hunter gatherer territoriality: the perspective from Australia. *Amer. Anthropol.*, **77**, 53–68.

Petursson, H. and Gudjonsson, G.H. (1981) Psychiatric aspects of homicide. *Acta Psychiat. Scand.*, **64**, 363–72.

Pfeifer, S. (1982) Disappearance and dispersal of *Spermophilus elegans* juveniles in relation to behavior. *Behav. Ecol. Sociobiol.*, **10**, 237–44.

Phillips, D.P. (1986) Natural experiments on the effects of mass media violence on fatal aggression: strengths and weaknesses of a new approach. *Adv. Exp. Soc. Psychol.*, **19**, pp. 207–50.

Phillips, L.H. and Konishi, M. (1973) Control of aggression by singing crickets. *Nature*, **241**, 64–65.

Phillips, R.E. (1964) 'Wildness' in the Mallard duck: effects of brain lesion and stimulation on 'escape behaviour' and reproduction. *J. Comp. Neurol.*, **122**, 139–56.

Phillips, R.E. and Youngren, O.M. (1971) Brain stimulation and species-typical behaviour: activities evoked by electrical stimulation of the brains of chickens (*Gallus gallus*). *Anim. Behav.*, **19**, 757–79.

Pienkowski, M.W. and Evans, P.R. (1985) The role of migration in the population dynamics of birds. In *Behavioural Ecology* (eds R.M. Sibly and R.H. Smith) Blackwell Scientific Publications, Oxford, pp. 331–51.

Poirier, F.A. and Smith, E.O. (1974) Socialising functions of primate play. *Amer. Zool.*, **14**, 275–87.

Polis, G.A. (1980) The effect of cannibalism on the demography and activity of a natural population of desert scorpions. *Behav. Ecol. Sociobiol.*, **7**, 25–35.

Polis, G.A. (1981) The evolution and dynamics of intraspecific predation. *Ann. Rev. Ecol. Syst.*, **12**, 225–51.

Polis, G.A. and Farley, R.D. (1979) Behaviour and ecology of mating in the cannibalistic scorpion, *Pararoctonus mesaensis*. *J. Arachnol.*, **7**, 33–46.

Poole, J.H., Kasman, L.H., Ramsay, E.C. and Lasley, B.L. (1984) Musth and urinary testosterone concentrations in the African elephant *Loxodonta africana*. *J. Rep. Fert.*, **70**, 255–60.

Poole, T.B. (1966) Aggressive play in polecats. *Symp. Zool. Soc. Lond.*, **18**, 23–44.

Poole, T.B. and Fish, J. (1975) An investigation of playful behaviour in *Rattus norvegicus* and *Mus musculus*. *J. Zool.*, **175**, 61–71.

Poshivalov, V.P. (1981) Some characteristics of the aggressive behaviour of mice after prolonged isolation: intraspecific and interspecific aspects. *Agg. Behav.*, **7**, 195–204.

Post, D., Hausfater, G. and McCluskey, S. (1980) Feeding behavior of yellow baboons (*Papio cynocephalus*): relationship to age, gender, and dominance rank. *Folia. Primatol.*, **34**, 170–95.

Potegal, M. (1978) The reinforcing value of several types of aggressive behaviour: a review. *Aggr. Behav.*, **5**, 353–73.

Potegal, M. and ten Brink, L. (1984) Behaviour of attack primed and attack-satiated female golden hamsters (*Mesocricetus auratus*). *J. Comp. Psychol.*, **93**, 66–75.

Potegal, M., Blau, A. and Glusman, M. (1981) Inhibition of intraspecific aggression in male hamsters by septal stimulation. *Physiol. Psychol.*, **9**, 213–18.

Potter, D.A. (1981) Agonistic behaviour in male spider mites: factors affecting frequency and intensity of fights. *Ann. Ent. Soc. Amer.*, **74**, 138–43.

Power, H.W. and Doner, C.G.P. (1980) Experiments on cuckoldry in the mountain bluebird. *Amer. Nat.*, **116**, 689–704.

Power, H.W., Litovich, E. and Lombardo, M. (1981) Male starlings delay incubation to avoid being cuckolded. *Auk*, **98**, 386–89.

Preer, J.L., Jr, Preer, L.B. and Jurand, A. (1974) Kappa and other endosymbionts in *Paramecium aurelia*. *Bacteriol. Rev.*, **38**, 113–63.

Pressley, P.H. (1981) Parental effort and the evolution of nest guarding tactics in the threespine stickleback, *Gasterosteus aculeatus*. *Evolution*, **35**, 282–98.

Prestwich, G.D. (1983) The chemical defenses of termites. *Sci. Amer.*, **249**, 68–75.

Pucilowski, O., Plaznik, A. and Kostowski, W. (1985) Aggressive behaviour inhibition by serotonin and quipazine injected into the amygdala in the rat. *Behav. Neur. Biol*, **43**, 58–68.

Pugesek, B.H. (1983) The relationship between parental age and reproductive effort in the Californian gull (*Larus californicus*). *Behav. Ecol. Sociobiol.*, **13**, 161–71.

Puleo, J.S. Jr (1978) Acquisition of imitative aggression in children as a function of the amount of reinforcement given the model. *Social Behav. Pers.*, **6**, 67–71.

Pulliam, H.R. and Millikan, G.C. (1982) Social organisation in the non-reproductive season. In *Avian Biology* (eds D.S. Farner and J.R. King), Academic Press, New York.

Purcell, J.E. (1977) Aggressive function and induced development of catch tentacles in the sea anemone *Metridium senile* (Coelenterata: Actiniana). *Biol. Bull.*, **153**, 355–68.

Pyle, D.W. and Gromko, M.H. (1981) Genetic basis for repeated mating in *Drosophila melanogaster*. *Amer. Nat.*, **117**, 133–46.

Raab, A., Seizinger, B.R. and Herz, A. (1985) Continuous social defeat induces an increase of endogenous opiates in discrete brain areas of the Mongolian gerbil. *Peptides*, **6**, 387–91.

Racey, P.A. and Skinner, J.D. (1979) Endocrine aspects of sexual mimicry in spotted hyaenas *Crocuta crocuta*. *J. Zool.*, **187**, 315–26.

Radesater, T. and Ferno, A. (1979) On the function of the 'eye-spots' in agonistic behaviour in the fire-mouth cichlid (*Cichlasoma meeki*). *Behav. Proc.*, **4**, 5–13.

Raisbeck, B. (1976) An aggression stimulating substance in the cockroach

Periplaneta americana. Ann. Ent. Soc. Amer., **69**, 793–96.

Rajecki, D.W., Nerenz, D.R., Hoff, S.J., Newman, E.R. and Volbrecht, V.J. (1981) Early development of aggression in chickens: the relative importance of pecking and leaping. *Behav. Proc*, 6, 239–48.

Ralls, K. (1977) Sexual dimorphism in mammals: avian models and unanswered questions. *Amer. Nat.*, **111**, 917–38.

Ramenofsky, M. (1984) Endogenous plasma hormones and agonistic behaviour in male Japanese quail, *Coturnix coturnix. Anim. Behav.*, **32**, 698–708.

Ramirez, M.J. and Delius, J.D. (1979) Nucleus striae terminalis lesions affect agonistic behaviour in pigeons. *Physiol. Behav.*, **22**, 871–75.

Rand, A.S. and Rand, W.M. (1978) Display and dispute settlement in nesting iguanas. In *Behaviour and Neurology of Lizards* (eds N. Greenberg and P.O. MacLean), National Institute Mental Health, Rockville, Maryland, pp. 245–52.

Rand, A.S. and Ryan, M.J. (1981) The adaptive significance of a complex vocal repertoire in a neotropical frog. *Zeit. Tierpsychol.*, **57**, 209–14.

Rappaport, J. and Holden, K. (1981) Prevention of Violence: the case for a nonspecific social policy. In *Violence and the Violent Individual* (ed J.R. Hay, T.K. Roberts and K.S. Solway), MTP Press, New York, pp. 409–40.

Rasa, O.A.E. (1969) Territoriality and the establishment of dominance by means of visual cues in *Pomacentrus jenkinsi. Zeit. Tierpsychol.*, **26**, 825–45.

Rasa, O.A.E. (1971) Appetance for aggression in juvenile damselfish. *Zeit. Tierpsychol., Suppl.*, **7**, 8–69.

Rasa, O.A.E. (1979) The effects of crowding on the social relationships and behaviour of the dwarf mongoose (*Helogale undulata rufula*). *Zeit. Tierpsychol.*, **49**, 317–29.

Rasa, O.A.E. and Vandenhoovel, H. (1984) Social stress in the fieldvole – differential causes of death in relation to behaviour and social structure. *Zeit. Tierpsychol.*, **65**, 108–33.

Ray, A., Sharma, K.K. and Sen, P. (1983) Morphine and naloxone in aggressive and non-aggressive mice: role of opiate peptides. In *Current Status of Centrally Acting Peptides* (ed B.N. Dhawan), Pergamon Press, Oxford, pp. 127–34.

Rebach, S. and Dunham, D.W. (1983) *Studies in Adaptation: the Behaviour of Higher Crustacea.* John Wiley, New York.

Redfield, J.A., Taitt, M.J. and Krebs, C.J. (1978) Experimental alteration of sex ratios in populations of *Microtus townsendii*, a field vole. *Can. J. Zool.*, **56**, 17–27.

Reed, T.M. (1982) Interspecific territoriality in the chaffinch and great tit on islands and the mainland of Scotland: playback and removal experiments. *Anim. Behav.*, **30**, 171–81.

Reeves, P. (1972) *The Bacteriocins*, Chapman and Hall, London.

Reid, J.B., Taplin, P.S. and Lorber, R. (1981) A social interactional approach to the treatment of abusive families. In *Violent Behaviour* (ed R.B. Stuart), Brunner-Mazel, New York, pp. 83–101.

Reinisch, J.M. (1981) Prenatal exposure to synthetic progestins increases potential for aggression in humans. *Science*, **211**, 1171–73.

Reis, D.J. and Fuxe, K. (1969) Brain norepinephrine; evidence that neuronal release is essential for sham rage behaviour following brainstem transection in the cat. *Proc. Nat. Acad. Sci.*, **64**, 108–11.

Rendell, J.M. (1951) Mating of ebony, vestigial and wildtype *Drosophila melanogaster* in light and dark. *Evolution*, **5**, 226–30.

Renfrew, J.W. and Hutchinson, R.R. (1983) The motivation of aggression. In *The Handbook of Behavioural Neurobiology, Vol. 6, Motivation* (eds S.E. Satinoff and P. Teifelbaum) Plenum Press, New York, pp. 511–41.

Reyer, H-U., Dittami, J.P. and Hall, M.R. (1986) Avian helpers at the nest are they psychologically castrated? *Ethology*, **71**, 216–28.

Richner, H.Z. (1985) One adult grey heron killing another on a feeding territory. *British Birds*, **78**, 297.

Ricklefs, R.E. (1977) Reactions of some Panamanian birds to human intrusion at the nest. *Condor*, 79, 376–79.

Ridley, M. (1978) Paternal care. *Anim. Behav.*, **26**, 904–32.

Ridley, M. and Thompson, D.J. (1979) Size and mating in *Asellus aquaticus*. *Zeit. Tierpsychol.*, **51**, 380–97.

Riechert, S.E. (1982) Spider interaction strategies: communication versus coercion. In *Spider Communication. Mechanisms and Ecological Significance* (eds P.N. Witt and J.S. Rovner). Princeton University Press, pp. 282–315.

Riechert, S.E. (1984) Games spiders play III. Cues underlying context-associated changes in agonistic behaviour. *Anim. Behav.*, **32**, 1–15.

Rissman, E.F. and Wingfield, J.C. (1984) Hormonal correlates of polyandry in the spotted sandpiper, *Actitis macularia*. *Gen. Comp. End.*, **56**, 401–05.

Ritzmann, R.E. (1974) Mechanisms for the snapping behaviour of two Alpheid shrimps, *Alpheus californiensis* and *Alpheus heterochelis*. *J. Comp. Physiol.*, **95**, 217–36.

Roberts, D.F. and Bachen, C.M. (1981) Mass communication effects. *Ann. Rev. Psychol.*, **32**, 307–56.

Roberts, E.P. and Weigl, R.D. (1984) Habitat preference in the dark-eyed junco (*Junco hyemalis*) the role of photoperiod and dominance. *Anim. Behav.*, **32**, 709–14.

Roberts, W.W., Steinberg, M.C. and Meais, L.W. (1967) Hypothalamic mechanisms for sexual, aggressive and other motivational behaviour in the opossum, *Didelphis virginiana*. *J. Comp. Physiol. Psychol.*, **64**, 1–15.

Robinson, B.W., Alexander, M. and Browne, G. (1969) Dominance reversal resulting from aggressive responses evoked by brain stimulation. *Physiol. Behav.*, **4**, 749–52.

Robinson, J.G. (1982) Intrasexual competition and mate choice in primates. *Amer. J. Prim. Suppl.*, 1, 131–44.

Robinson, S.K. (1986) Benefits, costs and determinants of dominance in a polygynous oriole. *Anim. Behav.*, **34**, 241–55.

Roche, K.E. and Leshner, A.I. (1979) ACTH and vasopressin immediately after a defeat increase future submissiveness in mice. *Science*, **204**, 1343–44.

Rodgers, R.J. and Hendrie, C.A. (1984) On the role of the endogenous opiate mechanisms in offense, defense and nociception. In *Ethopharmacological Aggression Research* (eds K.A. Miczek, M.R. Kruk and B. Olivier) Alan Liss, New York, pp. 27–42.

Rohrs, W-H. (1977) Veranderungen der sexuellen und aggressiven Handlungsbereitschaft des Schwerttragers *Xiphophorus helleri* (Pisces, Poeciliidae) unter dem Einfluss sozialer Isolation. *Zeit Tierpsychol.*, **44**, 402–22.

Rohwer. S. (1982) The evolution of reliable and unreliable badges of fighting ability. *Amer. Zool.*, **22**, 531–46.

Rohwer, S. (1985) Dyed birds achieve higher social status than controls in Harris' sparrows. *Anim. Behav.*, **33**, 1325–31.

Rohwer, S. and Ewald, P.W. (1981) The cost of dominance and advantage of subordination in a badge signalling system. *Evolution*, **35**, 441–54.

Rohwer, S. and Wingfield, J.C. (1981) A field study of social dominance, plasma levels of luteinizing hormone and steroid hormones in wintering Harris' sparrows. *Zeit Tierpsychol.*, **57**, 173–83.

Rollo, L.D. and Wellington, W.G. (1979) Intra- and inter-specific agonistic behaviour among terrestrial slugs (Pulmonata, Stylommatophora). *Can. J. Zool.*, **57**, 846–51.

Rood, J.P. (1980) Mating relationships and breeding suppression in the dwarf mongoose. *Anim. Behav.*, **28**, 143–50.

Ropartz, P. (1977) Chemical signals in agonistic and social behaviour of rodents. In *Chemical Signals in Vertebrates* (eds D. Müller-Schwartze and M.M. Mozell), Plenum Press, New York, pp. 169–84.

Roper, M.K. (1975) Evidence of warfare in the Near East from 10,000 to 4,300 B.C. In *War, its Causes and Correlates* (eds M.A. Nettleship) Mouton, The Hague, pp. 299–343.

Rose, F.L. (1970) Tortoise chin gland fatty acid composition: behavioural significance. *Comp. Biochem. Physiol.*, **32**, 577–80.

Rose, R.K. (1979) Levels of wounding in the meadow vole *Microtus pennsylvanicus*. *J.Mammal.*, **179**, 37–45.

Rose, R.M., Holaday, J.W. and Bernstein, I.S. (1971) Plasma testosterone, dominance rank and aggressive behaviour. *Nature*, **231**, 366–68.

Roseler, P.F., Roseler, I. and Stramboi, A. (1980) The activity of corpora allata in dominant and subordinate females of the wasp (*Polistes gallicus*). *Ins. Soc.*, **27**, 97–107.

Røskaft, E. (1983) Male promiscuity and female adultery by the rook *Corvus frugilegus*. *Orn. Scand.*, **14**, 175–79.

Røskaft, E., Järvi, T., Bakken, M., Beeh, C., and Reinertsen, R.E. (1986) The relationship between social status and resting metabolic rate in great tits

(*Parus major*) and pied flycatchers (*Ficedula hypoleuca*). *Anim. Behav.*, **34,** 838–42.

Rosvold, H.E., Mishkin, M. and Pribram, K.H. (1954) Influence of amygdalectomy on social behaviour in monkeys. *J. Comp. Physiol. Psychol.*, **47,** 173–78.

Rothstein, S.I. (1979) Gene frequencies and selection for inhibiting traits with special emphasis on the adaptiveness of territoriality. *Amer. Nat.*, **113,** 317–31.

Rotton, J., Frey, J., Barry, T., *et al* (1979) The air pollution experience and physical aggression. *J. Appl. Soc. Psychol.*, **9,** 397–412.

Rovner, J.S. (1967) Acoustic communication in a lycosid spider (*Lycosa rabida*). *Anim. Behav.*, **15,** 273–81.

Rowe, R.J. (1985) Intraspecific interactions of New Zealand damselfly larvae I. *Xanthocremis zealandica, Ischnura aurora* and *Austrolestes colensonis*. *New Zealand J. Zool.*, **12,** 1–15.

Rowley, I. (1965) The life history of the superb blue wren *Malurus cyaneus*. *Emu,* **64,** 251–97.

Rowley, I. (1981) The communal way of life in the splendid wren *Malurus splendens*. *Zeit. Tierpsychol.*, **55,** 228–68.

Rouse, E.F., Coppenger, C.J. and Barnes, P.R. (1977) The effect of an androgen inhibitor on behaviour and testicular morphology in the stickleback, *Gasterosteus aculeatus*. *Horm. Behav.*, **9,** 8–18.

Rudnai, J. (1973) Reproductive biology of lions (*Panthera leo massaica* Neumann) in Nairobi National Park. *East Afr. Wild. J.*, **11,** 241–53.

Ryan, M.J. (1980) Female mate choice in a neotropical frog. *Science,* **209,** 523–25.

Sachser, N. and Prove, E. (1984) Short-term effects of residence in the testosterone responses to fighting in male guineapigs. *Aggr. Behav.*, **10,** 285–92.

Sackett, G.P. (1967) Some persistent effects of different rearing conditions on preadult social behaviour of monkeys. *J. Comp. Physiol. Psychol.*, **64,** 363–65.

Sailly, E.P. (1983) The behavioural ecology of competition and resource utilization among hermit crabs. In *Studies in Adaptation* (eds S. Rebach, and D.W. Dunham) John Wiley, New York, pp. 23–55.

Sale, P.F. (1980) The ecology of fishes on coral reefs. *Ocean. Marine Biol. Ann. Rev.*, **18,** 367–421.

Salmon, M. and Horsch, K.W. (1972) Acoustic signalling and detection by semiterrestrial crabs of the family Ocypodidae. In *Behaviour of Marine Animals* (eds H.E. Winn and B.L. Olla) Plenum Press, New York, pp. 60–96.

Salmon, M. and Hyatt, G.W. (1982). Communication. In *The Biology of Crustacea* (eds F.J. Vernberg and W.B. Vernberg) Vol. 7, Academic Press, New York, pp. 1–40.

Sammarco, P.W. (1983) Effects of fish grazing and damselfish territoriality on coral reef algae. I. Algal community structure. *Marine Ecol.*, **13**, 1–14.

Sammarco, P.W. and Williams, A.N. (1982) Damselfish territoriality: influence on *Diadema* distribution and implications for coral community structure. *Marine Ecol.*, **8**, 53–59.

Samuels, A., Silk, J.B. and Rodman, P.S. (1984) Changes in the dominance rank and reproductive behaviour of male bonnet macaques (*Macaca radiata*). *Anim. Behav.*, **32**, 994–1003.

Sanders, D., Warner, P., Backstrom, T. and Bancroft, J. (1983) Mood, sexuality, hormones and the menstrual cycle: changes in mood and physical state. Descriptions of subjects and methods. *Psychosom. Med.*, **45**, 487–502.

Sapolsky, R.M. (1982) The endocrine stress-response and social status in the wild baboon. *Horm. Behav.*, **16**, 279–92.

Sarter, M. and Markowitsch, H.J. (1985) Involvement of the amygdala in learning and memory; a critical review, with emphasis on anatomical relations. *Behav. Neurosci.*, **99**, 342–80.

Savin-Williams, R.C. (1977) Dominance in a human adolescent group. *Anim. Behav.*, **25**, 400–06.

Savin-Williams, R.C. (1979) Dominance hierarchies in groups of early adolescents. *Child. Dev.*, **50**, 923–35.

Sawyer, T.E. (1980) Androgen effects on responsiveness to aggression and stress-related odors of male mice. *Physiol. Behav.*, **25**, 183–87.

Scallett, A.C., Suoni, S.J. and Bowman, R.F. (1981) Sex differences in adrenocortical response to controlled agonistic encounters in Rhesus monkeys. *Physiol. Behav.*, **26**, 385–90.

Scaramella, T.J. and Brown, W.A. (1978). Serum testosterone levels and aggressiveness in hockey players *Psychosom. Med.*, **40**, 262–63.

Schechter, D. and Gandelman, R. (1981) Intermale aggression in mice: influence of gonadectomy and prior fighting experience. *Aggr. Behav.*, **7**, 187–93.

Schechter, D., Howard, S.M. and Gandelman, R. (1981) Dihydrotestosterone promotes fighting behaviour in female mice. *Horm. Behav.*, **15**, 233–37.

Schein, H., Bock, W., Banke, D., *et al.* (1983) Acoustic communication in the guineafowl (*Numida meleagris*). In *Advances in Vertebrate Neuroethology* (eds J.P. Ewert, R.R. Caprianica and D. Ingle), Plenum Press, New York, pp. 731–82.

Schiavi, R.C., Theilgaard, A., Owen, D.R. and White, D. (1984) Sex chromosome anomalies, hormones, and aggressivity. *Arch. Gen. Psychiatr.*, **41**, 93–99.

Schmidt, W.J. (1984). L. dopa and apomorphine disrupt long but not short behavioural trains. *Physiol. Behav.*, **33**, 671–80.

Schneider, K.J. (1984) Dominance, predation and optimal foraging in white-throated sparrow flocks. *Ecology*, **65**, 1820–27.

Schoener, T.W. (1983a) Field experiments on interspecific competition. *Amer. Nat.*, **122**, 240–85.

Schoener, T.W. (1983b) Simple models of optimal territory feeding size: a reconciliation. *Amer. Nat.*, **121**, 608–629.

Schröder, J.H. (1980) Morphological and behavioural differences between BB/OB and B/W colour morphs of *Pseudotropheus zebra*. *Z.Zool. Syst. Evol. Forsch*, **18**, 69–76.

Schulsinger, F. (1977) Psychopathy: heredity and environment. In *Biosocial Bases of Criminal Behaviour* (eds S.A. Mednick and K.O. Christianson), Gardner Press, New York, pp. 109–25.

Schuster, I. (1983) Women's aggression: an African case study. *Aggr. Behav.*, **9**, 319–31.

Schuurman, T. (1980) Hormonal correlates of agonistic behaviour in adult male rats. *Prog. Brain. Res.*, **53**, 415–20.

Schwank, E. (1980) The effect of size and hormonal state in the establishment of dominance in *Tilapia*. *Behav. Proc.*, **5**, 45–53.

Schwartz, A. (1974a) Sound production and associated behaviours in a cichlid fish *Cichlasoma centrarchus*. *Zeit. Tierpsychol.*, **35**, 147–56.

Schwartz, A. (1974b) The inhibition of aggressive behaviour by sound in the cichlid fish, *Cichlasoma centrarchus*. *Zeit. Tierpsychol.*, **35**, 508–77.

Scott, J.P. (1976) Individual aggression as a cause of war. In *Discussions of War and Human Aggression* (ed. R.D. Givens and M.A. Nettleship), Mouton, The Hague, pp. 69–86.

Scott, J.P. and Fredericson, E. (1951) The causes of fighting in mice and rats. *Physiol. Zool.*, **24**, 273–309.

Scott, J.P. and Payne, L.F. (1934) The effect of gonadectomy on the behaviour of the bronze turkey *J. Exp. Zool.*, **69**, 123–36.

Searcy, W.C. and Wingfield, J.C. (1980). The effects of androgen and anti-androgen on dominance and aggressiveness in male redwinged blackbirds. *Horm. Behav.*, **14**, 126–35.

Seastedt, T.R. and MacLean, S.F. (1979) Territory size and composition in relation to resource abundance in Lapland longspurs breeding in Arctic Alaska. *Auk.*, **96**, 131–42.

Sebastian, R.J., Buttino, A.J., Burzynski, M. H. and Moore, S. (1981) Dynamics of hostile aggression: influence of anger, hurt, instructions and victim pain feedback. *J. Res. Pers.*, 15, 343–58.

Selmanoff, M.K. and Ginsburg, B.E. (1981) Genetic variability in aggression and endocrine function in inbred strains of mice. In *Multidisciplinary Approaches to Aggression* (eds P.F. Brain and D. Benton) Elsevier, North Holland Biomedical Press, Amsterdam, pp. 247–67.

Selmanoff, M.K., Jumonville, J.E., Maxson, S.C. and Ginsberg, B.E. (1975) Evidence for a Y chromosome contribution to aggressive phenotype in inbred mice. *Nature*, **253**, 529–30.

Selmanoff, M.K., Abreu, E., Goldman, B.D. and Gindsburg, B.E. (1977)

Manipulation of aggressive behaviour in adult DBA/2/Bg and C57BL/10/ Bg male mice implanted with testosterone in silastic tubing. *Horm. Behav.*, **8**, 177–90.

Selmanoff, M.K., Goldman, B.D. and Ginsburg, B.E. (1977) Serum testosterone and dominance in inbred strains of mice. *Horm. Behav.*, **8**, 107–19.

Sempere, A.J. and Lacroix, A. (1982) Temporal and seasonal relationships between LH, testosterone and antlers in fawn and adult male roe deer. (*Capreolus capreolus*): a longitudinal study from birth to 4 years of age. *Acta Endocrinol.*, **99**, 295–307.

Serri, G.A. and Ely, D.L. (1984) A comparative study of aggression related changes in brain serotonin in CBA, C57BC and DBA mice. *Behav. Brain Res.*, **12**, 283–89.

Sevenster, P. (1973) Incompatibility of response and reward. In *Constraints on Learning* (eds R.A. Hinde and J. Stevenson-Hinde). Academic Press, London.

Seyfarth, R.M. (1978) Social relationships among adult male and female baboons. I. Behaviour during sexual consortships. *Behaviour*, **64**, 205–26.

Sherman, K.J. (1983) The adaptive significance of postcopulatory mate guarding in a dragonfly, *Pachydiplax longipennis*. *Anim. Behav.*, **31**, 1107–15.

Sherman, P.W. (1981) Reproductive competition and infanticide in Belding's ground squirrels and other animals. In *Natural Selection and Social Behaviour: Recent Research and New Theory* (eds R.D. Alexander and D.W. Tinkle), Chiron Press, New York, pp. 311–31.

Shideler, S.E., Lindberg, D.G. and Lasley, B.L. (1983) Estrogen–behaviour correlates in the reproductive physiology and behaviour of the ruffed lemur (*Lemur variegatus*). *Horm. Behav.*, **17**, 249–63.

Shields, W.M. (1984) Barn swallow mobbing: self defence, collateral kin defence, group defence or parental care? *Anim. Behav.*, **32**, 132–49.

Shields, W.M. and Shields, L.M. (1983) Forcible rape: an evolutionary perspective. *Ethol. Sociobiol.*, **4**, 115–35.

Shine, R. (1978) Sexual selection and sexual dimorphism in the amphibia. *Copeia*, 297–306.

Shivers, M. and Edwards, D.A. (1978) Hypothalamic destruction and mouse aggression. *Physiol. Psychol.*, **6**, 485–87.

Shoham, S.G. (1982) The interdisciplinary study of sexual violence. *Dev. Behav.*, **3**, 245–74.

Sibly, R. and Calow, P. (1983) An integrated approach to life-cycle evolution using selective landscapes. *J. Theoret. Biol.*, **102**, 527–47.

Sibly, R.M. and Smith, R.H. (1985) Behavioural ecology and population dynamics: towards a synthesis. In *Behavioural Ecology* (eds R.M. Sibly and R.H. Smith), Blackwell Scientific Publications, Oxford, pp. 577–91.

Siegel, H.I. (1985) Aggressive behaviour. In *The Hamster: Reproduction and Behaviour*, (ed H.I. Siegel) Plenum Press, New York, pp. 261–86.

Siegfried, B., Frischnecht, H.R. and Waser, P.G. (1982) A new learning model for submissive behavior in mice: effects of naloxone. *Aggr. Behav.*, **8,** 112–15.

Siegfried, B., Frischnecht, H.R. and Waser, P.G. (1984) Vasopressin impairs or enhances retention of learned submissive behavior in mice depending on the time of application. *Behav. Brain Res.*, **11,** 259–69.

Silk, J.B., Clark-Wheatley, C.B., Rodman, P. and Samuels, A. (1981) Differential reproductive success and facultative adjustment of sex ratios among captive female bonnet macaques (*Macaca radiata*). *Anim. Behav.*, **29,** 1106–20.

Silverin, B. (1980) Long-acting testosterone injections given to a population of wild male pied flycatchers. *Anim. Behav.*, **28,** 906–12.

Simon, C.A. (1975) The influence of food abundance on territory size in the iguanid lizard, *Sceloporus jarrovi. Ecology*, **56,** 993–98.

Simon, N.G. (1979) The genetics of inter-male aggressive behaviour in mice: recent research and alternative strategies. *Neurosci. Biobehav. Rev*, 3, 97–106.

Simon, N.G. (1983) New strategies for aggression research. In *Aggressive Behaviour* (ed. E.C. Simmel, M.E. Hahn and J,K. Waters), Lawrence Erlbaum, Hilldale, New Jersey, pp. 19–36.

Sinclair, A.R.E. (1977) *The African Buffalo: A Study of Resource Limitation of Populations*, University of Chicago Press, Chicago.

Singer, J.L. and Singer, D.G. (1981) *Television, Imagination and Aggression: A Study of Preschoolers.* Erlbaum, Hillsdale, New Jersey.

Singer, M.C. (1982) Sexual selection for small size in male butterflies. *Amer. Nat.*, **119,** 440–45.

Singleton, G.R. and Hay, D.A. (1982) A genetic study of male social aggression in wild and laboratory mice. *Behav. Gen.*, **12,** 435–48.

Singleton, G.R. and Hay, D.A. (1983) The effect of social organisation on reproductive success and gene flow in colonies of wild house mice, *Mus musculus. Behav. Ecol. Sociobiol.*, **12,** 49–56.

Sipes, R.G. (1973) War, sports and aggression: an empirical test of two rival theories. *Amer. Anthropol.*, **75,** 64–86.

Slaney, P.A. and Northcote, T.G. (1974) Effects of prey abundance on density and territorial behaviour of young rainbow trout *Salmo gairdneri* in laboratory stream channels. *J. Fish. Res. Board Can.*, **31,** 1201–09.

Slater, P.J.B. (1981) Chaffinch song repertoires: observations, experiments and a discussion of their significance. *Zeit. Tierpsychol.*, **56,** 1–24.

Sluckin, A. and Smith, P. (1977) Two approaches to the concept of dominance in preschool children. *Child Dev.*, **48,** 917–23.

Smart, J.L. (1981) Undernutrition and aggression. In *Multidisciplinary Approaches to Aggression Research* (eds P.F. Brain and D. Benton), Elsevier, Amsterdam, pp. 179–91.

Smith, A.T. and Ivins, B.L. (1983) Colonisation in a pika population: dispersal versus philopatry. *Behav. Ecol. Sociobiol.*, **13,** 37–42.

Smith, C.C. (1968) The adaptive nature of the social organisation in the genus of tree squirels *Tamiasciurus. Ecol. Mon.*, **38**, 31–63.

Smith, D.G. (1976) An experimental analysis of the function of redwinged blackbird song. *Behaviour*, **56**, 136–56.

Smith, P.K. (1984) *Play in Animals.* Blackwell Scientific Publications, Oxford.

Smith, R.J.F. (1969) Control of prespawning behaviour of sunfish (*Lepomis megalotis* and *L. gibbosus*) i. gonadal androgens. *Anim. Behav.*, **17**, 279–85.

Smith, S.M. (1982) Raptor 'reverse' dimorphism revisited: a new hypothesis. *Oikos*, **39**, 118–22.

Smith, W.J., Pawlukiewicz, J. and Smith, S. (1978) Kinds of activity correlated with singing patterns of the yellow throated vireo. *Anim. Behav.*, **26**, 862–84.

Snow, B.K. and Snow, D.W. (1984) Long-term defence of fruit by mistle thrushes *Turdus viscivorus. Ibis*, **126**, 39–49.

Spassov, N.B. (1979) Sexual selection and the evolution of horn-like structures of Ceratopsian dinosaurs. *Palent. Stratig. Lithol.*, **11**, 37–43.

Spellacy, F. (1978) Neuropsychological discrimination between violent and nonviolent men. *J. Clin. Psychol.*, **34**, 49–52.

Spencer, S.R. and Cameron, G.N. (1983) Behavioural dominance and its relationship to habitat patch utilization by the hispid cotton rat (*Sigmodon hispidus*). *Behav. Ecol. Sociobiol.*, **13**, 27–36.

Spieth, H. (1982) Behavioural biology and evolution of the Hawaiian picture-winged species group of *Drosophila. Evol. Biol.*, **14**, 351–437.

Spiliotis, P.H. (1974) The effect of thyroxine and thiourea on territorial behaviour in cichlid fish. *Diss. Abs. Int.*, **35B**, 1030–31.

Stacey, P.B. and Chiszar, D. (1978) Body colour pattern and aggressive behaviour of male pumpkinseed sunfish (*Lepomis gibbosus*). *Behaviour*, **64**, 271–304.

Stamps, J.A. (1977) The relationship between resource competition, risk, and aggression in a tropical territorial lizard. *Ecology*, **58**, 349–58.

Stamps, J.A. (1978) A field study of the ontogeny of social behaviour in the lizard *Anolis aeneus. Behaviour*, **66**, 1–31.

Stamps, J.A. (1983a) The relationship between ontogenetic habitat shifts, competition and predator avoidance in a juvenile lizard (*Anolis aeneus*). *Behav. Ecol. Sociobiol*, **12**, 19–33.

Stamps, J.A. (1983b) Territoriality and the defence of predator refuges in juvenile lizards. *Anim. Behav.*, **31**, 857–70.

Stamps, J.A. and Metcalf, R.A (1978) Parent–offspring conflict. In *Sociobiology: Beyond Nature/Nurture?* (eds G.W. Barlow and J. Silverberg). Westview Press, Boulder, Colorado, pp. 589–618.

Stamps, J., Clark, A., Arrowood, P. and Kus, B. (1985) Parent–offspring conflict in budgerigars. *Behaviour*, **94**, 1–40.

Stanley, S.M. (1974) Relative growth of the titanothere horn: a new approach to an old problem. *Evolution*, **28**, 447–57.

Stanley, S.M. and Newman, W.A. (1980) Competitive exclusion in evolutionary time: the case of the acorn barnacles. *Palaeobiology*, **6**, 173–83.

Stark-Adamec, C. and Adamec, R.E. (1982) Aggression by man against women – adaptation or aberration. *Int. J. Women*, **5**, 1–21.

Stay, B. Friedel, T., Tobe, S.S. and Mundall, E.C. (1980) Feedback control of juvenile hormone synthesis in cockroaches; possible role for ecdysone. *Science*, **207**, 898–900.

Steger, R. and Caldwell, R.L. (1983) Intraspecific deception by bluffing: a defense strategy of newly moulted stomatopods. *Science*, **21**, 558–60.

Steinmetz, S.K. (1977) *The Cycle of Violence*, Praeger Publishers, New York.

Steinmetz, S. and Straus, M.A. (1974) *Violence in the Family*, Harper and Row, New York.

Stellar, E. (1954) Physiology of motivation. *Psychol. Rev.*, **61**, 5–22.

Stephens, M.L. (1982) Mate takeover and possible infanticide by a female northern jacana (*Jacana spinosa*). *Anim. Behav.*, **30**, 1253–54.

Stimpson, J. (1973) The role of territory in the ecology of the intertidal limpet. *Ecology*, **54**, 1020–30.

Stoeker, R.E. (1972) Competitive relations between sympatric populations of voles (*Microtus montanus* and *M. pennsylvanicus*). *J. Anim. Ecol.*, **41**, 311–29.

Stoessinger, J.G. (1982) *Why Nations go to War*, St Martins Press, New York.

Stokkan, K.-A., Howe, K. and Carr, W.R. (1980) Plasma concentrations of testosterone and luteinizing hormone in rutting reindeer bulls (*Rangifer tarandus*). *Can. J. Zool.*, **58**, 2081–83.

Straus, M.R. (1978) Sexual inequality, cultural norms and wife beating. In *The Victimization of Women* (eds J.R. Chapman and M. Gates) Sage, Beverly Hills.

Straus, M.R., Gelles, R. and Steinmetz, S. (1980) *Behind Closed Doors* Doubleday, New York.

Strayer, F.F. and Trudel, M. (1984) Developmental changes in nature and function of social dominance among young children. *Ethol. Sociobiol.*, **5**, 279–95.

Strong, D.R., Lawton, J.H. and Southwood, T.R. (1983) *Insects on Plants: Community Patterns and Mechanisms*, Blackwell Scientific Publications, Oxford.

Strong, D.R. Jr., Simberloff, D., Abele, L.G. and Thistle, B. (eds) (1984) *Ecological Communities: Conceptual Issues and the Evidence*, Princeton University, Press.

Stuart, R.J. and Allaway, T.M. (1983) The slave-making ant (*Harpagoxenus canadensis*) and its host species (*Leptothorax muscorum*): slave raiding and territoriality. *Behaviour*, **85**, 58–90.

Studd, M.V. and Robertson, R.J. (1985) Evidence for reliable badges of status in territorial yellow warblers (*Dendroica petechia*). *Anim. Behav.*, **33**, 1102–13.

Struhsaker, T.T. (1977) Infanticide and social organisation in the redtail monkey (*Cercopithecus ascanius schmidti*) in the Kibale forest. *Zeit. Tierp-sychol.*, **45**, 75–84.

Stumphauzer, J.S., Veloz, E.V. and Aiken, T.W. (1981) Violence by street gangs: East side story? In *Violent Behaviour* (ed R.B. Stuart) Brunner/Mazel, New York, pp. 68–82.

Sugerman, R.A. and Demski, L.S. (1978) Agonistic behavior elicited by electrical stimulation of the brain in western collared lizards, *Crotaphytus collaris*. *Brain. Behav Evol.*, **15**, 446–69.

Surwillo, W.W. (1980) The electroencephalogram and childhood aggression. *Aggr. Behav.*, **6**, 9–18.

Suter, R.B. and Keiley, M. (1984) Agonistic interactions between male *Fratrella pyramitela*. *Behav. Ecol. Sociobiol.*, **15**, 1–8.

Sutherland, W.A. and Parker, G.A. (1985) Distribution of unequal competitors. In *Behavioural Ecology* (eds R.M. Sibly and R.H. Smith), Blackwell Scientific Publications, Oxford, pp. 255–73.

Svare, B. (1979) Steroidal influences on pup-killing behavior in mice. *Horm. Behav.*, **13**, 153–64.

Svare, B.B. (1981) Maternal aggression in mammals. In *Parental Care in Mammals* (eds D.G. Gubernick and P.H. Klopfer), Plenum Press, New York, pp. 179–210.

Svare, B. (1983) *Hormones and Aggression*, Plenum Press, New York.

Svare, B.B., Mann, M. and Samuels, O. (1980) Suckling stimulation but not lactation important for maternal aggression. *Behav. Neur. Biol.*, **29**, 453–62.

Symons, D. (1974) Aggressive play and communication in rhesus monkeys *Macacca mulatta*. *Amer. Zool.*, **14**, 317–22.

Symons, D. (1978) *Play and Aggression: a Study of Rhesus Monkeys*, Columbia University Press, New York.

Symons, P.E.K. (1971) Behavioural adjustment of population density to available food by juvenile Atlantic salmon. *J. Anim. Ecol.*, **40**, 569–87.

Symons, P.E.K. (1974) Territorial behaviour of juvenile Atlantic salmon reduces predation by brook trout. *Can. J. Zool.*, **52**, 677–79.

Taitt, M.J. (1985) Experimental analysis of spacing behaviour in the vole *Microtus townsendii*, In *Behavioural Ecology* (eds R.M. Sibly and R.H. Smith), Blackwell Scientific Publications, Oxford, pp. 313–18.

Taitt, M.J. and Krebs, C.J. (1982) Manipulation of female behaviour in field populations of *Microtus townsendii*. *J. Anim. Ecol.*, **51**, 681–90.

Taitt, M.J. and Krebs, C.J. (1983) Predation cover and food manipulations during a spring decline of *Microtus townsendii*. *J. Anim. Ecol.*, **52**, 837–48.

Takahata, Y. (1982) The sociosexual behaviour of Japanese monkeys. *Zeit. Tierpsychol.*, **59**, 89–108.

Tallamy, D.W. (1982) Age-specific maternal defense in *Gargaphis solani* (Hemiptera, Tingidae). *Behav. Ecol. Sociobiol.*, **11**, 7–11.

Tamm, G.R. and Cobb, S. (1978) Behavior and the crustacean molt cycle: changes in aggression of *Homarus americanus*. *Science.*, **200**, 79–81.

Tarr, R.S. (1977) Role of the amygdala in the intraspecies aggressive behaviour of the iguanid lizard, *Sceloporus occidentalis*. *Physiol. Behav.*, **18**, 1153–58.

Tarr, R.S. (1982) Species typical display behaviour following stimulation of the reptilian striatum. *Physiol. Behav.*, **29**, 615–20.

Taylor, G.T. (1977) Affiliation in highly aggressive domesticated rats. *J. Comp. Physiol. Psychol.*, **91**, 1128–35.

Taylor, G.T. (1979) Reinforcement and intraspecific aggressive behaviour. *Behav. Neur. Biol.*, **27**, 1–24.

Taylor, G.T. (1980) Sex pheromone (produced by female) and aggressive behaviour in male rats. *Anim. Learn. Behav.*, **8**, 485–90.

Tazi, A., Dautzer, R., Mormede, P. and le Moal, M. (1983) Effects of post-trial administration of naloxone and beta-endorphin on shock-induced fighting in rats. *Behav. Neur. Biol.*, **39**, 192–202.

Tepedino, V.J. and Frohlich, D.R. (1980) Fratricide in *Megachile rotundata* a non-social megachilid bee: impartial treatment of sibs and non-sibs. *Behav. Ecol. Sociobiol.*, **15**, 19–24.

Theodor, J.L. (1970) Distinction between 'self' and 'not-self' in lower invertebrates. *Nature*. **277**, 690–92.

Thor, D.H. and Flannelly, K.J. (1978) Intruder gonadectomy and elicitation of territorial aggression in the rat. *Physiol. Behav.*, **17**, 725–27.

Thornhill, R. (1980) Competition and coexistence in *Panorpa* species. *Ecol. Monogr.*, **50**, 179–97.

Thornhill, R. (1981) *Panorpa* scorpionflies; systems for understanding resource defence polygyny and alternative male reproductive efforts. *Ann. Rev. Ecol. Syst.*, **12**, 355–68.

Thornhill, R. and Thornhill, N.W. (1983) Human rape: an evolutionary analysis. *Ethol. Sociobiol.*, **4**, 137–73.

Thorpe, W. (1963) *Learning and Instinct in Animals*, 2nd edn, Methuen, London.

Thresher, R.E. (1976) Field analysis of the territoriality of the 3-spot damselfish *Eupomacentrus planifrons*. *Copeia*, 266–75.

Thresher, R.E. (1979) The role of individual recognition in the territorial behaviour of the threespot damselfish, *Eupomacentrus planifrons*. *Marine Behav. Physiol.*, **6**, 83–93.

Tilson, R.L. and Hamilton, W.J. (1984) Social dominance and feeding patterns of spotted hyaenas. *Anim. Behav.*, **32**, 715–24.

Tinbergen, N. (1951) *The Study of Instinct*. Oxford University Press, London.

Tinbergen, N. (1953) *Social Behaviour in Animals*, Methuen, London.

Tinbergen, N. (1959) Comparative studies of the behaviour of gulls, Laridae, *Behaviour*, **15**, 1–70.

Tizabi, Y., Massari, V.J. and Jacobowitz, D.M. (1980) Isolation-induced aggression and catecholamine variations in discrete brain areas of the mouse. *Brain Res. Bull.*, **5**, 81–86.

Tizabi, Y., Thoa, N.B., Maengwyn-Davies, G.D. *et al.* (1979) Behavioural correlation of catecholamine concentration and turnover in discrete brain areas in three strains of mice. *Brain Res.*, **166**, 199–205.

Toates, F. (1983) Models of motivation. In *Animal Behaviour* (eds P.J. Slater and T.R. Halliday) Blackwell Scientific Publications, Oxford, pp. 168–96.

Tonkiss, J. and Smart, J.L. (1982) Interactive effects of genotype and early life undernutrition on the development of behaviour in rats. *Devel. Psychobiol*, **16**, 287–301.

Tooker, C.P. and Miller, R.J. (1980) The ontogeny of agonistic behaviour in the blue gourami (*Trichogaster trichogaster*). *Anim. Behav.*, **28**, 973–88.

Townshend, D.J. (1985) Decisions for a lifetime: establishment of spatial defence and movement patterns by juvenile grey plovers (*Pluvialis squatarola*). *J. Anim. Ecol.*, **54**, 267–74.

Tristram, D.A. (1977) Intraspecific olfactory communication in the terrestrial salamander *Plethodon cinereus*. *Copeia*, 597–600.

Trivers, R.L. (1972) Parental investment and sexual selection. In *Sexual Selection and the Descent of Man* (ed B. Campbell), Aldine Press, Chicago, pp. 136–79.

Trivers, R.L. (1974) Parent-offspring conflict. *Amer. Zool.*, **14**, 249–65.

Troisi, A. and D'Amato, F.R. (1984) Ambivalence in monkey mothering. Infant abuse combined with maternal possessiveness. *J. Nerv. Ment. Disease*, **172**, 105–08.

Trubec, R.J. and Oring, L.W. (1972) Effects of testosterone propionate implantation on lek behaviour of the sharp-tailed grouse. *Amer. Midl. Nat.*, **57**, 531–36.

Tschinkel, W. (1978) Dispersal behavior of the tenebrionid beetle *Zophobas rugipes*. *Physiol. Zool.*, **51**, 300–13.

Tsutsui, K. and Ishii, S. (1981) Effects of sex steroids on aggressive behaviour of adult male Japanese quail. *Gen. Comp. Endocrinol.*, **44**, 480–86.

Turner, B.W., Iverson, S.L. and Severson, K.L. (1980) Effects of castration on open field behaviour and aggression in male meadow voles (*Microtus pennsylvanicus*). *Can. J. Zool.*, **58**, 1927–32.

Turner, G. (1986) Teleost mating: systems and strategies. In *Teleost Behaviour* (ed T.J. Pitcher) Croom Helm, London. pp. 236–51.

Turner, G. and Huntingford, F.A. (1986) A problem for game theory analysis; assessment and intention in male mouthbrooder contests. *Anim. Behav.*, **34**, 961–70.

Tutin, C.E.G. (1979) Mating patterns and reproductive strategies in a community of wild chimpanzees (*Pan troglodytes schweinfurthii*). *Behav. Ecol. Sociobiol.*, **6**, 29–38.

Tuttle, M. and Ryan, M.J. (1981) Bat predation and the evolution of frog vocalizations in the neotropics. *Science.* **214**, 677–78.

Tyack, P. and Whitehead, M. (1983) Male competition in large groups of wintering hump-backed whales. *Behaviour*, **83**, 132–54.

Ukegbu, A.A. (1986) Life history patterns and reproduction in the three-spined stickleback (*Gasterosteus aculeatus*). Ph.D. Thesis, Glasgow University.

Ulrich, R.E. and Azrin, N.H. (1962) Reflexive fighting in response to aversive stimulation. *J. Exp. Anal. Behav.*, **5**, 511–20.

Ursin, H. (1981) Neuroanatomical basis of aggression. In *Multidisciplinary Approaches to Aggression Research* (eds P.F. Brain and D. Benton), Elsevier, Amsterdam, pp. 269–93.

Valenti, G., Vescovi, P.P., Volpi, R., *et al.* (1981) Thyroid hormone pattern and aggressiveness. *La Ricerca Clin. Lab.*, **11**, 117–72.

Valzelli, L. (1981) *Psychobiology of Aggression and Violence*, Raven Press, New York.

Van den Assem, J. (1967) Territory in the three-spined stickleback. *Gasterosteus aculeatus* L. Behaviour. *Suppl.*, **16**, 1–164.

Van den Berg, M.J. ter Horst, G.J. and Koolhaas, J.M. (1983) The nucleus premammillaris ventralis and aggressive behaviour in the rat. *Aggr. Behav.*, **9**, 41–47.

Van de Poll, N. E., Smeets, J. van Dyen, H. and Van der Zwai, S.M. (1981) Behavioural consequences of agonistic experience in rats: sex differences and the effects of testosterone. *J. Comp. Physiol. Psychol.*, **96**, 893–903.

Van der Bercken, J.H.L. and Cools, A.R. (1980) Information-statistical analysis of social interaction and communication: an analysis of variance approach. *Anim., Behav.*, **28**, 172–88.

Van Hoof, J.A.R.A.M. (1967) The facial displays of catarrhine monkeys and apes. In *Primate Ethology* (ed D. Morris) Weidenfeld and Nicolson, London, pp. 7–88.

Van Hoof, J.A.R.A.M. (1972) A comparative approach to the phylogeny of laughing and smiling. In *Non-Verbal Communication* (ed R.A. Hinde) Cambridge University Press, Cambridge, pp. 209–38.

Van Lawick H. (1974) *Solo*, Collins, London

Van Lawick-Goodall, J. (1971) *In the Shadow of Man*, Houghton Mifflin, Boston.

Vannini, M. (1981) Notes on some factors affecting aggressive behaviour in *Caranus mediterraneus*. *Marine Biol.*, **61**, 235–41.

Van Oortmersson, G.A. (in press) Studies in wild house mice: II Testosterone and aggression. *Horm. Behav.*

Van Oortmerssen, G.A. and Bakker, T.C.M. (1981) Artificial selection for short and long attack latencies in wild *Mus musculus domesticus*. *Behav. Genet.*, **11**, 115–26.

Van Rhijn, J.G. (1973) Behavioural dimorphism in male ruffs *Philomachus pugnax*. *Behaviour*, **47**, 153–229.

van Valen, L. (1965) Morphological variations and width of ecological niche. *Amer Nat.*, **99**, 377–90.

Van Zegeren, K. (1980) Variations in aggressiveness and the regulation of numbers in house mouse populations. *Neth. J. Zool.*, **30**, 635–77.

Vaughn, B.E. and Waters, E. (1980) Social organisation among preschool peers. In *Dominance Relationships* (ed D.R. Omark, F.F. Strayer and D.G. Freedman) Garland STPM Press, New York, pp. 359–79.

Vehrencamp, S.L. (1978) The adaptive significance of communal nesting in groove-billed anis (*Crotophaga sulcirostris*). *Behav. Ecol. Sociobiol.*, **4**, 1–33.

Vehrencamp, S. (1983) A model for the evolution of despotic versus egalitarian societies. *Anim. Behav.*, **31**, 667–82.

Vehrencamp, S.L. and Bradbury, J.W. (1984) Mating systems and ecology. In *Behavioural Ecology* (J.R. Krebs and N.B. Davies), Blackwell Scientific Publications, Oxford, pp. 251–78.

Verner, J. (1977) On the adaptive significance of territoriality. *Amer. Nat.*, **111**, 769–75.

Verrell, P.A. (1986) Wrestling in the red-spotted newt (*Nothophthalamus viridescens*): resource value and contestant asymmetry determine contest duration and outcome. *Anim. Behav.*, **34**, 398–402.

Villars, T.A. (1983) Hormones and aggressive behaviour in teleost fish. In *Hormones and Aggression* (ed B.B. Svare), Plenum Press, New York, pp. 407–33.

Vines, G. (1980) Spatial consequences of aggressive behaviour in flocks of oystercatchers (*Haematopus ostralegus*). *Anim. Behav.*, **28**, 1175–83.

Vodegal, N. (1978) A study of the underlying motivation of some communicative behaviors of *Pseudotropheus zebra* (Pisces, Cichlidae): a mathematical model (1). *Proc. Koninklijke Neder. Akad. van Wetenschappen. Amsterdam. series C*, **81**, 211–40.

Vollrath, F. (1980) Male body size and fitness in the web-building spider *Nephila clavipes. Zeit. Tierpsychol.*, **53**, 61–78.

Vom Saal, F.S. (1983) Models of early hormonal effects in intrasexual aggression in mice. In *Hormones and Aggressive Behaviour* (ed. B.B. Svare) Plenum Press, New York, pp. 197–222.

Vom Saal, F.S. and Howard, L.S. (1982) The regulation of infanticide and parental behavior: implications for reproductive success in male mice. *Science*, **215**, 1270–72.

Vowles, D.M. and Beasley, L.D. (1974) The neural substrate of emotional behaviour in birds. In *Birds: Brain and Behaviour* (eds I.J. Goodman and M.W. Schein) Academic Press, New York, pp. 221–58.

Wada, K. (1984) Barricade building in *Ilyoplax pusillus. J. Exp. Marine Biol. Ecol.*, **83**, 1773–88.

Walker, L.E. (1981) A feminist perspective in domestic violence. In *Violent Behavior* (ed R.B. Stuart) Brunner-Mazel, New York.

Walker, M.L., Wilson, M.E. and Gordon, T.P. (1983) Female rhesus monkey aggression during the menstrual cycle. *Anim. Behav.*, **31**, 1247–54.

Walkowiak, W. (1983) Neuroethological studies on intraspecific call discrimination in the grass frog. In *Advances in Vertebrate Neurobiology* (eds

J.P. Ewert, R.R. Caprianca and D.J. Ingle), Plenum Press, New York, pp. 833–888.

Walletschek, H. and Raab, A (1982) Spontaneous activity of dorsal raphe/ neurons during defensive and offensive encounters in the tree shrew. *Physiol. Behav.*, **28**, 697–705.

Wapler-Leang, D.C.Y. and Reinboth, R. (1974) The influence of androgenic hormone on the behaviour of *Haplochromis burtoni*. *Fortsch. Zool.*, **22**, 334–39.

Watson, A. (1977) Population limitation and the adaptive value of territorial behaviour in Scottish Red Grouse, *Lagopus lagopus scoticus* In *Evolutionary Ecology* (eds B. Stonehouse and C. Perrins) MacMillan, London, pp. 18–25.

Watson, A., and Parr, R. (1981) Hormone implants affecting territory size and aggressive and sexual behaviour in red grouse. *Ornis Scand.*, **12**, 55–61.

Watt, D.J. (1984) The role of plumage polymorphism in dominance relationships of the white-headed sparrow. *Auk*, **101**, 116–20.

Watt, D.J. (1986) A comparative study of status signalling in sparrows (Genus *Zonotrichia*) *Anim. Behav.*, **34**, 1–15.

Weis, C.S. and Coughlin, J.P. (1979) Maintained aggressive behaviour in gonadectomized male Siamese fighting fish (*Betta splendens*). *Physiol. Behav.*, **23**, 173–77.

Welch, B.L. and Welch, A.S. (1969) Aggression and the biogenic amine neurohumors. In *Aggressive Behaviour* (eds S. Garattini and E.B. Sigg) Wiley, New York, pp. 188–202.

Wells, K.D. (1977) The social behaviour of anuran amphibians. *Anim. Behav.*, **25**, 666–93.

Weygoldt, P. (1980) Complex brood care and reproductive behavior in captive poison arrow frogs, *Dendrobates pumilio* Schmidt. *Behav. Ecol. Sociobiol.*, **7**, 329–32.

Weygoldt, R. (1977) Communication in insects and crustacea. In *How Animals Communicate* (ed T.A. Sebeok), Indiana University Press, pp. 303–37.

Whitham, T.G. (1986) Costs and benefits of territoriality: behavioural and reproductive release by competing aphids, *Ecology*, **67**, pp. 139–47.

Whitsett, J.M. (1975) The development of aggression and marking behaviour in intact and castrated male hamsters. *Horm. Behav.*, **6**, 47–57.

Whitten, P.L. (1983) Diet and dominance among female vervet monkeys. (*Cercopithecus aethiops*). *Amer. J. Primatol.*, **5**, 139–59.

Whitworth, M.R. (1984) Maternal care and behavioural development in pikas *Ochotona princeps*. *Anim. Behav.*, **32**, 743–52.

Wiepkema, P.R., Koolhaus, J.G.H. and Olivier-Aardema, R. (1980) Adaptive aspects of neuronal elements in agonistic behaviour. *Progr. Brain Res.*, **53**, 369–84.

Wilcox, R.S. and Ruckdeschel, T. (1982) Food threshold territoriality in a water strider (*Gerris remiges*). *Behav. Ecol. Sociobiol.*, **11**, 85–90.

Wilkinson, G.S. (1984) Reciprocal food sharing in vampire bats. *Nature*, **309**, 181–84.

Wilkinson, P.F. and Shank, C.C. (1976) Rutting fight mortality among musk oxen on Banks Island, Northwest Territories, Canada. *Anim. Behav.*, **24**, 756–58.

Williams, G.C. (1966) *Adaptation and Natural Selection*. Princeton University Press.

Wilson, A.P. and Vessey, S.H. (1968) Behaviour of free-ranging castrated rhesus monkeys. *Folia. Primatol.*, **9**, 1–14.

Wilson, E.O. (1971) *The Social Insects*, Harvard University Press, Cambridge, Mass.

Wilson, E.O. (1975) *Sociobiology*. Harvard University Press, Cambridge, Mass.

Wilson, M.C. (1981) Social dominance and female reproductive behaviour in rhesus monkey (*Macaca mulatta*). *Anim. Behav.*, **29**, 472–82.

Wilson, M. and Daly, M. (1985) Competitiveness, risktaking and violence. *Ethol. Sociobiol.*, **6**, 59–73.

Wingfield, J.C. (1985) Short-term changes in plasma levels of hormones during establishment and defense of a breeding territory in male song sparrows, *Melospiza melodia*. *Horm. Behav.*, **19**, 174–87.

Wingfield, J.E. and Farner, D.S. (1978) The endocrinology of a natural breeding population of the white-crowned sparrow. *Physiol. Zool.*, **51**, 188–205.

Wingfield, J.C. and Ramenofsky, M. (1985) Testosterone and aggressive behaviour during the reproductive cycle in male birds. In *Neurobiology* (eds R. Giles and J. Balthazart), Springer-Verlag, Berlin, pp. 92–104.

Winn, H.E. (1954) Comparative reproductive behavior of 14 species of darter. *Ecol. Mon.*, **28**, 151–91.

Wirtz, P. and Diesel, R. (1983) The social structure of *Inachus phalanguim*, a spider crab associated with the sea anemone *Anemonia sulcata*. *Zeit. Tierpsychol.*, **62**, 209–34.

Wise, D.A. and Prior, K. (1977) Effects of prolactin on aggression in the post-partum hamster. *Horm. Behav.*, **8**, 36–39.

Wolcott, D.L. and Wolcott, T.G. (1984) Food quality and cannibalism in the red land crab *Gecareinus lateralis*. *Physiol. Zool.*, **57**, 318–24.

Wolf, L.L. and Watts, E.C. (1984) Dominance and site-fixed aggressive behaviour in breeding male *Leucorrhina intacta* (Odonata). *Behav. Ecol. Sociobiol.*, **14**, 107–18.

Woodman, D.D. (1979) Urinary catecholamine and test habituation in maximum security hospital patients. *J. Psychosom. Res.*, **23**, 263–67.

Wootton, R.J. (1976) *The Biology of Sticklebacks*, Academic Press, London.

Worchel, S. (1978) Aggression and power restoration. The effects of identifiability and timing of aggressive behaviour. *J. Exp. Psychol.*, **14**, 43–52.

Wourms, J.P. (1977) Reproduction and development in chondrichthyan fishes. *Amer. Zool.*, **17**, 379–410.

Wrangham, R. (1981) Drinking competition in vervet monkeys. *Anim. Behav.*, **29**, 904–10.

Wrangham, R.W. and Waterman, P.G. (1981) Feeding behaviour of vervet monkeys on *Acacia tortilis* and *Acacia xanthopheloea* with special reference to reproductive strategies and condensed tannin production. *J. Anim. Ecol.*, **50**, 715–31.

Wright, N.G. (1982) Ritualised behaviour in a territorial limpet. *J. Exp. Marine. Biol. Ecol.*, **60**, 245–51.

Wuensch, K.L. and Cooper, A.J. (1981) Postweaning paternal presence and later aggression in male *Mus musculus. Behav. Neur. Biol.*, **32**, 510–15.

Wyman, R.L. and Ward, J.A. (1973) The development of behaviour of the cichlid fish *Etroplus maculatus. Zeit. Tierpsychol.*, **33**, 461–91.

Yagil, R. and Etzion, Z. (1980) Hormonal and behavioural patterns in the male camel (*Camelus dromedarius*). *J. Rep. Fert.*, **58**, 61–65.

Yahr, P. (1983) Hormonal influences on territorial marking behaviour. In *Hormones and Aggressive Behaviour* (ed B. Svare) Plenum Press, New York, pp. 145–75.

Yasukawa, K. (1979) Territory establishment in red-winged blackbirds: importance of aggressive behaviour and experience. *Condor*, **81**, 258–64.

Yasukawa, K. and Bick, E.T. (1983) Dominance hierarchies in dark-eyed juncos (*Junco hyemalis*): a test of a game theory model. *Anim. Behav.*, **31**, 439–48.

Yoburn, B.C., Cohen, P.G. and Campagnoni, P.R. (1981) The role of intermittent food in the induction of attack in pigeons. *J. Exp. Anal. Behav.*, **36**, 101–17.

Zack, S. (1975) A description and analysis of agonistic behaviour patterns in an opisthobranch mollusc *Hermissenda crassicornis. Behaviour*, **53**, 238–67.

Zahavi, A. (1975) Mate selection – a selection for a handicap. *J. Theoret. Biol.*, **53**, 205–14.

Zahavi, A. (1976) Cooperative nesting in Eurasian birds. *Proc. XVI Int. Orn. Congr. Canberra*, 685–93.

Zahavi, A. (1979) Ritualisation and the evolution of movement signals. *Behaviour*, **72**, 77–81.

Zeichner, A. and Phil, R.O. (1979) Effects of alcohol and behavioural contingencies in human aggression. *J. Abn. Psychol.*, **88**, 153–60.

Zillman, D. (1984) *Connections between Sex and Aggression.* Erlbaum, Hillsdale, New Jersey.

Zimbardo, P.G. (1978) The psychology of evil: on the perversion of human potential. In *Aggression, Dominance and Individual Spacing* (eds L. Frances, P. Krames and T. Allaway). Plenum Press, New York, pp. 155–69.

Zucker, N. (1983) Courtship variation in the neotropical fiddler crab *Uca deichmanni*: another example of female incitation to male competition. *Marine Behav. Physiol.*, **10**, 57–79.

Zwickel, F.C. (1980) Surplus yearlings and the regulation of breeding density in blue grouse. *Can. J. Zool.*, **58**, 896–902.

Species index

Entries are in alphabetical order by common name, or by scientific name when no common name is available. Where more than one example of a given kind of animal is included these are entered under the general name and then listed in alphabetical order. Common names not given in the text are bracketed.

Subject index

NB '. . . and aggression' to be taken as read in all entries.